# Advanced MATLAB for Engineers

Mehran Andalibi · Samuel A. Cherkauer

# Advanced MATLAB for Engineers

## Integrating Theory and Programming

Springer

Mehran Andalibi
Department of Mechanical Engineering
Embry-Riddle Aeronautical University
Prescott, AZ, USA

Samuel A. Cherkauer
Department of Aerospace Engineering
Embry-Riddle Aeronautical University
Prescott, AZ, USA

ISBN 978-3-032-10676-6 ISBN 978-3-032-10677-3 (eBook)
https://doi.org/10.1007/978-3-032-10677-3

This book also provides accompanying materials for lecturers. You can request them by going to sn.pub/lecturer-material and filling out the form.



This Springer imprint is published by the registered company Springer Nature Switzerland AG
The registered company address is: Gewerbestrasse 11, 6330 Cham, Switzerland

*Dedicated to*

*My wife Nooshin, my son Ryan and my parents, Mehri and Morteza*

*—Mehran Andalibi*

*My mom and dad, Laura and Keith*

*—Sam Cherkauer*

# Foreword by Damon M. Chandler

MATLAB is a powerful tool that can play many roles—it's a calculator, a systems solver, a simulator, a data acquisition tool, a visualization tool, a filter/system designer, a signal/image/video processing and computer vision framework, a deep learning tool, a programming environment, a GUI framework, and more. If you have an engineering task that needs to be implemented, simulated, or otherwise automated on the computer, there is a good chance that MATLAB is the right tool for you. But therein lies the challenge: Whereas engineering textbooks can provide the theory, and MATLAB's documentation can describe specific parameters, there is otherwise no guidance on how to transform the theory into a working MATLAB implementation.

This text, *Advanced MATLAB for Engineers: Integrating Theory and Programming*, serves an invaluable role in this regard. It helps bridge the gap between theory and implementation through thorough explanations and easy-to-follow, meaningful examples. Whether you are just starting MATLAB and want to learn the basics of the language, or you are trying to use MATLAB to solve a specific task, or you are a seasoned MATLAB developer aiming to deepen your understanding, this text will cater to your needs. The text is a massive undertaking to thoroughly explain MATLAB's features—including the advanced aspects not normally discussed elsewhere—in an approachable and practical manner.

I have worked with Prof. Andalibi for many years on a variety of topics in engineering, computer science, and education. Most of our projects have made use of MATLAB in one way or another. During our research discussions, Prof. Andalibi has never ceased to amaze me with his ability to use MATLAB to its fullest—and to even push it beyond its limits. Even though I consider myself a veteran MATLAB user, I always seem to learn something new about MATLAB from Prof. Andalibi. Through this text, you too can gain from his experience and insights to start using MATLAB more effectively for your own projects.

Damon M. Chandler, Ph.D.
Professor, College of Information
Science and Engineering
Ritsumeikan University
Kyoto, Japan

# Foreword by Ron Madler

I have worked with Dr. Mehran Andalibi for many years. He is a prolific educator with extensive experience and knowledge. Dr. Andalibi has created an advanced MATLAB book with extensive examples of solving real problems with MATLAB. These examples allow the reader to quickly solve a multitude of engineering problems. It is a book I look forward to having on my bookshelf.

Ron Madler, Ph.D.
Professor of Aerospace Engineering
Embry-Riddle Aeronautical University
Prescott, AZ, USA

# Foreword by Oyku Ozsoy

I had the chance to review this MATLAB book, and I can genuinely say it is one of the most complete and student-friendly resources I have come across. It covers a wide range of topics, with clear explanations and a strong focus on hands-on learning through exercises. What I liked most was how easy and practical the book feels. It is clearly made for students and will be a great help for anyone learning MATLAB. I am happy to have been part of the process and look forward to seeing how this book supports learners in the future.

Oyku Ozsoy, Ph.D.
Assistant Professor of Computer Engineering
Embry-Riddle Aeronautical University
Prescott, AZ, USA

# Foreword by Michael Puopolo

*Advanced MATLAB for Engineers: Integrating Theory and Programming* is a treasure trove for students and professionals who are serious about furthering their skills and knowledge in MATLAB. Within its pages, the reader finds practical applications, helpful visualizations, and detailed programming techniques that equip them with the most powerful tools in the MATLAB environment. Although Advanced MATLAB for Engineers is marked by depth of content, its chapters are organized in the style of an outline, whereby information is delivered like a handy reference manual. Topics are well chosen, and example problems at the end of each chapter are appropriately challenging for advanced MATLAB users. These refreshing qualities of the book do not come as a surprise, as we have here in book form what Prof. Andalibi crafts so masterfully in the classroom: a learning environment where truly motivated students find deep edification.

Michael Puopolo, Ph.D.
Engineer, Writer

**Competing Interests** The authors have no competing interests to declare that are relevant to the content of this manuscript.

# Purpose of the Book and Intended Users

Welcome to *Advanced MATLAB for Engineers: Integrating Theory and Programming*, a book designed to deepen your understanding and enhance your skills in using MATLAB, a powerful tool for numerical computing, visualization, and programming. This book aims to bridge the gap between basic user knowledge and advanced applications, specifically tailored for engineers who are already familiar with MATLAB's fundamental concepts.

MATLAB is indispensable in the engineering world due to its simplicity and flexibility in solving a wide range of engineering problems. Whether you are involved in electrical, mechanical, civil, or aerospace engineering, MATLAB has applications that can benefit your work. This book focuses on advanced topics, such as mathematical modeling, sophisticated data analysis, and the development of comprehensive graphical user interfaces. Each topic is presented with practical engineering applications in mind, demonstrating how MATLAB can be used to solve complex engineering challenges efficiently.

The goal of this book is not only to introduce advanced MATLAB features and techniques but also to provide insight into their practical implementation. We will explore advanced programming techniques, delve into the powerful features of Simulink, and examine how MATLAB interfaces with other software and hardware. Special attention is given to animation techniques, advanced data visualization, and the creation of user-friendly applications with App Designer.

*Advanced MATLAB for Engineers: Integrating Theory and Programming* is structured to be accessible to those who have a basic understanding of MATLAB but want to push their skills further. It is assumed that the reader has an introductory knowledge of MATLAB programming; hence, the book immediately dives into more complex subjects. Each chapter builds upon the previous one, ensuring a smooth progression that follows the logical development of skills and knowledge.

We invite you to explore each chapter, engage with the examples provided, and apply the lessons learned to your engineering projects. Whether you use this book as a reference for specific tasks or as a guide for systematic learning, it is our hope that it will be a valuable addition to your professional and academic endeavors.

Thank you for choosing *Advanced MATLAB for Engineers: Integrating Theory and Programming*. Let's embark on this journey to not just learn MATLAB, but to master it for innovative and efficient engineering solutions.

# Contents

# About the Authors

**Dr. Mehran Andalibi** is an Associate Professor of Mechanical Engineering at Embry-Riddle Aeronautical University. With over 18 years of experience in higher education, he has taught a wide range of engineering and software courses, including MATLAB programming. As a certified MATLAB MCMA professional, Dr. Andalibi has been using MATLAB for over 22 years in diverse research areas, such as Artificial Intelligence, Robotics, and Engineering Education. His motivation for writing this book stems from a desire to distill his extensive knowledge of MATLAB, including its latest tools, into a comprehensive and accessible textbook. Designed with engineering students in mind, the book features numerous real-world examples that can support their learning throughout their academic journey.

**Samuel A. Cherkauer** is an Embry-Riddle Aeronautical University alumnus who graduated in May 2025 with a major in Aerospace Engineering and a minor in Robotics. During his time at Embry-Riddle, Sam became proficient at programming in MATLAB, Python, and C, and was motivated to write this book to further his knowledge in advanced programming. Following graduation, Sam plans on returning to work in the aerospace industry and potentially pursuing a graduate degree. Writing this book was a great educational experience and will hopefully help future engineering undergraduates.

# Chapter 1
# Introduction to MATLAB for Engineering

## 1.1 Overview of MATLAB

MATLAB, short for Matrix Laboratory, is a high-level programming and numeric computing environment used extensively by engineers and scientists around the world for its robust functionality and ease of use. Developed by MathWorks, MATLAB was first released in 1984 and has evolved significantly over the years. The current stable release, MATLAB 2025b, as of this writing, continues to expand its capabilities, ensuring that MATLAB remains at the cutting edge of scientific computing.

**Core Features of MATLAB**:

- **Matrix-Based Language**: MATLAB is fundamentally a matrix-based language, making it an intuitive choice for computational mathematics. Every data type in MATLAB is a matrix, and matrix operations are seamlessly integrated, making it highly suitable for engineering and other technical fields.
- **Interpreted Language**: As an interpreted language, MATLAB facilitates quick iteration and exploration. Code can be run immediately after writing, which speeds up the process of algorithm development and adjustment.
- **Rich Built-In Libraries**: MATLAB comes equipped with a vast library of pre-built functions covering various areas, such as statistical analysis, Fourier analysis, filtering, optimization, and many others.
- **Toolboxes**: Extensive toolboxes expand the base functionality of MATLAB, each designed to solve specific problems in different areas of engineering. These toolboxes are collections of functions that extend the MATLAB environment to solve particular classes of problems more efficiently.
- **Graphical Capabilities**: MATLAB offers extensive capabilities for visualizing data. Functions for plotting vectors and matrices, annotating graphs, processing images, and 3D visualization tools are all available directly within MATLAB.

 
M. Andalibi and S. A. Cherkauer, *Advanced MATLAB for Engineers*,
https://doi.org/10.1007/978-3-032-10677-3_1

- **Development Environment**: The MATLAB desktop environment includes tools for managing files, variables, and applications, providing a powerful yet simple interface for organizing large coding projects. Its integrated debugger and profiler make MATLAB an excellent environment for designing robust applications.
- **Simulink**: MATLAB integrates seamlessly with Simulink, a graphical environment for simulation and Model-Based Design of multidomain dynamic and embedded systems.

**Community and Support**

MATLAB benefits from a strong support network. MathWorks provides extensive documentation and user support, including tutorials and user forums. The MATLAB Central community, which includes the File Exchange, offers a platform where users can share code, applications, and ideas. This community-driven platform enhances the capabilities of MATLAB users by allowing them to share solutions and benefit from the innovations of others.

**Why is MATLAB Preferred in Engineering?**

Engineers prefer MATLAB because it allows them to perform intensive mathematical calculations with relatively simple code. This efficiency is crucial in fields where time and accuracy are of the essence, such as control systems, telecommunications, signal processing, and image processing. The ability to rapidly test algorithms without recompilation saves valuable time in the iterative process of design.

Furthermore, MATLAB's widespread use in academia and industry makes it a valuable skill for engineers, enhancing their employability and facilitating collaborative projects across disciplines and industries.

## 1.2 Importance in Engineering

MATLAB's significance in the engineering world cannot be overstated. Its comprehensive suite of computational tools and intuitive programming environment make it indispensable for today's engineers. Here are several key reasons why MATLAB holds such pivotal importance in engineering disciplines:

**Facilitates Complex Numerical Analysis**

Engineers frequently face challenges that require complex numerical analysis and mathematical computations. MATLAB provides an efficient platform for handling large data sets and performing complex calculations, including matrix operations, optimization tasks, and numerical integration. These capabilities are crucial for developing models that simulate real-world phenomena and for creating algorithms that improve technological processes.

### Accelerates the Design Process

With its high-level language, extensive function libraries, and interactive environment, MATLAB significantly accelerates the design process. Engineers can quickly prototype and iterate their designs, testing and refining systems without the need for lengthy compilations. This rapid prototyping capability is especially beneficial in control systems engineering, electronics, and mechanical design, where time-to-market and responsiveness to change are critical.

### Enhances Interdisciplinary Collaboration

The versatility of MATLAB allows it to serve as a bridge between various engineering domains, facilitating interdisciplinary collaboration. Engineers can easily share code and algorithms, making collaborative projects more streamlined and effective. Whether it's integrating mechanical designs with electrical systems or combining chemical processes with control logic, MATLAB's comprehensive environment supports a multidisciplinary approach to engineering solutions.

### Supports Advanced Engineering Applications

MATLAB is at the forefront of advanced engineering applications, including signal processing, image processing, communications, and control systems. It supports engineers in implementing complex algorithms that are integral to developing next-generation technologies such as autonomous vehicles, smart grids, and IoT devices. MATLAB's toolboxes further enhance its capabilities, providing specialized functions for nearly every engineering field.

### Enables Data Visualization and Analysis

Effective data visualization and analysis are crucial for engineers to interpret data, make decisions, and communicate findings. MATLAB excels in offering tools that simplify these tasks, providing a wide array of plotting functions and customizable diagrams to visualize complex data intuitively. These visualizations help engineers to uncover underlying patterns, assess the performance of systems, and present data compellingly to stakeholders.

### Educational and Professional Development

MATLAB is not only a tool for professional engineers but also a staple in engineering education. It is used extensively in universities around the world to teach fundamental and advanced engineering concepts. Familiarity with MATLAB enhances a student's educational outcomes and increases their marketability in the engineering job market. For professional engineers, continual learning and adaptation to new MATLAB features and toolboxes keep them at the cutting edge of technology and innovation.

### Conclusion

In summary, MATLAB is a linchpin in the engineering toolkit, respected not just for its computational capabilities but also for its role in enhancing productivity, fostering innovation, and bridging different technical disciplines. Its ongoing development ensures that it remains relevant to both current and emerging engineering challenges, making it a perpetual asset in the engineering community.

## 1.3 Basic MATLAB Syntax and Operations

MATLAB's environment is designed to be intuitive and efficient, making it ideal for both simple computations and complex engineering tasks. Understanding the basic syntax and operations in MATLAB is crucial for leveraging its full capabilities. Here are key elements of MATLAB's syntax and fundamental operations:

**MATLAB Interface**

The MATLAB default GUI includes several key components (shown in Fig. 1.1):

1. **Command Window**: Where you can execute commands directly.
2. **Workspace**: Displays the variables created during the current session.
3. **Current Folder**: Is the working directory where the software looks for files, scripts, and data, and where it saves new files by default.
4. **Home Tab**: Where you go to do general purpose operations like creating new files, importing data, managing your workspace, and setting your Desktop layout.
5. **Plot Tab**: Displays a gallery of plotting tools.
6. **Apps Tab**: Shows you the apps that are currently installed.

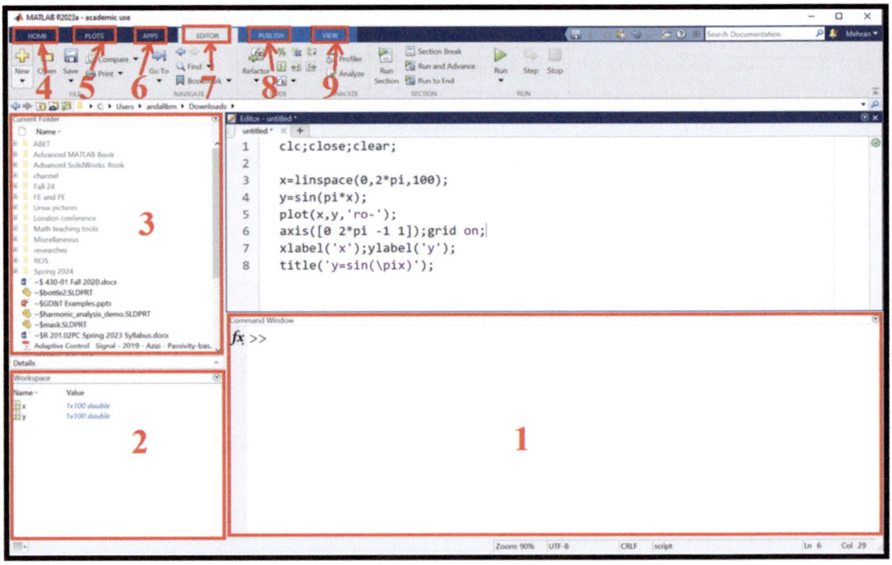


**Fig. 1.1** MALTAB GUI components

7. **Editor Tab**: Allows you to write, edit, and run MATLAB scripts and functions.
8. **Publish Tab**: Publishing tools for the Editor tab.
9. **View Tab**: View options for the Editor tab.

**Basic Syntax**

- **Variables**: Variables in MATLAB do not require explicit declaration (MATLAB is an untyped programming language). A variable is created the moment you assign it a value and the type of variable is determined from the given value automatically. For example, **x = 5;** creates a variable **x** of type double and assigns it the value 5, **y=true;** creates a variable y of type logical, and **z='hello world!';** creates a variable of type char. To determine the types of variables in the workspace, use the command "**whos**".
  - Note that only letters, numbers, and underscore are allowed to be used in the name of a variable and the name must only start with a letter.
  - Names of the variables must not be reserved keywords (such as, **pi, nan, eps, inf**) or the names of build-in functions. If you are not sure the name you chose for a variable is reserved, search that name in MATLAB help or in the command window using "**lookfor keyword**".
  - MATLAB is case-sensitive, and a and A are two different variables.

- **Operators**: MATLAB operators include the following categories:

1- **Arithmetic Operators**: They are +, −, *, /, and ^ (for exponentiation). Matrix operations use the same symbols but operate according to linear algebra rules.
   - Note that in arrays, the indices start at 1 as opposed to most of other programming languages, such as C++ or Python where indices start at zero.
   - The operator backward division \ between variables a and b, i.e., a\b is the same as b/a, and for matrices A and B, i.e., A\B is the same as inv(A)*B. A\B is typically recommended as an alternative to inv(A)*B due to the speed of operation.
   - Note that when a dot is added to the *, /, and ^ for matrices, (**.*,./,.^**), it indicates element-wise operation, e.g., $A. * B = \left[a_{ij} * b_{ij}\right]$.

2- **Relational Operators**: There are 6 operators (**==, ~=, >, >=, <, <=**) that are used to compare values and create results of type **logical**.
3- **Logical Operators**: There are 5 operators (**&, |, &&, ||, ~**) that are used to apply to logical variables and expressions. Note that && is called a logical and uses short-circuiting as opposed to the standard &. This means that **a&&b** will not evaluate b if the value of a is false. Similarly, in the logical or with short-circuiting (**a||b**), the value of b will not be evaluated if a is true.
4- **Special Characters**:

   4-1- Character @ is used for calling a function handle (**[T,Y]=ode45(@myfun,tspace, init_cond)**) and superclass methods, which will be seen in Object-Oriented-Programming (**disp@MySuper(obj)**).

4-2- Character period or dot (**.**) is used for decimal point (**2.5**), element-wise matrix operations (**A.*B**), structure field access (**students.names**), and object property or method specifications (**myobj.PropertyName**).

4-3- Character dot dot dot (**...**) is used to indicate line of code continuation when the line is too long (**if a==2 & b~=3 |...**).

4-4- Character comma (**,**) is used to separate elements of a matrix in a row. Whitespace can alternatively be used for this purpose (**B=[1,2;3 4]**).

4-5- Character colon (**:**) is used to create equally-space arrays (**initial_value:spacing:final_value**).

4-6- Character semicolon (**;**) is used to indicate the end of a row in a matrix (**A=[1 2;3 4]**), and to suppress displaying the results of code execution in the command window.

4-7- Parentheses **( )** are used to keep the operator precedence, where the highest priority of execution always belongs to the innermost parentheses (**A*(B+C)/(E-F)**). Parentheses are also used to enclose function arguments/inputs (**outp=myfun(inp1,inp2)**) and for array indexing (**A(3,:)**).

4-8- Brackets **[ ]** are used for array construction (**A=[1 2]**), array concatenation (**[A B]** or **[A;B]**), creating empty matrix and element deletion (**D=[], C(1,:)=[]**), and to assign multiple outputs for a function (**[out1,out2, out3]=myfun(inp1,inp2)**).

4-9- Curly Brackets **{ }** are used for cell array assignment and contents (**A={'a',2,true}, b=A{3}**).

4-10- Percent (**%**) is used to add comments as discussed in the following section and also in the **springf** and **fprintf** commands to show the conversion specifier (**fprinf('a=%4.6f',a)**).

4-11- Single quotes (**' '**) are used to declare character arrays (**a= 'Hello All'**).

4-12- Double quotes (**" "**) are used to declare string arrays (**a= "Hello All"**).

4-13- Tilde (**~**) is used as a logical NOT (**~(A&B)**) and for an argument placeholder when that argument is not needed (**[~,maxval]=max(A)**).

4-14- Left Angle Bracket (**<**) is used to specify superclasses in Object-Oriented-Programming (**classdef MyClass < MySuperclass**).

### Scripts and Functions

- **Scripts**: Scripts are files containing a sequence of MATLAB commands. Scripts run all commands as if they were typed into the command window, making them useful for automating repetitive tasks.
  - Scripts share their variables with the workspace, and they can be executed using the button "**run**" in the Editor tab.
  - The **.m** file containing scripts can have any name if it follows the same convention for naming the variables.

- **Functions**: Unlike scripts, functions can accept input arguments and return outputs. Functions are declared using the **function** keyword, followed by the output and input variables, e.g., **function [out1,out2] = myfun(inp1,inp2,inp3).**
  - Functions typically end with the keyword **end.**
  - Functions are used to make the code modular, reusable, and create data encapsulation, i.e., functions have their own workspace and the variables declared within a function will not be shared with the MATLAB workspace. Different types of functions will be discussed in Chap. 2.
  - Functions need to be called using their syntax (declaration line) and not executed using the "**run**" button.
  - The name of the **.m** file containing the code for a function must be the same as the name of the function; otherwise, MATLAB will issue an error. Furthermore, names of built-in functions (e.g., **sin, cos, log**) must not be used when creating the **.m** file for a new custom function to avoid overwriting them.

**Data Structures**

- **Matrices and Arrays**: MATLAB is particularly known for its powerful matrix and array operations. A matrix is defined by rows and columns, e.g., **A = [1, 2; 3, 4];** creates a matrix.
- **Cell Arrays**: Hold data of varying types and sizes. Defined using curly braces, e.g., **C = {'cat', 'dog'; 1, [2, 3]};**
- **Structures**: Allow you to store different types of data under the same variable with a specific structure provided by fields, e.g., **S.name = 'John'; S.age = 28;**
  - More on advanced data structures will be provided in Chap. 2.

**Conditional Statements and Loops**

- **Conditional Statements**: Use **if/elseif/else** and **switch** to execute code based on conditions.
- **Loops**: **for** and **while** loops are used to repeat operations a certain number of times or while a condition is true, respectively.
  - More on logic control and programming tools will be provided in Chap. 2.

**Input and Output**

- **Input**: Use **input('Prompt:')** to get input from the user.
- **Output**: **disp('Message')** displays a message in the command window.
  - More advanced I/O commands will be explained in Chap. 2.

**Commenting and Execution**

- Comments are marked using % for single-line comments or %{ and %} for block comments. You can alternatively use **Ctrl+R** and **Ctrl+T** for commenting and uncommenting lines of code or use the associated buttons in the Editor tab. Comments are essential for explaining and organizing your code.

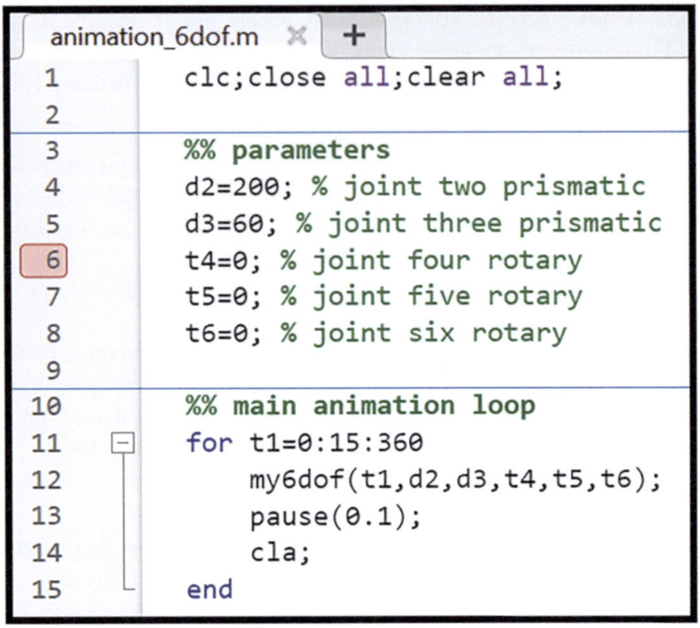


**Fig. 1.2** Section lines and break points in scripts

- When, using **%%** in the beginning of a line in the Editor, it indicates a separate section of the code that can be executed individually using **Ctrl+Enter** or the "**Run Section**" button in the Editor tab. Sections are separated using lines as can be seen in Fig. 1.2. Button "**Run and Advance**" can be used to move to the next section and button "**Run to End**" can be used to run from the current section to the end of the script.
- Any line or highlighted lines of code can be executed using the **F9** key.
- Break points can be inserted anywhere in a script by simply clicking on the line number (a red square will appear around the line number as shown in Fig. 1.2). They will break the code execution into pieces to facilitate debugging. Buttons "**Continue**" and "**Step**" can be used to move between the break points while running the code. Break points can be removed by clicking on them again.

**Meaning of Default Colors in MATLAB Syntax**:

Different colors of the code/syntax in MATLAB along with their meanings are provided in Table 1.1.

While in general MATLAB errors cannot be suppressed, warnings that are not blocking the code executions can be deactivated using **"warning off"**.

Also, the default colors in Table 1.1 can be altered under Home tab/Preferences/Color.

**Table 1.1** Meaning of default syntax colors in MATLAB

| Color | Meaning | Example |
|---|---|---|
| | Regular Code | x=5; |
| | Syntax Error | A(3,4)=[] |
| | Warning | Min value set to 0 |
| | Strings | C='Hello World!' |
| | Comments | % Parameters |
| | Keywords, Hyperlink | for ii=1:5 link |
| | System Commands | ! Touch testFile.txt |

## 1.4 Logic Control

In MATLAB, "logic controls" generally refer to constructs that allow you to control the flow of execution in a program based on certain conditions or logic. These are fundamental components in any programming language, enabling decision-making, looping, and control over the execution sequence. In MATLAB, the primary logic control structures are:

### *1.4.1 if, elseif, else*

These statements allow you to execute different sections of code based on certain conditions. The **if** statement evaluates a logical condition and executes a group of statements when the condition is true. The **elseif** and **else** statements extend this functionality to handle multiple conditions. Typically, **else** is reserved for cases not covered in previous conditions.

```
x=input('enter an integer number: ')
if x > 0
        disp('x is positive');
elseif x == 0
        disp('x is zero');
elseif x<0
        disp('x is negative');
else
        disp('entered value is not integer');
end
```

### 1.4.2 *switch, case, otherwise*

The **switch** statement provides an alternative to multiple **if** statements when you have several conditions to check against a *single* variable or expression. Each case specifies a different condition to check against the expression evaluated by **switch**. Typically, **otherwise** is reserved for cases not covered in previous conditions.

```
x=input('enter the number for a month (1-12): ')
switch x
        case {1,3,5,7,8,10,12}
                disp('31 days');
        case {4,6,9,11}
                disp('30 days');
        case 2
                disp('28 or 29 days');
        otherwise
                disp('entered number is wrong! Try again.');
end
```

**If/elseif/else** can check any logical condition as opposed to **switch/case/otherwise** that only acts based on finite values of a variable.

### 1.4.3 *return*

This command is used to exit a function prematurely. This command is particularly useful when you want to stop executing a function before reaching its end, often based on a specific condition. Using **return** allows for more readable and efficient code by preventing unnecessary computations or actions.

For instance, in the following function that returns the combination of n and r, i.e., $C_r^n$, if any of these two numbers are not positive, the **return** command stops the execution of the rest of the code and returns the control back to the user:

```
function y=returncombinations(n,r)
        if (n <= 0)|(r<0)
                disp('The numbers provided are not positive. Exiting the function.');
                return;
        end
        num=factorial(n);
        den=factorial(r)*factorial(n-r);
        y=num/den;
end
```

### 1.4.4 *for loop*

A **for** loop runs a group of statements a *fixed* number of times. It iterates over elements in a vector or matrix, executing the loop body once for each element.

**For** loops can be used inside each other (**nested for loops**) as well as with the logic controls mentioned previously to perform more complicated tasks.

For instance, the code below plots a chessboard using two **for** loops and the **fill** command as can be seen in Fig. 1.3:

```
n = 8;
figure;
hold on;
axis equal;
colors = {'w', 'k'};
for row = 1:n
        for col = 1:n
                colorIndex = mod(row + col, 2) + 1;
                fill([col-1, col, col, col-1], [row-1, row-1, row, row], colors{colorIndex});
        end
end
axis off;
hold off;
```

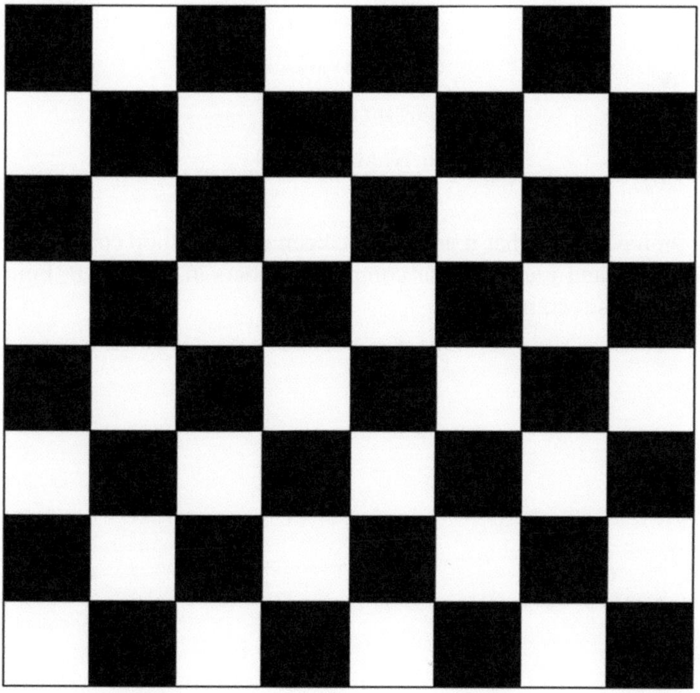

**Fig. 1.3** Chessboard plotted using nested for loops

- One of the common mistakes when using **for** loops is changing the loop counter during the iterations. The example below is an attempt to avoid displaying the value of 6:

```
for ii = 1:10
    if ii == 5
        ii = 7; % Attempt to change the loop counter to skip 6
    end
    disp(['Current ii value: ', num2str(ii)]);
end
```

However, when the code is executed, all numbers from 1 to 10 are displayed and number 6 is not skipped. Six is not skipped since the **for** loop continuously updates **ii** and effectively cancels the effect of the **if** statement:

```
Current ii value: 1
Current ii value: 2
Current ii value: 3
Current ii value: 4
Current ii value: 5
Current ii value: 6
Current ii value: 7
Current ii value: 8
Current ii value: 9
Current ii value: 10
```

- You might have noted that **ii** and **jj** have been used for loop counters instead of **i** and **j** as both **i** and **j** are used for complex numbers in MATLAB. Following the same practice is recommended.

### 1.4.5 *while loop*

A **while** loop repeatedly executes a target statement as long as a given condition is true. The **while** loop is useful when the number of iterations is unknown before the loop starts and makes the **while** loop more powerful than the **for** loop.

The function below is a guessing game for a random integer between 1 and 100 and it stops when the number is guessed correctly. It returns the number of guesses used and provides hints each time a guess is entered:

```
function guessNumberGame()
    % Generate a random number between 1 and 100
    targetNumber = randi(100);
    % Initialize attempt count
    attempts = 0;
    % Game introduction
    disp('Guess the number game!');
    disp('I am thinking of a number between 1 and 100.');
    % Game loop
    while true
        userGuess = input('Enter your guess: '); % Get user guess
        attempts = attempts + 1; % Increment attempts
        % Check guess and provide feedback
        if userGuess < targetNumber
            disp('Too low. Try again.');
        elseif userGuess > targetNumber
            disp('Too high. Try again.');
        else
            fprintf('Congratulations! You guessed the number in %d
        attempts.\n',  attempts);
            break; % Exit the loop if the guess is correct
        end
    end
end
```

- Note that since the syntax of a **while** loop does not initiate the loop counter and does not change it, loop counter initialization has to be done before the **while** loop starts and loop counter change must be done within the loop.
- Using **while true** constructs an **infinite loop**. To avoid being trapped in an infinite loop, there must be a stopping/breaking strategy. This example uses the break command, which is explained next.

### 1.4.6 *break*

This command is used within any looping structure (**for** or **while**). **break** immediately exits the loop, regardless of the loop's normal progression or condition as shown in the example of the **while** loop.

### 1.4.7 *continue*

This command is used within looping structures. **continue** skips the remaining part of the loop body and proceeds with the next iteration of the loop.

In the example below, the **continue** command is used to skip displaying the current loop counter/number if it is even:

```
for ii = 1:10
    if mod(ii, 2) == 0
        continue; % Skip the current iteration if ii is even
    end
    disp(['ii = ', num2str(ii)]);
end
```

It will generate the following results:

```
ii = 1
ii = 3
ii = 5
ii = 7
ii = 9
```

### 1.4.8 *try/catch*

Although this structure is not directly used as a logic control tool, it is a powerful tool for handling exceptions and errors during the execution of code. It allows you to try a block of code and catch any errors that occur, enabling graceful handling of unexpected situations without stopping the entire program. This mechanism is particularly useful for debugging and for ensuring that your program can handle different types of errors smoothly.

The function below used the **try/catch** structure to divide any two given numbers even if the denominator is zero. If the denominator is not zero, it proceeds with no error; otherwise, it catches the error message and displays it without generating an error message and breaking the flow of the code:

```
function result = safeDivide(num, denom)
    try
        % Attempt to divide two numbers
        result = num / denom;
        disp(['Result of division: ', num2str(result)]);
    catch ME
        % Error handling
        disp('An error occurred:');
        disp(ME.message); % Display the error message
        result = NaN; % Return NaN if an error occurs
    end
end
```

## 1.5 Functions for Matrices and Sets

Since there are hundreds of built-in functions for matrices, we focus on the most common functions and show their applications with brief examples.

- Note that the indices for matrices and arrays start from 1 as opposed to C++, Python, or some other programming languages that start array indices with 0.
- **Forming a Matrix**: entries in a row are separated by a comma or whitespace and rows are separated by a semicolon:

  **A= [1 3;2, 4; 5 6]**: forms a $3 \times 2$ matrix
  **B= [1 3 2 4]**: forms a $1 \times 4$ matrix (row vector)
  **C= [1; 3; 2; 4]**: forms a $4 \times 1$ matrix (column vector)

- **Multi-dimensional matrices**:

  **X = rand(3,4,2)**: returns a 3-dimensional matrix **X** with 3 rows, 4 columns, and 2 layers

```
X(:,:,1) =

0.0462  0.6948  0.0344  0.7655

0.0971  0.3171  0.4387  0.7952

0.8235  0.9502  0.3816  0.1869

X(:,:,2) =

0.4898  0.7094  0.6797  0.1190

0.4456  0.7547  0.6551  0.4984

0.6463  0.2760  0.1626  0.9597
```

- **Accessing entries, row, columns, or parts of a matrix**: We can use the colon operator.

  **A(2,3)**: entry at row 2 and column 3
  **A(1,:)**: first row
  **A(:,2)** second column
  **A(1:2,1:2)** first 2 rows and 2 columns of **A**
  **A([1,3],1)**: first and third rows column one of **A**
  **B(1:3)**: values of **B** from index 1 to index 3 (the first 3 entries)
  **B(end)**: last entry of **B** using the function **end**; **B(end-1)** gives the second to last entry of **B**
  **B(end:-1:1)**: reverse order of entries in **B** from the last one back to the first one

- **Converting a matrix to a vector (vectorization), reshaping, and repeating matrices:**

  **A(:)**: Converts **A** to a column vector by placing columns of **A** below each other starting from the first column
  **A=reshape(A,[2,3])**: converts the $3 \times 2$ matrix **A** to a $2 \times 3$ matrix
  **repmat(A,m,n)**: repeats matrix **A** in a $m \times n$ pattern

- **Deleting portions of a matrix:**

  **A(1,:)=[ ]**: deletes the first row of **A**
  **A(3,1)=[ ]**: generates an error as the result cannot be a matrix anymore (matrices cannot have empty elements)

- **Size of a matrix and length of a vector:**

  **size(A)**: returns the number of rows and columns of **A** as a $1 \times 2$ vector
  **length(B)**: returns the length of **B** as a single number

- **Arithmetic of Matrices**

  **A= [1 3; 2, 4; 5 6]**;
  **D= [1 3; 2, 4; 5 6]**;

**A+3**: adds a value of 3 to each entry of **A**.
**3*A**: multiplies each entry of **A** by 3.
**A+D**: adds corresponding entries of **A** and **D**.
**A-D**: subtracts corresponding entries of **D** from **A**
**A*D**: performs the standard matrix multiplications of matrices **A** and **D**. It generates an error since the number of columns of the left matrix (**A**) is not the same as the number of rows of the right matrix (**D**)
**A/D**: although matrix forward division is not mathematically defined, there will be no error message for these 2 matrices as MATLAB returns $AB^{+} = AB^{T}\left(BB^{T}\right)^{-1}$.
**A\D**: this is the same as $A^{-1}D$ which is the solution to the linear set of equations of form $AX = D$. This method is preferred over **inv(A)*D** in MATLAB, especially for a larger matrix **A**, as this backward division solves the system of equations using the Gauss-Jordan elimination method as opposed to calculating the matrix inverse, which is computationally demanding.

- **Transpose, Inverse, Trace, Determinant, eigenvalue and singular value decompositions of a Matrix:**

**transpose(A) or A'**: returns the transpose of **A**
**inv(A)**: returns the inverse of **A**
**trace(A)**: returns the trace of **A**
**det(A)**: returns the determinant of **A**
**[V,D]=eig(A)**: returns the eigenvalues of **A** in a diagonal matrix **D** and the eigenvectors of **A** as columns of matrix **V** for the corresponding eigenvalues in **D**, i.e., **A*V = V*D**
**[U,S,V] = svd(A)**: performs a singular value decomposition of matrix **A**, such that **A = U*S*V'**.

- **Special matrices**:

**zeros(m,n)**: generates an $m \times n$ matrix with all entries equal to zero
**ones(m,n)**: generates an $m \times n$ matrix with all entries equal to one
**eye(n)**: generates an $n \times n$ identity matrix
**D = diag(v)**: returns a square diagonal matrix with the elements of vector **v** on the main diagonal.
**X = diag(A)**: returns a column vector of the main diagonal elements of **A**.
**diag(v,k)**: places the elements of vector **v** on the kth diagonal.
**rand(m,n)**: generates an $m \times n$ matrix of random entries from a uniform distribution in the range [0 1]
**randi([-5,5],m,n)**: generates an $m \times n$ matrix of random integer entries in the range [-5 5]
**randn(m,n)**: generates an $m \times n$ matrix of random entries from a Gaussian distribution with a mean of zero and a variance of one
Note that these random-generator commands create pseudo-random numbers, meaning they use deterministic algorithms with an initial value called "**seed**". If we use the same seed with the same algorithm, it will generate the same random numbers.

**rand('seed', 42)**: creates a random scalar with a seed of 42.
MATLAB has also the command "**rng**", that can set the seed using the following syntax:
**rng(seed, 'generatorType')**
Using the "**rng**" command, we can save the current seed, and use it later to regenerate the same sequence of random number using the following commands:
**s = rng;**
**rng(s);**
Also, the command "**rng('default')**" restores the default settings for this command.
There are 3 generator algorithms in MATLAB: 'twister': Mersenne Twister (default), 'philox': (Philox 4x32-10), and 'threefry' (Threefry 4x64-20). Table 1.2 shows the attributes of these **Pseudo-Random Number Generation (PRNG)** algorithms and when to use each one.

- **Norms of Vectors and Matrices**:

  **norm(x,p)**: returns the p-norm of vector x: $|x|_p = \left(x_1^p + x_2^p + \cdots + x_n^p\right)^{\frac{1}{p}}$
  **p=2** is the Euclidean norm that is most used.
  **norm(A,p)**: returns the p-norm of matrix **A**, where
  **p=1**: is the maximum absolute column sum of the matrix.
  **p=2**: is defined by $\sqrt{\lambda_{max}\left(A^T A\right)}$
  $p = \infty$: is the maximum absolute row sum of the matrix.
  **norm(A, "fro")**: returns the **Frobenius norm** of matrix or array **A**: $\sqrt{\sum_{i=1}^{m}\sum_{j=1}^{n} a_{ij}^2}$

- **Dot product and cross product of vectors**:

  **dot(A,B)**: performs the dot product between vectors **A** and **B**: $A \cdot B$
  **cross(A,B)**: performs the cross product between vectors **A** and **B**: $A \times B$

**Table 1.2** PRNG Algorithms in MATLAB

| Feature | Twister | Philox | Threefry |
|---|---|---|---|
| Type | Pseudorandom number generator | Counter-based RNG | Counter-based RNG |
| Period | $2^{19937} - 1$ | $2^{128}$ | $2^{128}$ |
| State size | 624 words (19968 bits) | Stateless (based on counter) | Stateless (based on counter) |
| Parallel support | Limited | Excellent | Excellent |
| Cryptographic security | No | No | No (cryptographically inspired) |
| Speed | Fast for sequential use | Faster for parallel | Slower than Philox |

- **Elementwise arithmetic operations**:

  **A.*D**: performs the element-wise multiplication of matrices **A** and **D**: $\left[a_{ij} \times d_{ij}\right]$
  **A./D**: performs the element-wise division of matrices **A** and **D**: $\left[\frac{a_{ij}}{d_{ij}}\right]$
  **A.^2**: performs the square of elements for matrices **A**: $\left[a_{ij}^2\right]$

- **Maximum and minimum values in a vector or a matrix and their locations**:

  **[minval,minind]=min(A)**: finds the minimum value of vector **A** and the index of this value. A similar structure can be used for the maximum of a vector.
  For matrix **A**, the command above will not return the minimum value of the entire matrix, but rather the minimum values of each column and their indices. In order to find the minimum value of matrix **A** and its location, we use the vectorized form of **A**:
  **[minval,minind]=min(A(:))**: finds the minimum value of matrix **A** and the linear index of this value. To convert this linear index into a row and a column index, we use the following command:
  **[rowind,colind]=ind2sub(size(A),minind)**:finds the row index and column index of the minimum value of **A**.

- **Average and Standard Deviation of entries in a vector and in a matrix**:

  **mean(A)**: returns the average value of entries in a vector or of each column in a matrix
  **mean2(A)**: returns the average value of all entries in a matrix
  **std(A)**: returns the standard deviation of entries in a vector or of each column in a matrix
  **mode(A)**: returns the mode of entries in a vector or of each column in a matrix
  **median(A)**: returns the median of entries in a vector or of each column in a matrix

- **Sum, product, and difference of entries in a vector**:

  **sum(A)**: returns the sum of entries in a vector or of each column in a matrix
  **sum(A(:))**: returns the sum of all entries in a matrix
  **cumsum(A)**: returns the cumulative sum of entries in a vector or of each column in a matrix
  **prod(A)**: returns the product of entries in a vector or of each column in a matrix
  **cumprod(A)**: returns the cumulative product of entries in a vector or of each column in a matrix
  **diff(A)**: calculates differences between adjacent elements of **X** along the first array dimension. This command can be used for backward finite differences, such as for numerical differentiation:
  **diff(y)./diff(x)**: performs the numerical differentiation: $\frac{dy}{dx} \approx \frac{\Delta y}{\Delta x} \approx \frac{y_i - y_{i-1}}{x_i - x_{i-1}}$

- **Finding entries with specific conditions in a matrix/array and their locations (find)**:

  **k = find(X)**: returns a vector containing the linear indices of each nonzero element in array **X**
  **k = find(X>0)**: returns a vector containing the linear indices of positive element in array **X**
  **k = find((X>m)&(X<n))**: returns a vector containing the linear indices of elements in array **X** that are in the range [m n].

- **Logical Commands for arrays**:

  **isempty(A)**: returns logical 1 (true) if **A** is empty.
  **isnan(A)**: returns a logical array containing 1 (true) where the elements of **A** are **NaN**.

- **Colon operator, "linspace" and "logspace" for creating equally-space data**:

  **x=a:n:b** returns an equally-space vector **x** with entries starting at **a** incrementing a value of **n > 0** (or decrementing if **n < 0**) and ending at **b** if **mod(b - a, n) = 0**. We use this colon operator when the spacing (**n**) is *known* but the number of entries in **x** is *unknown.*
  **x=linspace(a,b,N)**: returns an equally-space vector **x** with **N** entries in the range [a b]. We use this command when the spacing is *unknown* but the number of entries (**N**) is *known.*
  **x=logspace(a,b,N)**: returns a vector **x** with **N** entries in the range [$10^a$ $10^b$] where the logarithms of entries are equally-space. This command is useful for logarithmic plots.

- **Shuffling entries in a vector**:

  **flipud(A)**: returns **A** with its rows flipped in the up-down direction.
  **fliplr(A)**: returns **A** with its columns flipped in the left-right direction.
  **circshift(A,K)**: circularly shifts the elements in array **A** by **K** positions

```
A = [5   7   0   9   -1   6   8]

circshift(A,2)

                        ans = [6   8   5   7   0   9   -1]

circshift(A,-3)

                        ans =[9   -1   6   8   5   7   0]
```

- **Sorting the values in a vector**

  **sort(x,'ascend')**: sorts the values in vector **x** in an ascending order (from smallest to largest).
  **sort(x,'descend')**: sorts the values in vector **x** in a descending order (from largest to smallest).
  If **x** is a matrix, then **sort(x)** treats the columns of **x** as vectors and sorts each column.

- **Frequency calculations in a vector**:

  **tabulate(x)**: displays a frequency table of the data in vector **x**.

```
x={'M','M','F','F','F','F'}
tabulate(x)
```

| Value | Count | Percent |
|---|---|---|
| M | 2 | 33.33% |
| F | 4 | 66.67% |

**count=hist(x,nbins)**: sorts **x** into the number of bins specified by the scalar **nbins** and returns the frequencies of data in each bin as a vector
Results can be seen in Fig. 1.4 for 1000 randomly generated data using 20 bins:

- **Sparse Matrices**:

  Sparse matrices are a type of matrix in which most of the elements are zero (there is no fixed threshold but at least 50% of the entries must be zero). They are commonly encountered in engineering systems with large dimensions or specific physical constraints, such as Finite Element Analysis (e.g., in the stiffness matrix), power systems (e.g., grid network), and machine learning (e.g., Natural Language Processing).

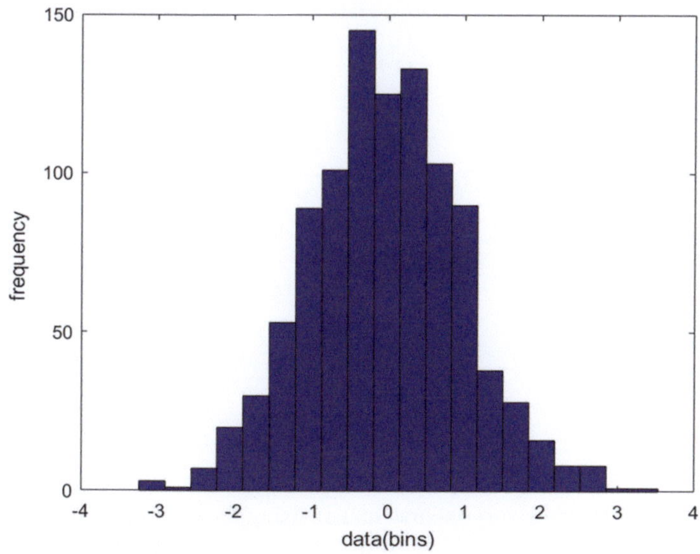


**Fig. 1.4** A sample histogram plot

In MATLAB, we can define a matrix as sparse, where only the nonzero entries along with their locations are stored and used in operations. This allows for efficient storage and computation by taking advantage of the sparsity of data. **"Sparsity"** is a measure of how many zero elements a matrix contains relative to its total number of elements. For instance, a sparsity of 0.2 (20%) for a $10 \times 10$ matrix means that only 20 out of 100 entries are nonzero. The complement value of sparsity is called **"density"**, i.e., $density = 1 - sparsity$. So, the matrix mentioned above has a density of 0.8 (80%).

To define a sparse matrix in MATLAB, there are 3 different methods:

(a) Directly define a sparse matrix using the command **sparse** and providing the nonzero entries withand their row and column locations:

```
ii = [1, 2, 3];
jj = [1, 3, 2];
v = [10, 20, 30];
S1 = sparse(ii, jj, v, 6, 4)
```

(b) Using the **spalloc** command with a syntax of **"spalloc(m,n,nz)"** to allocate a sparse matrix of an $m \times n$ dimension with **nz** nonzero entries and then adding the nonzero entries:

```
S1 =
(1,1)    10
(3,2)    30
(2,3)    20
```

```
S2 = spalloc(10,20,12)
S2(3,4)=6
S2(5,10)=2
S2(9,17)=-8
```

```
S2 =
(3,4)     6
(5,10)    2
(9,17)   -8
```

(c) Convert a “**full**” matrix to a sparse matrix using the command **sparse**. Let’s say we want to solve a set of $10 \times 10$ linear equations similar to those formed in FEA. The coefficient/stiffness matrix is given by:

```
K = [0 1 0 0 0 0 0 0 0 0;...      %    x1 = 4
     0 2 -3 0 0 0 0 0 0 0;...     %    2*x2 – 3*x3 = -9
     0 0 0 0 0 4 0 0 0 0;...      %    4*x6 = 8
     0 0 0 0 3 -2 0 0 0 0;...     %    3*x5 – 2*x6 = 6
     0 0 1 0 0 0 1 0 0 0;...      %    x3 + x7 = 5
     1 0 0 0 0 0 0 0 2 -1;...     %    x1 + 2*x9 – x10 = 0
     0 0 0 0 0 0 1 -3 5 0;...     %    x7 – 3*x8 + 5*x9 = 1
     0 0 0 1 -7 0 0 1 0 0;...     %    x4 – 7*x5 + x8 = -1
     0 0 0 0 0 1 0 0 0 -4;...     %    x6 – 4*x10 = 5
     2 0 5 0 0 0 0 0 3 0];        %    2*x1 + 5*x3 + 3*x9 = 7

K_sp = sparse(A)
```

```
K_sp =
(6,1)     1
(10,1)    2
(1,2)     1
(2,2)     2
(2,3)    -3
(5,3)     1
(10,3)    5
(8,4)     1
(4,5)     3
```

```
(8,5)    -7
(3,6)     4
(4,6)    -2
(9,6)     1
(5,7)     1
(7,7)     1
(7,8)    -3
(8,8)     1
(6,9)     2
(7,9)     5
(10,9)    3
(6,10)   -1
(9,10)   -4
```

If we have the following Force matrix:

```
F = [4;-9;8;6;5;0;1;-1;5;7];
```

Then, we can find the solution (nodal displacements) using the backslash operator either using the original **K** matrix or the Sparse **K** matrix (**K_sp**):

```
del = K\F;
del_sp = K_sp\F;
```

Storing the **K** matrix required more memory than the **K_sp** matrix. You can check this using the command **whos**:

| Name | Size | Bytes | Class | Attributes |
|---|---|---|---|---|
| **F** | **10x1** | **80** | **double** | |
| **K** | **10x10** | **800** | **double** | |
| **K_sp** | **10x10** | **440** | **double** | **sparse** |

Also, note that linear algebra operations can be done with much fewer calculations as only nonzero entries are considered in operations with sparse matrices.

In the following, we will look at some commands in MATLAB for handling sparse matrices:

**I_Full = full(I_sp)**: converts a sparse matrix to a full (dense) matrix.

**issparse(I_sp)**: returns a logical true if the matrix **I_sp** is defined as a sparse matrix.

**count = nnz(S_rand)**: returns the number of nonzero entries in a sparse matrix.

**v = nonzeros(S_rand)**: extracts the nonzero elements of a sparse matrix.

**[ival, jval, val] = find(S_rand)**: returns the row and column indices and values of nonzero elements.

**spy(S_rand)**: visualizes the sparsity pattern of a matrix as can be seen in Fig. 1.5 for the **K_sp** matrix introduced in the previous page. Each blue dot is a nonzero entry and the zero entries are shown with blank spaces.

**I_sp = speye(n)**: creates a sparse identity matrix of size $n \times n$.

**S_diag = spdiags(B, d, m, n)**: creates a sparse diagonal matrix of size $m \times n$, where **B** contains the diagonals' entries and **d** specifies their positions (on the main diagonal, above or below it).

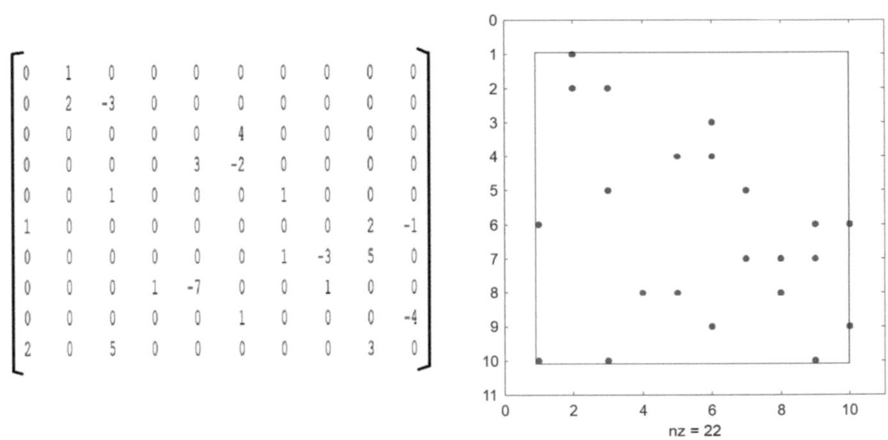


**Fig. 1.5** Sparsity pattern of sparse matrix

```
% Define the diagonals to insert
B = [1 2 3 4 5;     % Diagonal to place at position -1
    6 7 8 9 30;     % Main diagonal (position 0)
    10 11 12 13 14]; % Diagonal to place at position 1

d = [-1, 0, 1]; % -1: Below main diagonal, 0: Main diagonal, 1: Above main diagonal
m = 5;
n = 5;
S_diag = spdiags(B', d, m, n);
disp(full(S_diag));
```

```
6  11   0   0   0
1   7  12   0   0
0   2   8  13   0
0   0   3   9  14
0   0   0   4  30
```

**S_rand = sprand(m, n, density)**: creates a matrix of size $m \times n$ with all random entries in the range [0 1] with a given density. Note that the given density will be approximate, i.e., the number of nonzero entries will not exactly be equal to $m \times n \times density$ as this number might not be an integer value.

**spfun(f,s)**: applies the function **f** to nonzero entries of a sparse matrix **S**.

```
f = @(x)(sin(x).^2-2*cos(x)+5);
F3 = spfun(f,S2)
```

```
S2 =
(3,4)     6
(5,10)    2
(9,17)   -8
```

```
F3 =
(3,4)     3.1577
(5,10)    6.6591
(9,17)    6.2698
```

- **Useful Commands for Sets**:

  **intersect(A,B)**: returns the data common to both **A** and **B**.
  **union(A,B)**: returns the combined data from **A** and **B**.
  **setdiff(A,B)**: returns the data in **A** that is not in **B**.
  **unique(A)**: returns the same data as in **A**, but with no repetitions.
  **ismember(A,B)**: returns an array containing logical 1 where the data in **A** is found in **B**.

## 1.6 String Manipulations

- **Characters vs. Strings**: **Character Arrays**: In MATLAB, character arrays and strings serve similar purposes but have some differences in how they are handled and manipulated.

(I) **Character Arrays**

- They are essentially arrays of characters, like how strings are represented in many other programming languages.
- Character arrays are created using single quotes (') to enclose the characters. For example, **str1 = 'Hello'**.
- Each element of a character array is a single character, and you can access individual characters using indexing. For example, **str1(1)**.
- Character arrays have fixed lengths once created, meaning you cannot directly append characters to them without creating a new array.
- Operations on character arrays are typically performed elementwise.

(II) **Strings**

- Strings in MATLAB are a data type introduced in more recent versions (MATLAB R2016b and later).
- Strings are created using double quotes (") to enclose the characters. For example, **str2 = "Hello"**.
- Strings require more memory to store the same information:

```
str1='Hello';
str2="Hello";
whos
                         Name     Size     Bytes    Class

                            str1      1x5      10      char

                         str2     1x1      166     string
```

- Unlike character arrays, strings can hold sequences of characters of varying lengths, and you can concatenate strings using the **+** operator or functions like **strcat**.
- MATLAB provides a set of functions specifically designed for working with strings, such as **strlength**.

- **Character to double and double to character conversions based on ACSII table**: each keyboard character has an equivalent integer number of type **double** based on the ASCII table shown in Table 1.3 which can be used for conversions:

```
val='a'-'A'
val = 32
```

**Table 1.3** ASCII Table

| Decimal | Hex | Char | Decimal | Hex | Char | Decimal | Hex | Char | Decimal | Hex | Char |
|---|---|---|---|---|---|---|---|---|---|---|---|
| 0 | 00 | [NULL] | 32 | 20 | [SPACE] | 64 | 40 | @ | 96 | 60 | ‘ |
| 1 | 01 | [START OF HEADING] | 33 | 21 | ! | 65 | 41 | A | 97 | 61 | a |
| 2 | 02 | [START OF TEXT] | 34 | 22 | " | 66 | 42 | B | 98 | 62 | b |
| 3 | 03 | [END OF TEXT] | 35 | 23 | # | 67 | 43 | C | 99 | 63 | c |
| 4 | 04 | [END OF TRANSMISSION] | 36 | 24 | $ | 68 | 44 | D | 100 | 64 | d |
| 5 | 05 | [ENQUIRY] | 37 | 25 | % | 69 | 45 | E | 101 | 65 | e |
| 6 | 06 | [ACKNOWLEDGE] | 38 | 26 | & | 70 | 46 | F | 102 | 66 | f |
| 7 | 07 | [BELL] | 39 | 27 | ' | 71 | 47 | G | 103 | 67 | g |
| 8 | 08 | [BACKSPACE] | 40 | 28 | ( | 72 | 48 | H | 104 | 68 | h |
| 9 | 09 | [HORIZONTAL TAB] | 41 | 29 | ) | 73 | 49 | I | 105 | 69 | i |
| 10 | 0A | [LINE FEED] | 42 | 2A | * | 74 | 4A | J | 106 | 6A | j |
| 11 | 0B | [VERTICAL TAB] | 43 | 2B | + | 75 | 4B | K | 107 | 6B | k |
| 12 | 0C | [FORM FEED] | 44 | 2C | , | 76 | 4C | L | 108 | 6C | l |
| 13 | 0D | [CARRIAGE RETURN] | 45 | 2D | - | 77 | 4D | M | 109 | 6D | m |
| 14 | 0E | [SHIFT OUT] | 46 | 2E | . | 78 | 4E | N | 110 | 6E | n |
| 15 | 0F | [SHIFT IN] | 47 | 2F | / | 79 | 4F | O | 111 | 6F | o |
| 16 | 10 | [DATA LINK ESCAPE] | 48 | 30 | 0 | 80 | 50 | P | 112 | 70 | p |
| 17 | 11 | [DEVICE CONTROL 1] | 49 | 31 | 1 | 81 | 51 | Q | 113 | 71 | q |
| 18 | 12 | [DEVICE CONTROL 2] | 50 | 32 | 2 | 82 | 52 | R | 114 | 72 | r |
| 19 | 13 | [DEVICE CONTROL 3] | 51 | 33 | 3 | 83 | 53 | S | 115 | 73 | s |
| 20 | 14 | [DEVICE CONTROL 4] | 52 | 34 | 4 | 84 | 54 | T | 116 | 74 | t |
| 21 | 15 | [NEGATIVE ACKNOWLEDGE] | 53 | 35 | 5 | 85 | 55 | U | 117 | 75 | u |

(continued)

**Table 1.3** (continued)

| Decimal | Hex | Char | Decimal | Hex | Char | Decimal | Hex | Char | Decimal | Hex | Char |
|---|---|---|---|---|---|---|---|---|---|---|---|
| 22 | 16 | [SYNCHRONOUS IDLE] | 54 | 36 | 6 | 86 | 56 | V | 118 | 76 | v |
| 23 | 17 | [END OF TRANS. BLOCK] | 55 | 37 | 7 | 87 | 57 | W | 119 | 77 | w |
| 24 | 18 | [CANCEL] | 56 | 38 | 8 | 88 | 58 | X | 120 | 78 | x |
| 25 | 19 | [END OF MEDIUM] | 57 | 39 | 9 | 89 | 59 | Y | 121 | 79 | y |
| 26 | 1A | [SUBSTITUTE] | 58 | 3A | : | 90 | 5A | Z | 122 | 7A | z |
| 27 | 1B | [ESCAPE] | 59 | 3B | ; | 91 | 5B | [ | 123 | 7B | { |
| 28 | 1C | [FILE SEPARATOR] | 60 | 3C | < | 92 | 5C | \ | 124 | 7C | \| |
| 29 | 1D | [GROUP SEPARATOR] | 61 | 3D | = | 93 | 5D | ] | 125 | 7D | } |
| 30 | 1E | [RECORD SEPARATOR] | 62 | 3E | > | 94 | 5E | ^ | 126 | 7E | ~ |
| 31 | 1F | [UNIT SEPARATOR] | 63 | 3F | ? | 95 | 5F | _ | 127 | 7F | [DEL] |

In what follows, we will look at some of the most important commands for string/character array manipulation:

- **String length**: to find the length of a character array, we use the command **length** similar to arrays of numbers. For strings, however, we use the command **strlength**:

```
str1='Hello World!';
length(str1)
str2="Hello World!";
strlength(str2)
```

- **String concatenation**: character arrays can be concatenated horizontally using **[str1 str2]** or the **strcat(str1,str2)**. The first option does not eliminate trailing white spaces, while the second command does.

```
str1='Hello '
str2='World!'
str3=[str1 str2]
str4=strcat(str1,str2)
                              str3 = 'Hello World!'
                              str4 = 'HelloWorld!'
```

To concatenate character arrays vertically, use **[str1;str2]** or **strvcat(str1,str2)**. For the first method to work, however, the length of the character arrays have to be the same. When using **strvcat**, if the character arrays are of different lengths, it adds white spaces to the shorter array to make it the same length as the longer array.

```
str1='Hello'
str2='World!'
str5=[str1;str2] : issues an error
str6=strvcat(str1,str2)
                              str6 =  2×6 char array
                                    'Hello '
                                    'World!'
```

To join the string variables, we can use **+**:

```
str1="Hello ";
str2="World!";
str3=str1+str2
```

- **String comparison**: to compare character arrays, we use **strcmp**. The following commands will return a value of 0 (false) since MATLAB is case-sensitive (similar results can be obtained if **str1** and **str2** are defined as string variables instead of character arrays):

```
str1='MATLAB';
str2='matlab';
strcmp(str1,str2)
```

If lowercase and uppercase comparisons are to be avoided, we can use the command **strcmpi**, and if only the first **n** characters are to be compared, we use the command **strcmpn** and **strcmpni**.

- **Uppercase and lowercase conversions**: commands **upper** and **lower** can be used for this purpose:

```
str='MATLAB';
str1=upper(str)
str2=lower(str)
                              str1 = 'MATLAB'
                             str2 = 'matlab'
```

- **Categorization of characters**: command **isstrprop(str,category)** can be used to categorize different characters in a character array. The categories are **alpha, alphanum, cntrl, digit, graphic, lower, print, punct, wspace, upper,** and **Xdigit**. The result is a logical array with the same length as **str**:

```
str='Advanced MATLAB 2024!';
isstrprop(str,'alpha')
isstrprop(str,'digit')
isstrprop(str,'punct')
isstrprop(str,'wspace')
```

```
ans = 1 1 1 1 1 1 1 1 0 1 1 1 1 1 1 0 0 0 0 0 0

ans = 0 0 0 0 0 0 0 0 0 0 0 0 0 0 0 0 1 1 1 1 0
ans = 0 0 0 0 0 0 0 0 0 0 0 0 0 0 0 0 0 0 0 0 1

ans = 0 0 0 0 0 0 0 0 1 0 0 0 0 0 0 1 0 0 0 0 0
```

- **Removing white space from strings**: command **deblank** can be used to remove trailing whitespaces and command **strtrim** can be used to remove leading and trailing white spaces from character arrays. The following commands will return values of 18, 16, and 17, respectively:

```
str=' Hello Engineers! ';
length(str)
length(strtrim(str))
length(deblank(str))
```

- **String parsing**: command **strtok()** can be used to parse a character array into tokens separated by delimiters. The following commands will count the number of words in a sentence separated by white spaces:

```
str='Welcome to Advanced MATLAB for Engineers!';
[token,remainder]= strtok(str,' ');
cnt=1;
while ~isempty(remainder)
        cnt=cnt+1;
        [token,remainder]= strtok(remainder,' ');
end
fprintf('word count = %u\n',cnt)
```

- **Find and replace in strings**: commands **strfind** and **replace** can be used for this purpose. Command **strfind(str,template)** returns the indices of the first occurrence of **template** within **str**:

```
str='Find the starting indices of templates in a character array';
strfind(str,'in')
str=replace(str,'indices','locations')
                                  ans = 2    15    19    40
         str =  'Find the starting locations of templates in a character array'
```

- **String to number and number to string conversion, string to symbolic conversion, string evaluation**:

Commands **str2num** and **num2str** are used to convert strings to numbers and vice versa:

```
str='3.2';
```

```
str2num(str)+4
                                       ans =  7.2
a=6;
disp(['a=' num2str(a)])
```

Command **str2sym** converts strings to symbolic expressions, which can be useful when passing a function as an argument to another function, but not when using anonymous function handles:

```
syms x;
f='sin(x)';
diff(str2sym(f),x,2)
ans = -sin(x)
```

Command **eval** can be used to convert mathematical expressions in character arrays to the equivalent numbers:

```
b='sin(pi/2)';
eval(b)
```

```
ans = 1
```

- **Generating strings from text and variables**: command **sprintf** generates a new character array by inserting variables into an existing character array. This function uses a similar syntax to the command **fprintf** discussed in Chap. 2, but the output is a variable:

```
a=2;
f=@(x)(sin(x)-x*cos(x))
str=sprintf('output of function f(x)=sin(x)-x*cos(x) at %u is %4.2f\n',a,f(a))
str = 'output of function f(x)=sin(x)-x*cos(x) at 2 is 1.74'
```

**Practice Problems**

1 Write a script that simulates a simple temperature control system for a heating unit in a smart home. The system should turn the heating unit on or off based on the current temperature and a user-defined setpoint temperature.

**Inputs**: Ask the user to input the current temperature (in degrees Celsius) and the desired setpoint temperature (in degrees Celsius).

**Conditionals**: **Use if/elseif/else.**

- If the current temperature is more than 2 degrees below the setpoint temperature, the script should display the message: "**Heating ON**".
- If the current temperature is within 2 degrees of the setpoint temperature (including exactly at the setpoint), the script should display the message: "**Heating OFF, Temperature is OK**".
- If the current temperature is more than 2 degrees above the setpoint temperature, the script should display the message: "**Cooling ON**".

**Output**: Display the appropriate message based on the conditions.

2 Write the script for problem 1 using the **switch/case/otherwise** structure.

3 Write a script that multiplies two matrices using **nested for loops**. You will implement the matrix multiplication process manually without using MATLAB's built-in matrix multiplication (*) operator.

**Input**: Matrices **A** and **B** from the user. Ensure that the number of columns in **A** equals the number of rows in **B** for matrix multiplication to be valid. If dimensions don't match, exit the program using the **return** command.

**Output**: Display the resulting matrix **C**.

4 Write the code for problem 3 using nested while loops. Which code is more convenient to write? Why?

5 Write a script that searches for prime numbers within a given range. The script should find the first **N** prime numbers within the range, but if a certain threshold of non-prime numbers is encountered consecutively, the search should stop early. To determine if a number is prime, utilize a helper/utility function. Use **break** and **continue** commands in your script.

**Input**: The range limits as two numbers: **lower_limit** and **upper_limit**; the value of **N** (the number of prime numbers to find); and the threshold **T** (the maximum number of consecutive non-prime numbers allowed before stopping).

**Output**: If the loop completes successfully (i.e., **N** prime numbers are found), display those prime numbers. If the loop stops early due to reaching the threshold of consecutive non-prime numbers, display a message indicating the early exit and show the prime numbers found up to that point.

6 Write a script to analyze how different matrix norms change when all entries of a square matrix **A** are multiplied by a scalar constant **n** provided by the user. Specifically, you will generate a random square matrix **A** and compute the **Frobenius norm**, **1-norm**, **2-norm**, and **infinity-norm** of **A** and **nA** and display the ratio of these two norms for each type of matrix norm mentioned above. Is there any difference among these ratios?

7 Write a script that analyzes two sets of numerical data, representing the IDs of participants in two different surveys, Group A and Group B. Your objective is to use set operations to analyze the similarities and differences between the participant IDs in the two groups.

**Inputs**: Group A and Group B's participant IDs are represented as two vectors of integers. Each integer in the vector represents a unique participant ID. The lists may contain some common IDs as well as some that are unique to each group.

**Outputs**:

Find all unique participant IDs that are in either Group A or Group B.
Find the participant IDs that are common to both Group A and Group B.
Find the participant IDs that are unique to Group A.
Find the participant IDs that are in either Group A or Group B, but not in both.
Determine which group has more unique participant IDs.

8 Let's say you work for an IT company. You need to write a script that analyzes a text dataset containing multiple lines of information. Each line in the dataset contains a string representing a log entry from a server as shown below:

| [TIMESTAMP] | USER_ID | ACTION | DETAILS |
|---|---|---|---|
| [2024-08-22 14:55:01] | User123 | LOGIN | Success |
| [2024-08-22 15:10:05] | User456 | LOGIN | Failure: Invalid password |
| [2024-08-22 15:11:22] | User123 | LOGOUT | Success |
| [2024-08-22 16:00:00] | User789 | LOGIN | Success |

Your code must return/display the following items:

- **Extract Timestamps**: Extract the timestamps from each log entry and convert them into MATLAB datetime objects.
- **Login Failure Analysis**: Identify all log entries where the action was "**LOGIN**" and the result was a "**Failure**", and extract the user IDs associated with these login failures.
- **Identify Anomalous Patterns**: Identify users who logged in multiple times without a corresponding logout between logins. This may indicate a potential issue or anomaly.

9 Write a script that can process numerical data from a set of matrices provided as input. The script must be robust and able to handle various potential issues that could arise during data processing, such as attempting operations on incompatible matrix dimensions or invalid data types. To achieve this, you will use the **try/catch** structure to catch errors, handle them appropriately, and continue processing without crashing the entire script.

**Inputs**: A set of matrices stored in a cell array. Each matrix may have different dimensions and data types (e.g., numeric, logical, etc.). Some matrices may be incompatible for certain operations, such as matrix multiplication, element-wise addition, or inversion.

**Tasks**:

**Matrix Operations**: For each matrix in the cell array:

1. Attempt to compute the matrix inverse if it is square.
2. Attempt to add a predefined constant matrix to each matrix.
3. Attempt to multiply the matrix by a predefined vector.

**Error Handling**:

1. If a matrix cannot be inverted due to being non-square or singular, catch the error and display a message.

2. If a matrix cannot be added to the predefined matrix due to incompatible dimensions, catch the error and display a message.
3. If a matrix cannot be multiplied by the predefined vector due to incompatible dimensions, catch the error and display a message.

**Continue Processing**: Even if an error occurs with one matrix, the script should continue processing the remaining matrices in the cell array.

**Output**: For each operation attempted (inverse, addition, multiplication), display whether it was successful or if an error was caught. If successful, display the result of the operation.

# Chapter 2
# Advanced Programming Techniques

## 2.1 Scripts, Functions, and Live Scripts

### *2.1.1 Scripts*

A MATLAB script is essentially a file containing a series of MATLAB commands. Scripts are invaluable for automating repetitive tasks and can be executed simply by calling the script's name. They do not accept input arguments nor return output arguments but can manipulate and produce results using data in the workspace.

The name of the **.m** file containing a script can be arbitrary if it follows the naming convention for variables, so, **_myscript1.m** is not a valid name.

Scripts share their variables to the workspace and can access the variables in the workspace.

A common practice is to use the following commands in the beginning of each script to clean the workspace, clear the command window, and close all open windows:

```
clear;
clc;
close;
```

Adding comments and sectioning off code can organize your program to make it understandable by other users and make debugging more convenient.

Another common practice is to display appropriate messages at the top of a program to inform the user about the goal of the program, what they need to provide as inputs, and what they should expect as outputs. For instance, if the goal is to write a script that returns the area of a triangle given the x–y coordinates of its vertices, it might contain the following lines in the beginning:



M. Andalibi and S. A. Cherkauer, *Advanced MATLAB for Engineers*,
https://doi.org/10.1007/978-3-032-10677-3_2

```
disp('This program accepts the x-y vertices of a triangle as a 3x2 matrix and returns its area');
verts=input('enter the x-y vertices matrix: ');
```

Since scripts do not have input arguments and do not return outputs like functions, communications with the user must be done using Input/Output commands, including **disp**, **input**, and **fprintf**:

**disp('This is Mehran''s book')** displays a message that is only a character array (here **This is Mehran's book**). Note that if you need to use possessive apostrophes, you need to use 2 separate single quotes as shown here to avoid terminating the character array in the middle.

**A = [2 1; 3 5];**

**disp(A):** displays a variable entirely (here, matrix **A**).

**N = 5;**

**disp(['factorial of ' num2str(n) 'is' num2str(factorial(n))]):** displays a concatenated character array involving variables that are converted to strings using **num2str.**

**n = input('enter a positive integer: '):** displays a message in the command window, waits for the user to enter a value and then assigns it to variable **n**.

**txt = input('Do you want more? Y/N:','s'):** displays a message in the command window, waits for the user to enter a value, converts it to a character, and assigns it to variable **txt**.

**fprintf('X is %4.2f meters or %8.3f mm\n',A1,A2):** displays a text that can involve variables (here **A1** and **A2**) with specific display formats on the command window. Note that the % sign here is a conversion specifier. Table 2.1 shows **Conversion Characters** for the **fprintf** command:

**Field Width and Precision for fprintf:** In the command **fprintf('A = %4.2f\n',A)**, **%4.2f** means floating point number format with a filed width of 4 (minimum of 4 characters to print) and a 2-decimal precision. For instance, if **A = 6*pi**, then it will display:

**A = 18.85**

However, if we use **fprintf('A = %4.4f\n',A)**, it will display:

**A = 18.8496**

Note that for **%f**, **%e**, and **%E**, the precision is the number of digits to the right of the decimal point, but for **%g** and **%G** it is the total number of significant digits. For example, **%.04**f and **%.4g** for **A = pi** will display 3.1416 and 3.142, respectively.

**Text Before or After Formatting Operators:** Table 2.2 shows text before and after formatting options for the **fprintf** command.

One of the most important aspects of writing professional programs is to check for user inputs and issue warning or error messages when appropriate.

**warning('Division by zero'):** generates a warning message to attract the user's attention, but it does not stop the rest of code from running.

Automatic warnings by MATLAB can be turned off using **warning('off')**.

**error('Division by zero'):** generates an error message and stops the rest of code from running.

**Table 2.1** Conversion characters for fprintf

| Value Type | Conversion | Details |
|---|---|---|
| **Integer, signed** | %d or %i | Base 10 |
| **Integer, unsigned** | %u | Base 10 |
| | %o | Base 8 (octal) |
| | %x | Base 16 (hexadecimal), lowercase letters a–f |
| | %X | Same as %x, but uppercase letters A–F |
| **Floating-point number** | %f | Fixed-point notation (Use a precision operator to specify the number of digits after the decimal point.) |
| | %e | Exponential notation, such as 3.141593e+00 (Use a precision operator to specify the number of digits after the decimal point.) |
| | %E | Same as %e, but uppercase, such as 3.141593E+00 (Use a precision operator to specify the number of digits after the decimal point.) |
| | %g | The more compact of %e or %f, with no trailing zeros (Use a precision operator to specify the number of significant digits.) |
| | %G | The more compact of %E or %F, with no trailing zeros (Use a precision operator to specify the number of significant digits.) |
| **Characters or strings** | %c | Single character |
| | %s | Character vector or string array. The type of the output text is the same as the type of formatSpec. |

**Table 2.2** Text before or after formatting operators for fprintf

| Special character | Representation |
|---|---|
| Single quotation mark | '' |
| Percent character | %% |
| Backslash | \\ |
| Alarm | \a |
| Backspace | \b |
| Form feed | \f |
| New line | \n |
| Carriage return | \r |
| Horizontal tab | \t |
| Vertical tab | \v |
| Character whose Unicode numeric value can be represented by hexadecimal number, N | \xN<br>**Example: fprintf**('\x5A') returns 'Z' |
| Character whose Unicode numeric value can be represented by octal number, N | \N<br>**Example: fprintf**('\132') returns 'Z' |

**f = errordlg('File not found','File Error'):** generates an error dialogue box with a handle **f** that has the title **'File Error'** and the text **'File not found'** as shown in Fig. 2.1.

**Fig. 2.1** Error dialogue box

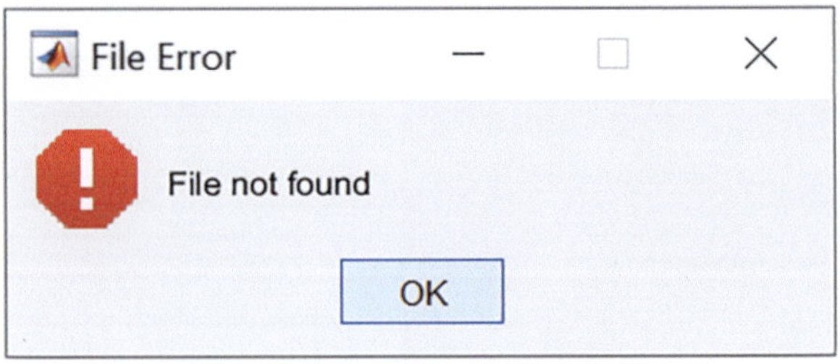


## *2.1.2 Functions*

Functions in MATLAB are more versatile than scripts. They enable encapsulation and reusability of code, accept input arguments, and can return one or more outputs. Here's the basic syntax for a MATLAB function with 3 inputs and 2 outputs (the first line is called the **function declaration line**):

```
function [out1, out2] = myfun(in1, in2, in3)
        out1 = in1 + in2;
        out2 = in1 * in3;
end
```

The name of the **.m** file storing this function must be **myfun.m** and nothing else.

Avoid using the names of existing functions or reserved keywords for a new function. If not sure whether a name has already been used in MATLAB, use **lookfor name** to search it.

Not all functions need to use the keyword **end** as will be discussed next.

A custom-written function can be called similar to MATLAB's built-in functions in a script if they are in the same folder, or if they are added to the MATLAB path.

**Types of Functions in MATLAB:**

I. **Simple Functions:** Perform specific tasks and return a result as illustrated earlier. They are stored as individual **.m** files.

To check the number of inputs and outputs to and from a function, we can use the commands **nargin** and **nargout**:

```
function [out1, out2] = myfunc(in1, in2, in3)
        if nargin<3
                error('You need to enter 3 inputs');
        end
        out1 = in1 + in2;
        out2 = in1 * in3;
end
```

Once a function is declared with a specific number of inputs, if it is called with a different number of inputs, it will generate an error, such as the following command for the function declared above:

```
[out1, out2] = myfunc(in1, in2)
```

If an output is not needed but is declared in the function's first line, we can use ~ as a placeholder:

```
[out1, ~ ] = myfunc(in1, in2)
```

To allow for a variable number of inputs and outputs and avoid generating errors, we can use the **varargin** and **varargout** cell arrays:

```
function [out1, out2] = myfunc(varargin)
    if nargin==3
        in1=varargin{1};
        in2=varargin{2};
        in3=varargin{3};
    elseif nargin==2
        in1=varargin{1};
        in2=varargin{2};
        in3=0;
    elseif nargin==1
        in1=varargin{1};
        in2=0;
        in3=0;
    else
        error('Invalid number of inputs');
    end
    out1 = in1 + in2;
    out2 = in1 * in3;
end
```

II. **Local Functions:** Also called **utility functions** or **subfunctions**, they are defined within another function file called the **primary function.** An example can be seen below, where the primary function **myboxoutlier** utilizes the **boxlimits** local function to find the minimum and maximum limits for the outlier box method.

```
function outliers = myboxoutlier(u)
% primary function to find the outliers in u using the box method
[a,b] = boxlimits(u);
outliers=u((u<a)|(u>b));
coef1=1.6;
% disp(coef2)

function [a,b] = boxlimits(u)
% utility function to calculate the upper and lower limits of the box method
coef2=1.5;
a=prctile(u,25)-coef2*iqr(u);
b=prctile(u,75)+coef2*iqr(u);
% disp(coef1)
```

Note that primary and utility functions can all be stored in the same **.m** file named with the primary function name. Only the primary function can be called from outside of the function file.

Primary and utility functions do not need to end with the keyword **end**.

Primary and utility functions can only access their own variables. For instance, in the example above, the commented **disp()** commands will generate an error.

III. **Nested Functions:** They are defined within another function's body (called the **host function**) as shown in the example below, where **myboxoutlier** is the host function and **boxlimits** is the nested function.

Nested functions have access to the host function's variables, if those variables are defined before calling the nested function. For example, variable **coef** is declared before calling the function **boxlimits** in the program below; otherwise, an error will be encountered (note the color of this variable, which indicates it acts like a global variable for nested functions). However, the host function does not have access to its nested functions' variables (the commented **disp(u_iqr)** line will generate an error).

```
function outliers = myboxoutlier(u)
        % host function to find the outliers in u using the box method
        coef=1.5;
        [a,b] = boxlimits(u);
        outliers=u((u<a)|(u>b));
        % disp(u_iqr)


        function [a,b] = boxlimits(u)
        % nested function to calculate the upper and lower limits of the box method
        u_iqr=iqr(u);
        a=prctile(u,25)-coef*u_iqr;
        b=prctile(u,75)+coef*u_iqr;
        end
end
```

Host and nested functions must end with the keyword **end** and the definitions of nested functions must end before the host function ends.

IV. **Anonymous Functions:** functions that are not stored in a separate **.m** file but are instead defined in the workspace, scripts, or other functions as a variable called a **function handle**. When the workspace is cleared, the function handle disappears. Therefore, the functions are called anonymous since no trace of them will be left behind.

Anonymous functions can accept inputs and return outputs, just as standard functions do. However, they can contain only a single executable statement:

```
myfun = @(x,y) (x.*y-x.^3);
z=myfun(25,20)
```

One of the most important applications of anonymous functions is when their handles are passed as an input to other functions. In the code below, the function handle for the anonymous function **myfun** is passed to the **integral, fminbnd**, and **fzero** functions, respectively, to find the integral of the function over the range [0 1], find the minimum of it over the range of [0 10], and find one of its roots starting with an initial guess of 0:

```
myfun = @(x) x.^2-sin(x)-1;
integral(myfun,0,1)
fminbnd(myfun,0,10)
fzero(myfun,0)
```

- **Global Variables:** As mentioned earlier, one way to share variables of a function with other functions is to write them as host and nested functions, where nested functions can access their own variables and the host function variables. Another common yet unpopular method of sharing variables among functions and scripts is to use global variables. In the example below, variable **coef** is a global variable declared in the script and used in the utility functions.

Note that the color of a global variable is cyan/green.

Note that a global variable must be declared in one line and a value to it must be assigned in a separate line. Also, wherever used, it must be declared again.

The main reason programmers avoid global variables is because any utility function can modify the variable. This modification could have adverse effects on the entire program and make debugging difficult (in the example below, both utility functions can change the value of **coef**).

```
clc;close;clear;
r=5;

global coef;
coef=3.1415;
surface_area = sphere_surfacearea(r)
volume = sphere_volume(r)

function surface_area = sphere_surfacearea(r)
        global coef;
        surface_area=4*coef*(r^2);
end

function volume = sphere_volume(r)
        global coef;
        volume=4*coef*(r^3)/3;
end
```

- **Persistent Variables:** variables that retain their values between calls to the function in which they are declared. Unlike global variables that are accessible from any function or the base workspace, persistent variables are visible only to the function in which they are defined. This makes them useful for storing information that must persist between function calls but should not be exposed or modified elsewhere.

In the example below, the function **callCounter** can keep track of the number of times it has been called using the persistent variable **numCalls**:

```
function count = callCounter()
% callCounter Counts the number of times the function is called.

% Declare the persistent variable
persistent numCalls;
```

```
% Initialize the persistent variable the first time the function is called
if isempty(numCalls)
        numCalls = 0;
end

% Increment the counter each time the function is called
numCalls = numCalls + 1;

count = numCalls;
end
```

To test, run the code below in the command window:

```
for ii = 1:5
    fprintf('Function has been called %d times.\n', callCounter());
end
```

It will generate the following results:

```
Function has been called 1 times.
Function has been called 2 times.
Function has been called 3 times.
Function has been called 4 times.
Function has been called 5 times.
```

Similar to how **nargin** and **nargout** can be used to alter the function's response to different numbers of inputs, simple and local functions can use argument blocks to set expected input sizes, types, and values. Additionally, the argument block can be used to set default values of an input so that it does not need to be provided and **nargin** does not need to be used. If an input does not match the expected size or data type, MATLAB will attempt to convert it or throw an error. If an input does not fall within expected values, MATLAB once again throws an error. Argument blocks are placed at the beginning of a function's contents and follow a similar format to functions, with an initial **arguments** keyword followed by contents and an **end** keyword.

```
arguments
    variableName (rows, columns) class {validator1, validator2, ...} = value
end
```

The only contents of argument blocks are variable definitions, where the name of the variable is followed by the variable dimensions, data type, and a cell array of requirements (called validators) separated by spaces. Any of these three objects can be left out, but those that remain must follow the order specified. The whole sequence of variable information can be set equal to a default value, which will set the input's value if one is not provided in the function call. If an input does not have the correct size or class type, MATLAB will attempt to convert it. If the input cannot be converted or does not comply with all of the validators, MATLAB raises an error. Argument blocks must contain all of the function's input variables in the same order.

An example of an argument block is shown below, where the function named **myfunc** accepts three variables, **x**, **y**, and **z**. Of these three variables, **x** must be a column vector of any height with numeric contents (**mustBeNumeric** is the validator), **y** must be a double, and **z** must be a string. Additionally, **y** and **z** have default values of 1 and '**z**', respectively.

```
function myfunc(x, y, z)
        arguments
                x (:, 1) {mustBeNumeric(x)};
                y double = 1;
                z string = 'z';
        end

        x
        y
        z
end

myfunc(2)
                                        x = 2
                                        y = 1
                                        z = "z"
myfunc([2; 4], 'z', 1)
                                        x = 2
                                            4
                                        y = 122
                                        z = "1"
```

All validator functions available for use are shown in Table 2.3. The **value** variable used in the table should be replaced by the name of the variable the validator is being applied to. Aside from the built-in validator functions, new validator functions can be created like any other functions. Validator functions must accept at least one argument, which is the variable being validated, must throw an error if its validation fails, and must not return anything.

### *2.1.3 Live Scripts*

MATLAB Live Scripts offer a unique and interactive way to create, document, and share your work directly within the MATLAB environment. Introduced to enhance the MATLAB experience, Live Scripts provide a rich format for combining code, output, formatted text, equations, and images in a single, executable document.

Here are several advantages of using Live Scripts in MATLAB:

1. **Interactive Documentation:** Live Scripts allow you to integrate code and documentation in one place. You can add explanations, mathematical descriptions, and formatted text right next to the code that performs the computation. This makes it easier to understand the purpose and functioning of the code.
2. **Rich Text Formatting:** Unlike traditional script files, Live Scripts support rich text formatting, including different styles and sizes of text, bold, italics, hyperlinks, and bullets. This feature helps in highlighting important information and organizing documentation more effectively.

**Table 2.3** All available validator functions

| Name | Effect |
|---|---|
| mustBePositive(value) | value > 0 |
| mustBeNonpositive(value) | value <= 0 |
| mustBeNonnegative(value) | value >= 0 |
| mustBeNegative(value) | value < 0 |
| mustBeFinite(value) | value has no NaN and no Inf elements |
| mustBeNonNan(value) | value has no NaN elements |
| mustBeNonzero(value) | value ~= 0 |
| mustBeNonsparse(value) | value is not sparse |
| mustBeSparse(value) | value is sparse |
| mustBeReal(value) | value has no imaginary part |
| mustBeInteger(value) | value == floor(value) |
| mustBeNonmissing(value) | value cannot contain missing values |
| mustBeGreaterThan(value, x) | value > x |
| mustBeLessThan(value, x) | value < x |
| mustBeGreaterThanOrEqual(value, x) | value >= x |
| mustBeLessThanOrEqual(value, x) | value <= x |
| mustBeA(value, classnames) | value must be one of the specified classes |
| mustBeNumeric(value) | value must be numeric |
| mustBeNumericOrLogical(value) | value must be numeric or logical |
| mustBeFloat(value) | value must be a floating-point array |
| mustBeUnderlyingType(value, typename) | value must have specified underlying type |
| mustBeNonempty(value) | value is not empty |
| mustBeScalarOrEmpty(value) | value must be a scalar or be empty |
| mustBeVector(value) | value must be a vector |
| mustBeMember(value, x) | value is an exact match to a member of x |
| mustBeInRange(value, lower, upper, flags) | value must be within the specified range |
| mustBeFile(stringValue) | value must be a path referring to a file |
| mustBeFolder(stringValue) | value must be a path referring to a folder |
| mustBeNonzeroLengthText(value) | value must be a string with nonzero length |
| mustBeText(value) | value must be a string array, character array, or cell array of character vectors |
| mustBeTextScalar(value) | value must be a single piece of text |
| mustBeValidVariableName(value) | value must be a valid variable name |

3. **Embedded Output:** Outputs from MATLAB commands, including plots and tables, can be embedded directly in the Live Script. This feature is particularly useful for visualization and data analysis tasks, as it keeps the results and the code that generated them together.
4. **Interactive Controls:** Live Scripts support interactive controls such as sliders, drop-downs, and buttons. These controls can be linked to variables in your script, allowing you to dynamically change the parameters and immediately see the results without re-running the entire script.
5. **Equation Editor:** An integrated equation editor allows you to include nicely formatted mathematical expressions using LaTeX syntax. This is particularly advantageous for teaching, academic research, and any situation where mathematical documentation is necessary.
6. **Ease of Sharing:** Live Scripts can be shared as is, or they can be exported to a variety of formats including PDF, HTML, and Microsoft Word. This makes it easy to share your findings with others who may not have MATLAB.
7. **Tool for Education:** Live Scripts are excellent educational tools. They allow educators to create interactive lessons that students can follow and modify themselves. This interactive component can enhance learning by allowing students to see the immediate impact of changes to parameters and code.
8. **Enhanced Debugging:** Live Scripts split the code into executable sections that can be run independently. This allows for easier debugging and testing of specific parts of a script without needing to run the whole script every time.

Figure 2.2 shows a demo of a live script that dynamically changes the period and amplitude of a sine wave using interactive controls and plots it immediately. It also has several code sections, some rich text, equations, and a picture. It also allows publication of this interactive education demo using pdf, HTML, Word, LaTeX, and other formats.

## 2.2 Advanced User Input Commands

### *2.2.1 Menu*

This advanced input command allows the user to create a GUI with a title and a selection of push buttons, and then it can assign the selected option to a variable. In the code below, the push buttons are defined in the cell array **clrs** and used as the second argument, while the title string is the first argument of the function. The selection is assigned as a number to the variable **val** (here, 1, 2, or 3) which can be used for further processing. The generated menu can be seen in Fig. 2.3.

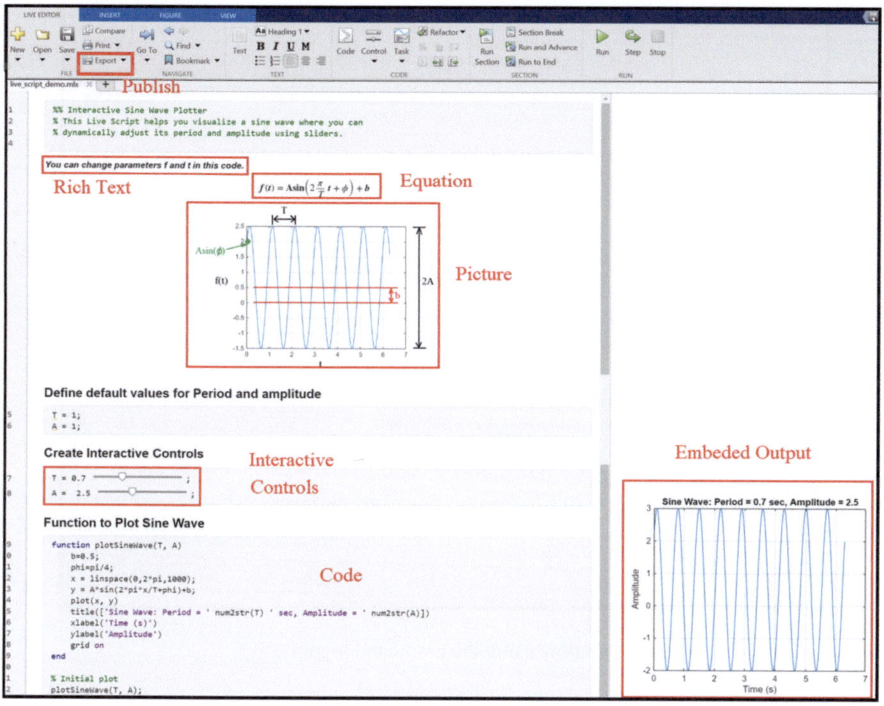


Fig. 2.2 Live script demo

**Fig. 2.3** Input GUI generated using the command menu

```
clrs={'red','yellow','gray'};
val=menu('color to choose',clrs);
disp('you chose');
disp(clrs{val})
```

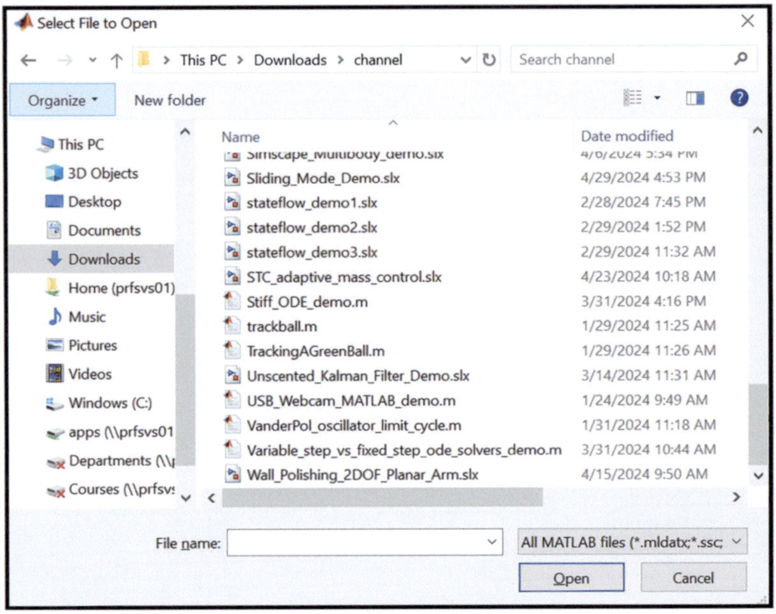


**Fig. 2.4** Generated browsing window using the command uigetfile

### *2.2.2 Uigetfile*

This advanced input command is used to create a GUI and browse for files. It returns the name of the file and its path as string variables. The code below creates a browsing window as shown in Fig. 2.4 so the user can find and click on a desired picture. Then MATLAB opens the selected image file.

```
[file,path]=uigetfile;
img=imread([path file]);
```

### *2.2.3 Ginput*

This advanced input command is used to obtain the coordinates of *n* points the user clicks on when a figure is open. It can be very useful to approximately recreate a plot when the raw data is not available and only the graph is accessible. In Fig. 2.5, **ginput** was used to record the locations of 20 clicks on the plot of the world population versus time, which was downloaded from the web (https://www.jdmoyer.com/2011/03/25/how-it-might-go-down/) and displayed in part (a). The following code used

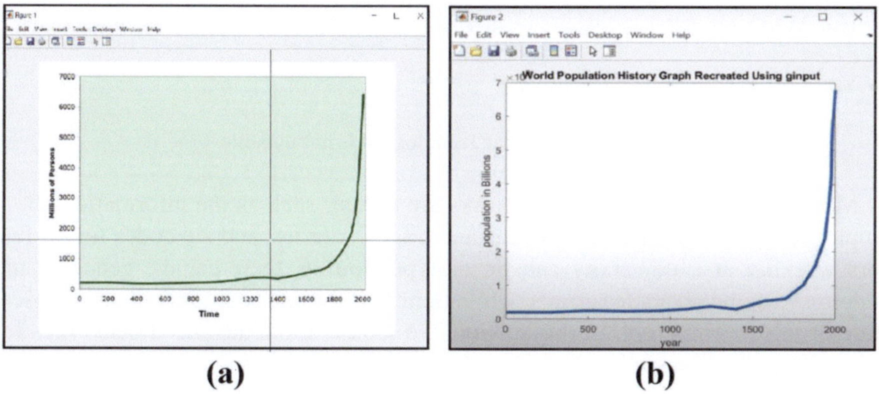


**Fig. 2.5** Recreated graph without access to raw data using command ginput **a** Original graph and clicking on the graph points; **b** Reconstructed graph

the **ginput** data to regenerate the population versus time plot, which can be seen in part (b):

```
[file,path]=uigetfile;
img=imread([path file]);
img=imresize(img,3,'bicubic');
imshow(img);
[x,y]=ginput(20);
xn=2000*(x(2:end)-x(1))/(x(end)-x(1));
yn=6.8e9*(y(2:end)-y(1))/(y(end)-y(1));
figure;
plot(xn,yn,'LineWidth',3);
xlabel('year');ylabel('population in Billions');
title('World Population History Graph Recreated Using ginput');
ax = gca;
ax.YGrid = 'on';
axis([0 2200 0 8e9]);
```

## 2.3 Advanced Data Structures (Cell Arrays, Structures, Tables)

While arrays are the basic data type in MATLAB and can be utilized for most routine tasks in programming, they have some limitations, such as the inability to store heterogenous types of data. Let's consider that you want to store an array (variable) containing a value of type **double**, one of type **logical**, and one of type **char** as shown below:

```
a=[3.2 'MATLAB' false]
```

You will get the following error:

```
Conversion to char from logical is not possible.
```

Many data encountered in daily tasks are mixed, such as the information about employees of an organization. For instance, ages, salaries, and zip codes for professors working at a university can be of type **double**, their names, genders, and addresses can be character arrays, while their tenure status could be of type **logical** (true or false). Therefore, in this section, we look at more advanced data types for different applications in MATLAB.

## *2.3.1 Cell Arrays*

Cell arrays can store a combination of numbers, strings, matrices, and even other cell arrays. Each element in a cell array is indexed by curly braces **{}**. This capability allows them to be used when a variable number of different inputs or outputs are used in functions (remember that **varargin** and **varargout** are cell arrays). Below are some commands related to creating and modifying cell arrays.

**b = cell(3,1):** creates a 3 × 1 cell array and assigns **{0 x 0 double}** to all of its entries.

**a = {3.2 6 'MATLAB' false}:** creates a 1 × 4 cell array and assigns the corresponding entries.

**a{1}:** accesses the content of the first entry in cell array **a** → ans = 3.2

**a(1):** accesses the first entry in cell array **a** as a cell:

```
ans = 1×1 cell array
      {[3.2000]}
```

**cell2mat(a(1)):** converts the first cell in cell array **a** to a matrix: ans = 3.2

**length(a):** gives the length of cell **a**, which is 4.

**a{1} = 3.1:** changes the first entry from its current value to 3.1

**a{1:2}:** accesses the first and second entries in cell array **a** but as two separate cells:

```
ans = 3.2000
ans = 6
```

**[a{1:2}]:** uses concatenation to merge the first and second entries of the cell array into one array:

```
ans =  3.2000    6.0000
```

**a{4} = []:** converts the fourth cell in **a** to **{0 × 0 double}** but does not delete it.
**a(4) = []:** deletes the fourth cell in **a**.
**2*a** or **2*b**: generates the error shown below:

```
Operator '*' is not supported for operands of type 'cell'.
```

**C1 = {2 True}**
**C2 = {3 6 'Hi'}**
**{c1 c2}:** merges cells **c1** and **c2** without merging their contents:

```
ans =  1×2 cell array
{1×2 cell}    {1×3 cell}
```

**[c1 c2]:** merges cells **c1** and **c2** and their contents:

```
ans =  1×5 cell array
{[2]}    {[1]}    {[3]}    {[6]}    {'Hi'}
```

While character arrays of different length cannot be concatenated vertically, this can be done using cell arrays:

```
str1=['ABC DE'];
str2=[str1; 'FG']
```

The commands above generate the following error, since **st1** has a length 6 and **'FG'** has a length of 2:

```
Dimensions of arrays being concatenated are not consistent.
```

**str2 = {str1; 'FG'}:** will generate a 2 × 1 cell array:

```
str2={'ABC DE'}
    {'FG'      }
```

### 2.3.2 *Structure Arrays*

Similar to cell arrays, structure arrays can store heterogenous data, but unlike in cell arrays, that data can only be stored and accessed via their indices. Similar categories of data are stored and accessed under the same branch of the structure, called **fields**.

Figure 2.6 shows a structure array called **Author** that has 5 fields (name, addr1, city, state, zip), four of which contain character arrays and one of which contains numbers.

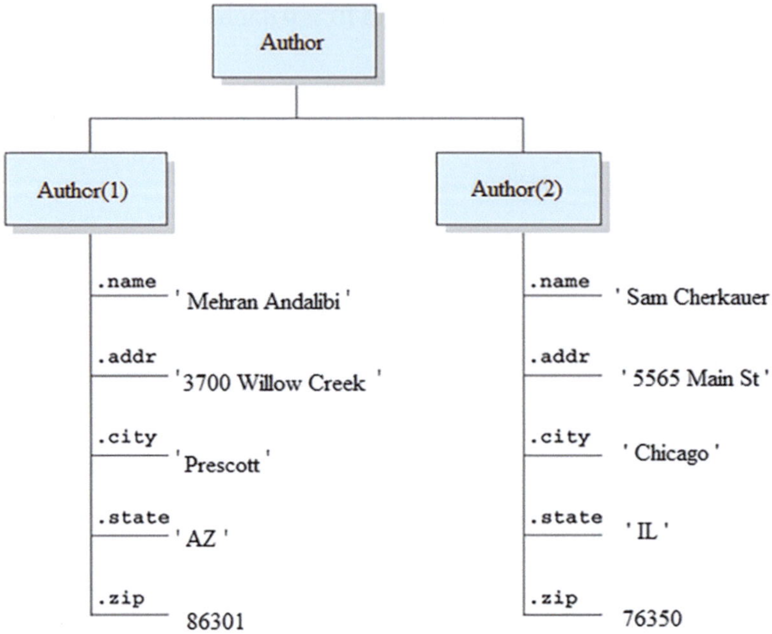


**Fig. 2.6** Structure array author with five fields

To create such a structure array, first we create the array of 2 structures with empty fields:

```
Author(2)=struct('name',[ ],'addr',[ ],'city',[ ],'state', [ ], 'zip',[ ]);
```

Then, we perform the data entry for all members:

```
Author(1).name='Mehran Andalibi';
Author(1).addr='3700 Willow Creek';
Author(1).city='Prescott';
Author(1).state='AZ';
Author(1).zip=86301;
```

```
Author(2).name='Sam Cherkauer';
Author(2).addr='5565 Main St';
Author(2).city='Chicago';
Author(2).state='IL';
Author(2).zip=76350;
```

As can be seen, fields are accessed using the dot operator (.).
Another way to set the value of a field for a structure is the command **setfield**:
**setfield(Author(1), 'name', 'Mehran Andalibi')**
To get the value stored in a field of a structure, there are two different methods:

```
Author(2).name

getfield(Author(2),'name')
```

To get the names of the fields (and the number of fields) for a structure, use the **fieldnames** command:

```
fieldnames(Author)
```

In this case, the **fieldnames** command returns a 5 × 1 cell array:

```
ans =  5×1 cell array
        {'name' }
        {'addr' }
        {'city' }
        {'state'}
        {'zip'  }
```

```
length(fieldnames(Author)):
```

```
5
```

To get the values stored in one field for all structures in a structure array use the following syntax:

```
Author.zip
```

Note that this command will create individual values and not a single matrix:

```
ans =  86301
ans =  76350
```

In order to combine all these values in a single matrix, use brackets and concatenation:

```
[Author.zip]
ans =  86301     76350
```

Note that this command will make all character arrays into a single character array:

```
[Author.city]
ans = 'PrescottChicago'
```

To remove a field from a structure, use the **rmfield** command and then reassign the result back to the same structure array:

```
Author=rmfield(Author,'state')
```

To add a field to the entire structure, use cells:

```
temp=num2cell([40 21])
[Author.age]=temp{:}
```

Note that you can always add new structures to the end of a structure array:

```
Author(3).name='John Doe'
```

### 2.3.3 *Tables*

Tables are a type of data structure specifically designed for storing tabular data. Each column can contain different types of data, but within each column, all elements must be of the same type. Tables are particularly useful for handling and analyzing structured data where each row represents an observation, and each column represents a variable. Therefore, tables are very similar to structure arrays, but they have more methods, especially for visualization, as will be shown:

```
Name = {'John Doe'; 'Jane Smith'; 'Emily Jones'};
Age = [28; 34; 40];
Height = [176; 163; 158];
Weight = [70; 60; 45];
Group = {'A'; 'B'; 'A'};
T = table(Name, Age, Height, Weight, Group);
disp(T);
```

Figure 2.7 shows the table generated by the above code.

Another advantage of using Tables is that their values can be accessed and modified both via the dot operator and indices.

| Name | Age | Height | Weight | Group |
|---|---|---|---|---|
| {'John Doe'   } | 28 | 176 | 70 | {'A'} |
| {'Jane Smith' } | 34 | 163 | 60 | {'B'} |
| {'Emily Jones'} | 40 | 158 | 45 | {'A'} |

**Fig. 2.7** Table array with five columns and 3 rows

```
% Access the age column
ages = T.Age;

% Access data for the first person
firstPerson = T(1,:);

% Modify the height for the second person
T.Height(2) = 165;

% Add a new column for BMI (Body Mass Index)
T.BMI = T.Weight ./ (T.Height/100).^2;
```

### *2.3.4 Categorical Arrays*

Categorical arrays are used to represent data that can take on one of a limited number of distinct values, known as **categories** or **levels**. These are particularly useful for representing data that falls into a fixed number of categories, such as gender, blood type, or experimental group.

To define a categorical array, we use the command **categorical.** Note that categories can be repeated.

```
% categorical 1D array
colors_cat = categorical({'red', 'blue', 'green', 'blue', 'red', 'green', 'green'})

% categorical matrix
colors_cat_mat = categorical({'red','blue';'pink','red';'blue','red'})
```

Categories of a categorical variable can be accessed using the command **categories:**

```
categories(colors_cat)
                                    ans =  3×1 cell array
                                          {'blue' }
                                          {'green'}
                                          {'red'  }
```

Entries of categorical arrays can be accessed, changed, and omitted similar to numerical arrays:

```
% Accessing the second entry
second_color = colors_cat(2)

% Modifying the third entry
colors_cat(3) = 'black';

% removing the second entry
colors_cat(end) = [ ]
```

Command **addcats** adds a new category to the categories of a categorical array:

```
colors_cat1 = addcats(colors_cat, 'yellow')
categories(colors_cat1)
                    ans = 5×1 cell array
                        {'blue'  }
                        {'green' }
                        {'red'   }
                        {'black' }
                        {'yellow'}
```

Command **removecats** removes an existing category from the categories of a categorical array and replace it with < **undefined** >:

```
colors_cat2 = removecats(colors_cat, 'green')
                    colors_cat2 = 1×7 categorical array
     red    blue    black    blue    red    <undefined>    <undefined>
```

Categories in a categorical array can be merged using the command **mergecats.** For instance, the second line converts any category named **'apple'** or **'banana'** into a merged category called **'fruit'**:

```
fruits = categorical({'apple', 'banana', 'orange', 'apple', 'grape', 'banana', 'orange'});
fruits = mergecats(fruits, {'apple', 'banana'}, 'fruit')
                    fruits =  1×7 categorical array
          fruit    fruit    orange    fruit    grape    fruit    orange
```

- Advantages of Categorical Variables:

1. **Memory Efficiency**: Categorical variables store data in a memory-efficient way, especially when there are many repetitions of the same category.

```
colors_cat = categorical({'red', 'blue', 'green', 'blue', 'red', 'green', 'green'});
colors_cell={'red', 'blue', 'green', 'blue', 'red', 'green', 'green'};
whos
```

Running the code above shows that for storing the same data, a categorical array only required 345 Bytes, while a cell array needs 786 Bytes:

```
Name            Size        Bytes  Class        Attributes
  colors_cat      1x7          345        categorical
colors_cell     1x7        786          cell
```

2. **Improved Readability:** They make the data more readable and interpretable, particularly in cases where numerical coding of categories (e.g., 1 for Male, 2 for Female) would be confusing.
   When plotting, categorical variables can be more legible and easier to understand than numerical values. For example, labeling the x axis as groups A through D rather than [1,2,3,4]. In the following code, the average age on the y axis is numerical, but the group number for the x axis is categorical, so we use a categorical array in the bar plot shown in Fig. 2.8.

```
categories = categorical({'Group A', 'Group B', 'Group C', 'Group D'});
average_age = [23, 45, 12, 67];
bar(categories, average_age);
xlabel('Groups');
ylabel('average age');
title('Bar Chart Example with Categorical Data');
```

3. **Facilitates Statistical Analysis**: Many statistical functions in MATLAB are designed to work with categorical data directly, which can simplify analyses that involve grouping and classification. One such command is **crosstab**. which

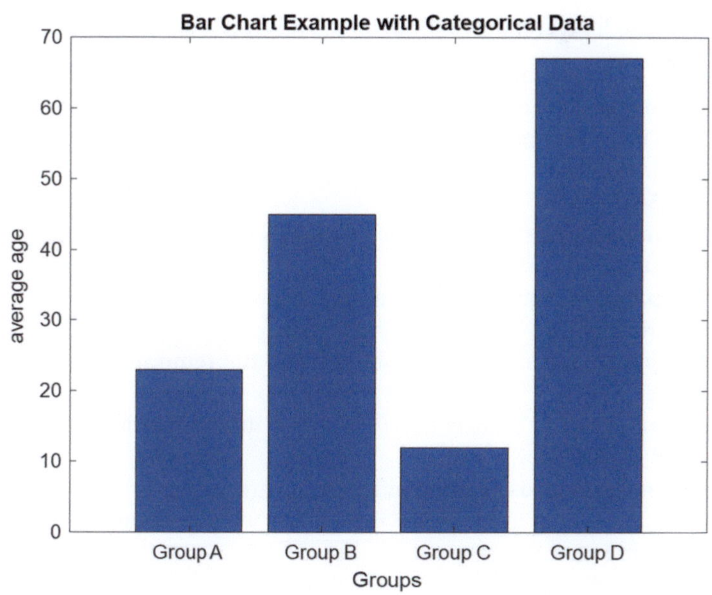


**Fig. 2.8** Using categorical arrays in plots

is used to create a cross tabulation among two variables. In the code below, both variables used are categorical (gender and drink type).

```
gender = categorical({ 'Male', 'Female', 'Female', 'Male', 'Female', 'Male', 'Male', 'Female',
'Male', 'Female'});
preference = categorical({'Tea', 'Coffee', 'Tea', 'Coffee', 'Tea', 'Tea', 'Coffee', 'Coffee',
'Tea', 'Tea'});

[table, chi2, p] = crosstab(gender, preference);
```

4. **Handling Missing Data:** MATLAB provides tools to handle missing data within categorical arrays, making them robust for data preprocessing.
   Here, we remove the missing data, i.e., **< undefined >** using the command **rmmissing**. We also use the command **isundefined** to find the indices of missing data and replace them with a category of **'purple'**.

```
colors_cat3=colors_cat2
colors_cat3 = rmmissing(colors_cat3) % removes the <undefined> entries
missingIdx = isundefined(colors_cat2) % returns a logical array
colors_cat2(missingIdx) = 'purple' % replaces the <undefined> entries with 'purple'
```

## 2.4 File I/O

In this section, we look at input and output to and from different types of files:

(a) **Mat files:** These are binary encoded files generated in MATLAB to store data, such as workspace variables. Commands **save** and **load** are used to write to and read from these files according to the following syntax:
    **X = '3.141592';**
    **Y = 250;**
    **save temp.mat**: saves all workspace variables to the mat file called **temp.mat** in the current directory.
    **save temp.mat 'x'**: saves only variable **x** to the mat file called **temp.mat** in the current directory.
    **load temp.mat**: loads all data inside the mat file **temp.mat** to the MATLAB workspace.
    **load temp.mat 'y'**: loads only variable **y** inside the mat file **temp.mat** to MATLAB workspace.
(b) **Excel files:** Commands **xlsread** and **xlswrite** are used to write to and read from Excel files according to the following syntax, with illustrations shown in parts (a) and (b) in Fig. 2.9:

| | A | B | C |
|---|---|---|---|
| 1 | x | y | correlation coefficient |
| 2 | 1 | 2 | |
| 3 | 4 | 9 | |
| 4 | 5 | 0 | |
| 5 | 7 | 6 | |
| 6 | 8 | 3 | |

**(a)**

| | A | B | C |
|---|---|---|---|
| 1 | x | y | correlation coefficient |
| 2 | 1 | 2 | 0.103279556 |
| 3 | 4 | 9 | |
| 4 | 5 | 0 | |
| 5 | 7 | 6 | |
| 6 | 8 | 3 | |

**(b)**

**Fig. 2.9** **a** Original excel file; **b** Modified excel file

```
myData = xlsread('mydata.xlsx','Sheet1','a2:b6');
x=myData(:,1);
y=myData(:,2);
z=corrcoef(x,y)
xlswrite('mydata.xlsx',z(1,2),'Sheet1','c2')
```

(c) **Text files:** Commands **textscan** and **fprintf** are used to write to and read from text files. First, the text file must be opened using the command **fopen**, which requires a file name and file access permission.

**[fileID,message] = fopen('mydata.txt', 'r'):** opens the file **mydata.txt** shown in Fig. 2.10a for reading only. If opening the file is successful, **fileID** is a going to be a positive integer and the **message** will be empty; otherwise, **fileID** will be -1 and **message** will contain the opening error message.

– If the file does not exist, **fopen** will create one.

Table 2.4 shows different access permissions for **fopen**.

Once open, the command **textscan** reads the file's formatted data. **textscan** requires a reading format for different columns of data and the **delimiter** that separates these columns:

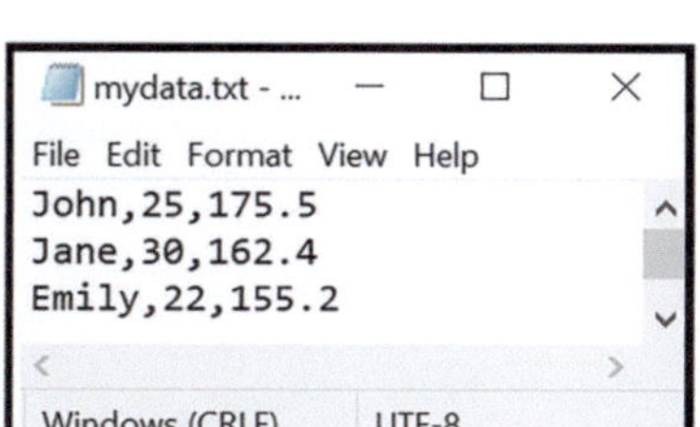


**(a)**

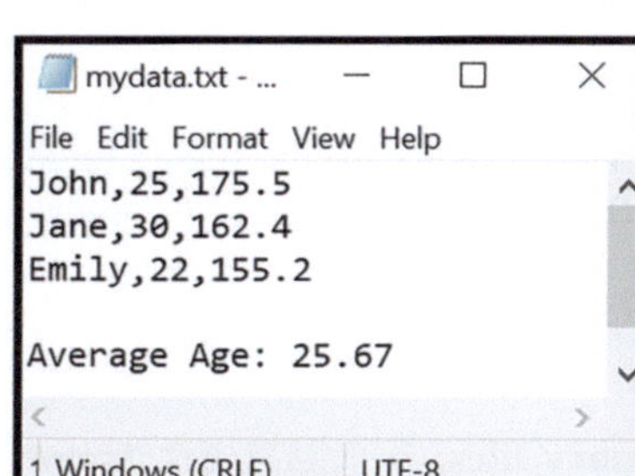


**(b)**

**Fig. 2.10** **a** Original text file; **b** modified text file

**Table 2.4** Access permission for fopen

| Permission | Description |
|---|---|
| 'r' | Open file for reading |
| 'w' | Open or create new file for writing. Discard existing contents, if any |
| 'a' | Open or create new file for writing. Append data to the end of the file |
| 'r+' | Open file for reading and writing |
| 'w+' | Open or create new file for reading and writing. Discard existing contents, if any |
| 'a+' | Open or create new file for reading and writing. Append data to the end of the file |
| 'A' | Open file for appending without automatic flushing of the current output buffer |
| 'W' | Open file for writing without automatic flushing of the current output buffer |

```
% reading format: string (name), integer (age), floating-point number (height)
formatSpec = '%s %d %f';
data = textscan(fileID, formatSpec, 'Delimiter', ',');
ages = data{2};
averageAge = mean(ages);
```

Finally, the file must be closed to avoid corrupting it:

```
fclose(fileID);
```

To write to the text file, we can use the **fprintf** command. First, the file has to be reopened for appending:

```
fileID = fopen('mydata.txt', 'a');
```

Next, the variable **averageAge** will be appended to the end of the text file with the **fprintf** command shown below. The cursor is moved to the beginning of the next line using **'\n'**:

```
fprintf(fileID, '\nAverage Age: %.2f\n', averageAge);
```

The modified text file is shown in Fig. 2.10b.

Don't forget to close the file again after writing the text to it. The variable status will be 0 if closing is successful; otherwise, it will be –1.

```
status=fclose(fileID);
```

(d) **Binary files:** In MATLAB, **fread** and **fwrite** are functions used for reading from and writing to binary files, respectively. These functions are particularly useful when you need to handle data in formats other than simple text, such as custom binary formats.

The code below, opens/creates the binary file **'test.txt'** with writing permission and writes the matrix **A** to it using **int8** format. Variable **count** returns the number of values written to the file (here 6):

```
A=[1 2;3 4;5 6]
fid=fopen('test.txt','w');
count=fwrite(fid,A,'int8')
fclose(fid)
```

The code below, opens the previous binary file with reading permission and reads from it 6 values using **int8** format. The code then assigns the values to the variable **Anew** and the number of successfully read values to the variable **count**:

```
fid=fopen('test.txt','r');
[Anew,count]=fread(fid,6,'int8')
fclose(fid)
```

(e) **Tables from text, Spreadsheets, HTML, or Word files:** Command **readtable** can be used to read the formatted data from a text or dat file, a csv or an Excel file, HTML, or even Microsoft Word file, and convert it to a table. Below are examples of **readtable** used with different file formats and different reading options:

**T = readtable("mydata.dat"):** reads the data in the **mydata.dat** file.

**T = readtable("mydata.txt",MissingRule = "omitrow"):** reads the data in **mydata.txt** and omits any row of data that has missing values.

**T = readtable("mydata.xlsx",Range = "a2:b6",ReadVariableNames = false):** reads the data from the Excel file but does not read the names of the variables/columns. Instead, the table uses substitute names such as **var1**, **var2**, etc.

**opts = detectImportOptions("mydata.csv"):** detect the import options of the csv file as shown in Fig. 2.11.

**opts.SelectedVariableNames = ["x","y"];** selecting only **x** and **y** variables to import.

**T = readtable("Mydata.csv",opts)**

**T = readtable("mydata.docx",TableIndex = 2):** reads the second table from the word document **"mydata.docx"**.

The following code reads the table in the given url on MathWorks.com that contains the word 'readtable':

```
url = "https://www.mathworks.com/help/MATLAB/text-files.html";
T = readtable(url,TableSelector="//TABLE[contains(.,'readtable')]", ...
ReadVariableNames=false)
```

```
opts =

  DelimitedTextImportOptions with properties:

   Format Properties:
                    Delimiter: {','}
                   Whitespace: '\b\t '
                   LineEnding: {'\n'  '\r'  '\r\n'}
                 CommentStyle: {}
    ConsecutiveDelimitersRule: 'split'
        LeadingDelimitersRule: 'keep'
       TrailingDelimitersRule: 'ignore'
                EmptyLineRule: 'skip'
                     Encoding: 'UTF-8'

   Replacement Properties:
                  MissingRule: 'fill'
              ImportErrorRule: 'fill'
             ExtraColumnsRule: 'addvars'

   Variable Import Properties: Set types by name using setvartype
                VariableNames: {'x', 'y', 'z'}
                VariableTypes: {'double', 'double', 'double'}
        SelectedVariableNames: {'x', 'y', 'z'}
              VariableOptions: Show all 3 VariableOptions
    Access VariableOptions sub-properties using setvaropts/getvaropts
           VariableNamingRule: 'modify'

   Location Properties:
                    DataLines: [2 Inf]
            VariableNamesLine: 1
               RowNamesColumn: 0
            VariableUnitsLine: 0
     VariableDescriptionsLine: 0
```

**Fig. 2.11** Import options of a csv file for readtable

## 2.5 Code Optimization Techniques and Timing

In this section, we look at techniques for improving the performance of the codes written in MATLAB, including reduction in execution time and memory allocation.

### *2.5.1 Vectorization*

Vectorization is a powerful technique in MATLAB for improving the performance of code by replacing explicit loops with matrix and vector operations. MATLAB is optimized for operations on arrays, making vectorized code generally faster and more concise than equivalent code using loops.

Let's say we want to perform an operation on a 1000 × 1000 matrix of integer numbers between 0 and 10 and those numbers bigger than or equal 6 will be set to 6 (e.g., simulating a saturation operation). First, we do it using **nested for loops** and use the **tic** and **toc** commands to find the run time.

- Since the run time is variable, we run the code 100 times and report the average to provide a better estimate of the run time:

```
for kk=1:100
A=randi(10,[1000 1000]);
        tic;
        for ii=1:1000
                for jj=1:1000
                        if A(ii,jj)>=6
                                A(ii,jj)=6;
                        end
                end
        end
        t(kk)=toc;
end
disp(mean(t))
```

The average run time for nested for loop is about 0.0178 s.

Next, we use logical arrays and vectorization to perform the same task:

```
for kk=1:100
        A=randi(10,[1000 1000]);
        tic;
        A(A>=6)=6;
        t(kk)=toc;
end
disp(mean(t))
```

The average run time for vectorization is about 0.0119 s which indicates a 33% reduction.

Note that the main line that performs vectorization **A(A >= 6) = 6;** and elegantly replaces the nested for loops with a single line of code is made of three steps:

First, a logical array is created using **A >= 6** which will be a logical matrix with the same size as **A**, but with values of 1 where $a_{ij \geq 6}$ and a value of 0 otherwise.

Next, by passing the logical matrix made above as the argument to the matrix **A**, i.e., **A(A >= 6)**, we target those entries in A that qualify for the specified conditions.

Finally, we assign a value of 6 to those entries by **(A >= 6) = 6**.

As a second example, let's work on the following practice problem:

Write a MATLAB code to calculate the following multi-criteria function for $-6\pi \leq t \leq 6\pi$ at intervals of $\pi/1000$ once with and once without Vectorization and compare the running times using **tic** and **toc**. Run each part 1000 times and report the average run time.

$$f(t) = \begin{cases} \sin(t) & \textit{if } \sin(t)>0 \\ 0 & \textit{others} \end{cases} \tag{2.1}$$

```
t=-6*pi:pi/1000:6*pi;
for kk=1:1000
        f=sin(t);
        tic;
        for ii=1:length(t)
                if f(ii)<=0
                        f(ii)=0;
                end
        end
        tm(kk)=toc;
end
disp(mean(tm))
```

The average run time is 41.2 microseconds, while for the vectorization method shown below it is 33.2 microseconds, which means a 20% reduction in the run time.

```
for kk=1:1000
```

```
        f=sin(t);
        tic;
        f(f<=0)=0;
        tm(kk)=toc;
end
disp(mean(tm))
```

### 2.5.2 Pre-allocation

Pre-allocation refers to the practice of allocating memory for an array or other data structure before entering a loop or a series of operations where the size of the data structure is known in advance. This technique can dramatically improve performance by reducing the number of memory allocation operations that MATLAB needs to perform as the array grows.

To demonstrate what happens without pre-allocation, first we create an empty array and keep filling it with the squares of positive integers from 1 to 1000, where the size of the array increases each time in the **for** loop:

```
tic;
N = 10000; % Size of the array
results = [ ]; % Start with an empty array
for ii = 1:N
        results(ii) = ii^2; % Append the square of ii to the array
end
toc
```

The average run time without pre-allocation was 0.011 s.

Next, since we know in advance that the size of the array results is going to be $1000 \times 1$, we create an array of this size with all zeros first, and then fill its entries with the squared values. This way, only computations for squares and assignment takes place and the size of the results array does not change in the loop, which saves run time:

```
tic;
N = 10000; % Size of the array
results = zeros(N, 1); % Pre-allocate a column vector of size N
        for ii = 1:N
results(ii) = ii^2; % Insert the square of ii into the pre-allocated array
end
toc
```

The average run time with pre-allocation was 0.0044 s which means a 60% reduction in the run time.

### 2.5.3 *Avoiding Redundant Calculations*

Redundant calculations are some of the most common mistakes when optimizing code, but can be avoided by moving calculations that do not change within a loop outside the loop.

Let's consider you want to create an array that contains the values of $f_k = \frac{\cos(\pi)}{k^2}$. Since the numerator does not change with **k**, it is not a wise idea to keep calculating it in the **for** loop:

```
N = 10000;
result = zeros(N,1);
for k = 1:N
        result(k) = cos(pi)/(k^2); % Compute numerator each time inside the loop
end
```

It is appropriate to calculate the numerator once outside the loop and use its value in the loop each time:

```
N = 10000;
result = zeros(N,1);
num = cos(pi); % Compute once outside the loop
for k = 1:N
        result(k) = num/(k^2);
end
```

### 2.5.4 *Using Built-In Functions*

Utilize MATLAB's built-in functions whenever possible, as they are highly optimized.

### 2.5.5 *Using MALTAB Profiler Tool*

Use this tool to identify bottlenecks in your code. Focus your optimization efforts on the parts of the code where they will have the most impact. The profiler tool can be found in the middle of the Editor tab.

The results of running the profiler tool on the code with redundant **for** loop calculations can be seen in Fig. 2.12.

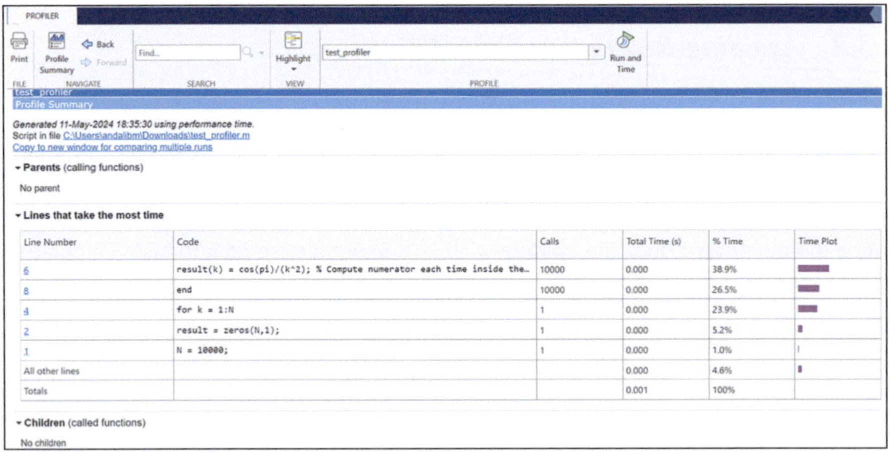


**Fig. 2.12** Running the profiler tool on a sample code

### 2.5.6 Minimize Use of Global Variables

Global variables can slow down MATLAB code because accessing global memory is slower than accessing local memory. Try to pass variables as arguments to functions instead of using global variables.

### 2.5.7 Function Handles and Anonymous Functions

Use function handles and anonymous functions judiciously; they can be slower than inline code in a script or within a loop for simple operations.

### 2.5.8 Using Appropriate Data Types

For instance, if you do not need the precision of **double** (which is MATLAB's default), consider using **single** or **integer** types to save memory and possibly speed up computations.

### 2.5.9 Using Parallel for Loops

Although this part is not about optimizing your code, it can reduce the run time by taking advantage of multi-core processors that can be found on most computers these days.

Note that, regularly, MATLAB only allocates one of your computer cores to perform the computations in a **for** loop. However, you can distribute the computations among all your computer cores. To use the parallel **for** loops (**parfor**), some conditions must be satisfied first:

(a) Parallel Computing Toolbox must be installed
(b) No loop iteration can depend on previous iterations, as iterations are performed in a nondeterministic manner.

To demonstrate this, let's look at a heavy computation for calculating the condition number of a system of linear equations with the coefficient matrix being a $400 \times 400$ matrix of random numbers. The condition number is the ratio of the maximum and minimum absolute values of eigenvalues of the coefficient matrix. We perform the operations 200 times and report the total run time.

First, we do it with a regular **for** loop:

```
tic;
n=200;
A=400;
a=zeros(1,n);
for ii=1:n
        res=abs(eig(rand(A)));
        a(ii)=max(res)/min(res);
end
toc
```

The total run time was about 23.67 s.

Next, we do it using the **parfor** command:

```
tic;
n=200;
A=400;
a=zeros(1,n);
parfor ii=1:n
        res=abs(eig(rand(A)));
        a(ii)=max(res)/min(res);
end
toc
```

The total run time was about 9.54 s which indicates a 60% reduction in the run time.

## 2.6 Operating System Commands

Using the commands below, we can create folders, delete folders, change the current folder, move, copy, and paste files, find the contents of a folder, and open files.

**d = dir:** returns a structure for the current folder with 6 fields, including the **name** (names of all folders and files), **folder** (full address of the current folder), **date** (date of last modification formatted as Day/Month/Year Hour:Minutes:Seconds), **bytes** (size of contents), **isdir** (logical variable indicating whether each component is a folder), and **datenum** (date of last modification as a serial data number).

```
d =  27×1 struct array with fields:

    name
    folder
    date
    bytes
    isdir
    datenum
```

**d = dir myfolder:** returns all the information mentioned above but for the folder named **"myfolder"**.

**sum([d.bytes]):** returns the total size of all files in the folder.

**dir *.m:** returns the information for all **.m** files.

**dir my*.*:** returns all files and folders that start with the first two letters **"my"**.

**datetime(d(1).datenum,ConvertFrom = "datenum"):** returns the date of last modification for the first subfolder.

**mkdir myfolder:** creates a subfolder named **"myfolder"** in the current folder.

**rmdir myfolder s:** removes the folder named **"myfolder"** and all its contents regardless of write permissions.

**cd newfolder:** changes the current folder to the folder named **"newfolder"**.

**cd:** displays the full address of the current folder.

**cd..:** goes up one level (folder).

**movefile file1.m mfiles:** moves the file **"file1.m"** from the current folder to the folder **"mfiles"**.

**copyfile file1.m mfiles:** copies the file **"file1.m"** from the current folder to the folder **"mfiles"**.

**delete *.mat:** deletes all **.mat** files from the current folder.

**winopen myfile:** opens the file **"myfile"** in the appropriate windows application.

## 2.7 Peripherals Control Commands

To control (simulate) the mouse, keyboard, and take snapshots from the screen, we need to use the **java.awt.Robot** class. This class is used to generate native system input events for the purposes of test automation, self-running demos, and other applications where control of the mouse and keyboard is needed.

First, we need to import the **robot** and **event.*** from this library and then create a **robot** object:

```
import java.awt.Robot;
import java.awt.event.*;
robot = java.awt.Robot;
```

Next, we can get the location of the cursor/mouse and move it to a desired location:

**loc = get(0,'PointerLocation'):** returns the (x, y) pixel location of the mouse.

**set(0,'PointerLocation',[600 300]):** moves the cursor pixel location with respect to the top left corner.

**robot.mouseMove(600,300):** moves the cursor with respect to the current location.

We can also simulate pressing the mouse right-click, and we can move the cursor down the options of the contextual menu:

```
robot.mousePress(InputEvent.BUTTON3_MASK);
robot.mouseRelease(InputEvent.BUTTON3_MASK);
loc=get(0,'PointerLocation');
set(0,'PointerLocation',[loc(1)-50 loc(2)]);
```

Then, we can use the mouse left-click to select an option:

```
robot.mousePress(InputEvent.BUTTON1_MASK);
robot.mouseRelease(InputEvent.BUTTON1_MASK);
```

Middle mouse button can also be simulated:

```
robot.mousePress(InputEvent.BUTTON2_MASK);
robot.mouseRelease(InputEvent.BUTTON2_MASK);
```

To take snapshots from the screen (similar to **prtsc**), we first need to specify the size and location of the capture rectangle:

```
pos = [0 0 1800 1000];   % [left top width height]
rect = java.awt.Rectangle(pos(1),pos(2),pos(3),pos(4));
cap = robot.createScreenCapture(rect);  % cap is a buffered image
```

To convert the buffered image **"cap"** to an RGB image, we need to perform a type-casting and matrix reshaping, and then we can show the resulting image:

```
rgb = typecast(cap.getRGB(0,0,cap.getWidth,cap.getHeight,[],0,cap.getWidth),'uint8');
imgData = zeros(cap.getHeight,cap.getWidth,3,'uint8');
imgData(:,:,1) = reshape(rgb(3:4:end),cap.getWidth,[])';
imgData(:,:,2) = reshape(rgb(2:4:end),cap.getWidth,[])';
imgData(:,:,3) = reshape(rgb(1:4:end),cap.getWidth,[])';

imshow(imgData)
```

Finally, to simulate the press and release of other keys on the keyboard, we can use the methods **keyPress** and **keyRelease** of the **robot** object. The first two lines below are the same as pressing the lowercase **"a"**:

```
robot.keyPress (java.awt.event.KeyEvent.VK_A);
robot.keyRelease (java.awt.event.KeyEvent.VK_A);
```

The 4 lines below are equivalent to typing the uppercase **"M"** using the shift key:

```
robot.keyPress (java.awt.event.KeyEvent.VK_SHIFT);
robot.keyPress (java.awt.event.KeyEvent.VK_M);
robot.keyRelease (java.awt.event.KeyEvent.VK_M);
robot.keyRelease (java.awt.event.KeyEvent.VK_SHIFT);
```

## 2.8 Techniques for Hiding Source Code

Some common techniques for hiding or obfuscating source code in MATLAB are:

1. **P-Coding**: MATLAB's **pcode** function converts **.m** files into obfuscated, binary **.p** files. These files can be executed in MATLAB but are not human-readable.
   **pcode('myfile.m'):** creates a **.p** file (**'myfile.p'**) from the **.m** file **'myfile.m'** which can be executed similar to the source **.m** file but cannot be opened or seen. The source file can then be removed.
2. **Compiling to Executables or Libraries:** Use MATLAB Compiler or MATLAB Compiler SDK to compile MATLAB code into standalone executables, shared libraries, or software components that can be integrated with other applications.
   **mcc -m myfile.m:** This generates an executable file which can run independently of MATLAB. The source file can then be removed.
3. **Packaging as Apps**: Package your code as MATLAB Apps using the MATLAB App Designer. Apps can be distributed as **.mlappinstall** files, which can be installed and run in MATLAB without exposing the underlying source code.
4. **Using third-party Encrypting or Obfuscation tools:** MATLAB does not have built-in encryption or obfuscation tools, so third-party tools can be used, but note that they might rename variables and make some unwanted changes in the code.

## 2.9 MEX Files

**MEX** files (**MATLAB Executable** files) in MATLAB are functions written in C, C++, or Fortran that are compiled into binary shared libraries. These files have a **.mex** extension, which varies depending on the operating system (e.g., **.mexw64** for Windows, **.mexa64** for Linux, **.mexmaci64** for MacOS). Once compiled, MEX files can be called directly from MATLAB as if they were built-in functions or scripts.

Applications of MEX Files in MATLAB:

i. **Performance Optimization:** MEX files are often used to accelerate computationally intensive operations by leveraging the speed of compiled languages like C/C++. For instance, if a MATLAB function is slow due to heavy looping or complex calculations, rewriting that function in C/C++ as a MEX file can significantly boost performance.
ii. **Integration with External Libraries:** MEX files allow MATLAB to interface with external libraries written in C/C++ or Fortran. This is particularly useful for leveraging existing codebases or functionalities not natively available in MATLAB.
iii. **Custom Algorithms:** Users can implement custom algorithms that are not easily achievable with MATLAB's built-in functions. By writing these algorithms in C/C++ and integrating them with MATLAB through MEX files, users can create highly specialized functions.

iv. **Memory Management:** MEX files can be used to handle large datasets more efficiently by managing memory explicitly, which can sometimes be a limitation when using MATLAB alone.
v. **Access to Hardware:** MEX files can be used to interact with hardware components directly, such as for custom data acquisition or control applications, where MATLAB's native functions might not provide direct access.
vi. **Parallel and Multithreaded Computing:** By writing MEX files, users can create multithreaded applications or use parallel computing techniques not directly supported by MATLAB, thereby improving the performance of operations that can benefit from concurrency.

A MEX file in MATLAB consists of several key components that work together to allow the execution of compiled C, C++, or Fortran code within the MATLAB environment.

Here are the main components:

I. **Source Code File (C, C++, or Fortran)**

- **C/C++**: The source code is typically written in C or C++, with functions that MATLAB will call. The main entry point for a MEX file is a function called **mexFunction**, which acts as the **gateway** between MATLAB and the compiled code.
- **Fortran**: Similarly, if using Fortran, the entry point function is called **mexFunction**.

II. **Header Files**

- For C/C++ MEX files, you need to include the **mex.h** header file. This header provides the necessary definitions, data types, and function prototypes needed to interface with MATLAB. You also need to include the **mexAdapter.hpp** file:

```
# include "mex.hpp"
# include "mexAdapter.hpp"
```

III. **The MexFunction Entry Point**

- This is the core function of the MEX file. It is a special function that MATLAB calls when the MEX file is invoked. The signature of the mexFunction in C/C++ is as follows:

```
void mexFunction(int nlhs, mxArray *plhs[], int nrhs, const mxArray *prhs[])
```

Arguments:

- **nlhs:** Number of left-hand side (output) arguments.
- **plhs[]:** Array of pointers to the left-hand side (output) arguments.
- **nrhs:** Number of right-hand side (input) arguments.

- **prhs[]:** Array of pointers to the right-hand side (input) arguments.

Note that input parameters (found in the **prhs** array) are read-only; do not modify them in your MEX file.

IV. **mxArray Data Type**

MEX functions operate on MATLAB data through the **mxArray** * type, which is a structure representing MATLAB arrays. You use **mxArray** pointers to access data passed between MATLAB and the MEX function. The **mxArray** is used for input and output arguments. The keyword **const**, which modifies **prhs**, means that your MEX file does not modify the input arguments.

V. **Managing Input and Output Parameters**

You can manipulate **mxArray** data using various utility functions provided by the MATLAB API, such as:

- **mxGetPr**: Get a pointer to the real data of a double array.
- **mxCreateDoubleMatrix**: Create a double-precision matrix.

To copy an input array into a locally defined variable, **myData**, call the **mxDuplicateArray** function to make of copy of the input array. For example:
**mxArray *MyData = mxCreateStructMatrix(1,1,nfields,fnames);**
**mxSetField(MyData,0,"myFieldName",mxDuplicateArray(Prhs[0]));**

VI. **Error Handling**

- MEX files can include error handling using the **mexErrMsgIdAndTxt** or **mexErrMsgTxt** functions to throw errors back to MATLAB if something goes wrong during execution.

```
if (error_condition) {
  mexErrMsgIdAndTxt("MATLAB:myMEXFile:error", "An error occurred.");
}
```

Figure 2.13 shows a MEX function that takes two input arguments, a scalar and a vector, multiplies the vector entries by the scalar, and returns the result. As can be seen, writing MEX files is more complicated than MATLAB files and the **computational routine** is only a small portion of the file (depicted in the red box). A large part of the code in a MEX file is for data and error handling.

To use MEX files, the first step is to use the MEX setup:

```
mex -setup
```

This command will set up the compiler for C++ or Fortran that can work with MATLAB. If the compiler is found (here **MinGW64**), you will get the following messages on the command window:

```
MEX configured to use 'MinGW64 Compiler (C)' for C language compilation.
```

```
class MexFunction : public matlab::mex::Function {
public:
    void operator()(matlab::mex::ArgumentList outputs, matlab::mex::ArgumentList inputs) {
        checkArguments(outputs, inputs);
        double multiplier = inputs[0][0];
        matlab::data::TypedArray<double> in = std::move(inputs[1]);
        arrayProduct(in, multiplier);
        outputs[0] = std::move(in);
    }

    void arrayProduct(matlab::data::TypedArray<double>& inMatrix, double multiplier) {

        for (auto& elem : inMatrix) {
            elem *= multiplier;
        }
    }

    void checkArguments(matlab::mex::ArgumentList outputs, matlab::mex::ArgumentList inputs) {
        std::shared_ptr<matlab::engine::MATLABEngine> matlabPtr = getEngine();
        matlab::data::ArrayFactory factory;

        if (inputs.size() != 2) {
            matlabPtr->feval(u"error",
                0, std::vector<matlab::data::Array>({ factory.createScalar("Two inputs required") }));
        }

        if (inputs[0].getNumberOfElements() != 1) {
            matlabPtr->feval(u"error",
                0, std::vector<matlab::data::Array>({ factory.createScalar("Input multiplier must be a scalar") }));
        }

        if (inputs[0].getType() != matlab::data::ArrayType::DOUBLE ||
            inputs[0].getType() == matlab::data::ArrayType::COMPLEX_DOUBLE) {
            matlabPtr->feval(u"error",
                0, std::vector<matlab::data::Array>({ factory.createScalar("Input multiplier must be a noncomplex scalar double") }));
        }

        if (inputs[1].getType() != matlab::data::ArrayType::DOUBLE ||
            inputs[1].getType() == matlab::data::ArrayType::COMPLEX_DOUBLE) {
            matlabPtr->feval(u"error",
                0, std::vector<matlab::data::Array>({ factory.createScalar("Input matrix must be type double") }));
        }

        if (inputs[1].getDimensions().size() != 2) {
            matlabPtr->feval(u"error",
                0, std::vector<matlab::data::Array>({ factory.createScalar("Input must be m-by-n dimension") }));
        }
    }
};
```

**Fig. 2.13** Structure of a C MEX function that multiplies a vector by a scalar

**To choose a different language, select one from the following:**

**mex -setup C++**

**mex -setup FORTRAN**

Now, we can compile a file, such as the C++ file **arrayProduct.cpp** using the **mex** command:

**mex arrayProduct.cpp**

If building the MEX file is successful, the following messages will appear in the command window:

**Building with 'MinGW64 Compiler (C++)'.**

**MEX completed successfully.**

The following code will generate a binary executable file titled **arrayProduct.mexw64** on a 64-bit Windows operating system. This file can be called and used directly from the command window:

```
a=-5;
b=[1 3 5 6 0 9 -4 -3 -10 -1]
c=arrayProduct(a,b)
```

The result shown on the command window is the proper product of ***a*** and ***b***:

```
c =   -5  -15  -25  -30    0  -45   20   15   50    5
```

If inside a **for** loop, you can use this MEX file to carry several array products. Then do the same task using the simple * operation in MATLAB and time both operations. You will find that the MATLAB built-in * works faster even in a **for** loop, which might be different from your expectations (you probably heard that MATLAB is slow in loops and C++ is faster than MATLAB):

```
a=-5;
b=[1 3 5 6 0 9 -4 -3 -10 -1];
tic;
for ii=1:1000
        b=arrayProduct(a,b);
end
toc;
```

```
a=-5;
b=[1 3 5 6 0 9 -4 -3 -10 -1];
tic;
for ii=1:1000
        b=a*b;
end
toc;
```

The run times you will get will be different each time if you run this code several times, but each time the MATLAB built-in operator will be faster:

```
Elapsed time is 0.048544 seconds.

Elapsed time is 0.010277 seconds.
```

So, **MEX files do not always make your MATLAB code run faster**. The reason is that most MATLAB functions are simple **wrappers** for the code written in C++

or Fortran, meaning that these MATLAB functions use compiled code and not interpreted code (**.m** file). Wrapper libraries consist of a thin layer of code which translates a library's existing interface into a compatible interface.

**Practice Problems**

1. Write a script that generates a detailed financial report for a business that sells multiple products. The report will display product IDs, units sold, unit prices, revenue, discount percentage, discount amount, and final revenue after applying the discount. Additionally, the report will calculate and display the overall total revenue before and after discounts at the end.

**Inputs**:

- Define the product IDs as a vector of integers.
- Define the number of units sold as a vector of integers.
- Define the price per unit as a vector of floating-point numbers.
- Define a discount percentage for each product as a vector of floating-point numbers.

**Output**:

- Use **fprintf** to display the data in a well-formatted table with aligned columns.
- Format the price, revenue, discount percentage, discount amount, and final revenue to two decimal places using **%n1.n2f** formatting.
- Ensure that the columns are aligned correctly, even with varying product ID lengths.
- At the end of the table, display the overall total revenue before and after discounts.

For instance, if the following inputs are provided:

```
product_ids = [101, 102, 103, 104, 105, 106];
units_sold = [150, 200, 350, 400, 250, 300];
price_per_unit = [29.99, 49.99, 79.95, 99.99, 59.99, 89.99];
discount_percentage = [5, 10, 7.5, 15, 0, 12];
```

The output must be:

```
Product ID   Units Sold   Unit Price   Revenue      Discount %   Discount Amount   Final Revenue
----------   ----------   ----------   ----------   -----------  ---------------   -------------
       101          150        29.99      4498.50         5.00            224.93         4273.57
       102          200        49.99      9998.00        10.00            999.80         8998.20
       103          350        79.95     27982.50         7.50           2098.69        25883.81
       104          400        99.99     39996.00        15.00           5999.40        33996.60
       105          250        59.99     14997.50         0.00              0.00        14997.50
       106          300        89.99     26997.00        12.00           3239.64        23757.36

Total Revenue Before Discount:  124469.50
Total Revenue After Discount:   111907.05
```

2. Write a function named **multiFunctionStats** that can calculate a range of statistics ('mean', 'median', 'std', 'var', 'min', 'max') for multiple datasets. The function should be able to handle a flexible number of datasets and allow the user

to specify which statistics to compute for each dataset. The function declaration line will be:

```
function varargout = multiFunctionStats(varargin)
```

**Input Arguments:** The function accepts a variable number of input arguments (**varargin**). The first input is always a string specifying which statistics to compute. Subsequent inputs are datasets (vectors) for which the statistics should be computed.

**Output Arguments:**

- The function returns the requested statistics for each dataset as separate outputs. If the user does not specify any datasets, the function should throw an error.
- If no specific statistic is requested, the function should compute the 'mean', 'median', and 'std' for each dataset by default.

**Validation:**

- Use **nargin** to check the number of input arguments and ensure that at least two inputs are provided (the statistic type and one dataset).
- Use **nargchk** to provide a custom error message if the number of inputs is incorrect.

3. Write a function to return the total normal stress in a beam that is fixed at both ends. The beam experiences two types of stress:

- **Thermal Stress**: Due to an increase in temperature.
- **Bending Stress**: Due to a time-varying concentrated load applied in the middle of the beam.

**Assumptions**:

- The beam is homogeneous and isotropic.
- The temperature is uniformly distributed along the length of the beam.
- The beam remains within the elastic limit (no plastic deformation).
- The load is applied vertically downward in the middle of the beam.

**Derivation of Equations:**

**Thermal Stress:** The thermal stress due to a uniform temperature change $\Delta T$ in a beam that is fixed at both ends is given by:

$$\sigma_{\text{thermal}}(T) = E\,\alpha\,\Delta T \tag{2.2}$$

where E is Young's Modulus, $\alpha$ is the coefficient of thermal expansion, and $\Delta T$ is the change in temperature.

**Bending Stress:** For a beam with a length L, fixed at both ends, and subjected to a concentrated load F(t) at the midpoint, the bending stress at a distance y from the neutral axis is given by:

$$\sigma_{\text{bending}}(y, t) = -M(t)\, y/I \tag{2.3}$$

where $M(t) = F(t) \cdot L/4$ is the maximum moment at the midpoint, $I = b \cdot h^3/12$ is the moment of inertia of the cross-section, b is the width of the beam, and h is the height (thickness) of the beam.

**Total Normal Stress:** The total normal stress at a distance y from the neutral axis is:

$$\sigma_{\text{total}}(y,t,T) = \sigma_{\text{thermal}}(T) + \sigma_{\text{bending}}(y,t) \tag{2.4}$$

Inputs must be time, temperature, and y for the main function and the output must be the total stress. Use anonymous functions to calculate the bending the thermal stress values.

4. Write the code for problem 3 with a main function that takes the 3 inputs using the **menu** command. The 3 separate menus must offer 6 time points, 4 temperatures, and 3 y values and return the total stress.
5. Write a function that manages a database of engineering materials. The database contains different types of materials: Metals, Polymers, and Ceramics. Each type has specific properties, and you will use **cell arrays** to store and manage this data. Initialize the database with a few materials and test all functions.

**Data Storage:**

- Create a cell array to store the materials.
- Each material is stored as a sub-cell array with attributes:

  **Metals:** {Name, Density, Young's Modulus, Yield Strength}

  **Polymers:** {Name, Density, Tensile Strength, Glass Transition Temp}

  **Ceramics:** {Name, Density, Fracture Toughness, Melting Point}

**Functions:**

- **addMaterial:** Add a new material to the database.
- **displayMaterials:** Display all materials, categorized by type.
- **searchByName:** Search for a material by name.
- **removeMaterial:** Remove a material by name.

6. Write a script to manage and analyze a fleet of electric vehicles (EVs) used by a transportation company. Each vehicle in the fleet has specific attributes, including model name, battery capacity, range per charge, and average energy consumption. Your code will store and manage this data using a **structure array**. Initialize the structure array using the following data:

**Vehicles:**

- "Tesla Model 3", 75 kWh, 500 km, 15 kWh/100 km
- "Nissan Leaf", 40 kWh, 240 km, 18 kWh/100 km
- "Chevy Bolt", 66 kWh, 380 km, 17 kWh/100 km

**Your Code Must Perform the Following Tasks:**

- Add a new vehicle: "BMW i3", 42 kWh, 260 km, 16 kWh/100 km
- Display all vehicles in the fleet.
- Identify the vehicle with the maximum range per charge.
- Calculate and display the average battery capacity and energy consumption.
- Calculate the total range of the fleet if all vehicles were fully charged.

7. Redo problem 6 using table data type.
8. Write a script to manage and analyze a fleet of electric vehicles (EVs) used by a transportation company. The vehicles are categorized by type (e.g., sedan, SUV, truck), and each vehicle has attributes such as battery capacity, range per charge, and average energy consumption. The vehicle types will be represented as **categorical variables**. The goal is to categorize these vehicles, analyze their distribution, and perform basic statistical analysis on their properties based on their categories as explained at the bottom of this page.

   **Data Storage:** Store the data in separate arrays for each attribute:

- **Model:** The model name of each vehicle (e.g., "Tesla Model 3", "Nissan Leaf").
- **Type:** The type of vehicle, represented as a categorical variable (e.g., "Sedan", "SUV", "Truck").
- **BatteryCapacity:** The battery capacity of the vehicle (in kWh).
- **RangePerCharge:** The range per full charge (in km).
- **EnergyConsumption:** The average energy consumption (in kWh per 100 km).

   **Data:** Vehicles are:

- "Tesla Model 3", "Sedan", 75 kWh, 500 km, 15 kWh/100 km
- "Nissan Leaf", "Hatchback", 40 kWh, 240 km, 18 kWh/100 km
- "Chevy Bolt", "Hatchback", 66 kWh, 380 km, 17 kWh/100 km
- "Tesla Model X", "SUV", 100 kWh, 450 km, 20 kWh/100 km
- "Ford Mustang Mach-E", "SUV", 88 kWh, 430 km, 19 kWh/100 km
- "Rivian R1T", "Truck", 135 kWh, 500 km, 23 kWh/100 km
- "Tesla Cybertruck", "Truck", 200 kWh, 800 km, 25 kWh/100 km
- "BMW i3", "Hatchback", 42 kWh, 260 km, 16 kWh/100 km
- "Audi e-tron", "SUV", 95 kWh, 350 km, 22 kWh/100 km
- "Porsche Taycan", "Sedan", 93 kWh, 400 km, 21 kWh/100 km

**Your Program Must Do the Following Tasks:**

- **Convert Vehicle Types:** Convert the vehicle type information to a categorical array.

- **Frequency Distribution:** Display the frequency distribution of each vehicle type.
- **Statistical Analysis:** Calculate and display the mean range per charge and mean energy consumption for each vehicle type.
- **Maximum Capacity:** Identify and display the vehicle with the highest battery capacity within each vehicle type category.

9. Write a script that reads data from a **text file**, processes the data, and writes the results to a new output file. The data are measurements from a stress–strain experiment for steel from a tensile test. The program should be able to:

**Read the Input File:** The input file contains columns of data representing the strain and corresponding stress values from the experiment.

**Process the Data:** Calculate the Young's modulus, yield strength, and ultimate tensile strength from the data.

**Write the Output File:** Write the results to a new text file, including the calculated values and a summary of the processed data.

**Test Data:** Create a text file with the following values:

| Strain | Stress |
|---|---|
| 0.0001 | 50 |
| 0.0002 | 100 |
| 0.0003 | 150 |
| 0.0004 | 200 |
| 0.0005 | 240 |
| 0.0006 | 270 |
| 0.0007 | 280 |
| 0.0008 | 290 |
| 0.0009 | 295 |
| 0.0010 | 300 |

10. Repeat the coding for problem 9 but this time, stress–strain data is stored in an **Excel file**. The goal is to calculate the strain rate and compare the performance (run time) of the calculations using a **for loop** and **vectorization**. The script will then write the results to a new Excel file and display a runtime comparison of both methods in the command window.

# Chapter 3
# Object-Oriented Programming

## 3.1 Class Definition

### *3.1.1 Introduction to OOP*

Most programming done using MATLAB is procedural, a coding paradigm focused on defining a step-by-step process to reach a desired state. This process is often performed using functions. MATLAB does, however, provide a second programming paradigm, known as object-oriented programming (**OOP**). OOP involves breaking sections of repeated code into classes, which themselves are collections of functions (called **methods**). Classes are useful for grouping together sections of code and functions that perform operations relating to the same concept or object. By grouping functions together, classes make it easier to hide and protect data from external functions and classes, which is known as **encapsulation**. Additionally, classes can be **instanced** (duplicated) to create multiple objects that have the same methods but unique instanced variables (**properties**), leading to a modular style of programming. The modularity of OOP becomes much more important and useful as the complexity and size of projects increase.

A simple example to help realize where OOP could be useful is in an engineering simulation. A robot simulation program could have a class representing a single link. This class could have properties for the link's mass, position, rotation, and velocity and have methods for updating the link's position and rotation. The program can then make multiple instances of the same link class and change the values of the properties to create multiple different links to form a complete robot. Additionally, subclasses could be made from this general link class that inherit all the link class's properties and methods while also adding unique properties and methods for dealing with rotary and prismatic joints, respectively, in a process known as **inheritance**. The rotary and prismatic subclasses might override a couple of the default link's properties and methods, such as the function that updates rotation, which is known as **polymorphism**.

 
M. Andalibi and S. A. Cherkauer, *Advanced MATLAB for Engineers*,
https://doi.org/10.1007/978-3-032-10677-3_3

### 3.1.2 Components of a Class in MATLAB

Like any other OOP programming language, MATLAB classes are made up of properties and methods, where the properties are variables owned by the class and methods are functions relating to the class. MATLAB classes tend to have a **constructor** and a **deconstructor**, which are methods that initialize and destroy the class object, respectively. The constructor is the function that is automatically called when a new instance of a class is created, and the inputs given to that new class are the inputs given to the constructor. This function's primary purpose is to assign the inputs to class properties and carry out any other operations to initialize the new class object. To work correctly, the constructor must have the exact same name as the class object and should be able to handle cases when the class is initialized without enough inputs. The deconstructor is the function implicitly called before MATLAB destroys that class instance's handle. This function must be named **delete**, must only have one input representing the class handle, and must have no outputs. The deconstructor cannot create any class instances, change any properties, call any methods, or throw any errors, or MATLAB will not call the **delete** function and could fail to destroy the object. Constructors and deconstructors will be discussed more in depth in the next section.

### 3.1.3 Creating a Class in MATLAB

To define a new class in MATLAB, type the keyword **classdef** and follow it with the name of the class. The class can only be defined at the start of a ***.m** file, and it must have the exact same name as the file (just like function files). Additional functions can be defined after the class in the same file, but these functions can only be accessed by the class and not externally. Aside from the basic class definition, two more optional settings can be applied to new classes, which are superclasses and class attributes. If this new class is a supposed to be a subclass, the name of the class is then followed by a less than sign (**<**) and the name of the superclass. Multiple superclasses can be specified by separating the names with **&**. If this new class needs attributes, these can be applied by listing the attribute keywords in parentheses between the **classdef** keyword and the class name. All available class attributes are listed and explained in Table 3.1. Attributes applied to a superclass are not inherited by the subclass.

```
classdef (attributeName = value, ...) ClassName < SuperclassName1 &
SuperclassName2
        ...
end
```

After creating a new class, it can be instanced using function notation. A new class instance is returned by calling the class name followed by parenthesis containing the inputs to the constructor function. If no constructor is defined or the constructor

**Table 3.1** All available class attributes

| Attribute name | Description |
|---|---|
| Abstract | If set to **true**, this class cannot be instanced |
| AllowedSubclasses | If set equal to one class object or a cell array of class objects, only those classes can be subclasses of this class. An empty cell array works the same as the **Sealed** attribute. Specify classes using a question mark operator before the name: **?classname** |
| ConstructOnLoad | If **true**, MATLAB calls this class's constructor when loading an object from a MAT-file. This class must have a no-argument constructor |
| HandleCompatible | If **true**, this class can be a superclass for handle classes even if it is not a handle class itself |
| Hidden | If **true**, this class does not appear in the output of the **help** or **superclasses** functions |
| InferiorClasses | If set equal to a cell array of class objects (specified using the **?** operator), the specified classes are inferior to this class. This class becomes dominant and is prioritized when overwriting other objects |
| Sealed | If **true**, this class cannot be subclassed |

takes no inputs, the parenthesis can be left empty or the parenthesis can be removed entirely.

```
ClassName(arg1, arg2, ...)
                    ans = ClassName with no properties.
```

### 3.1.4 Adding Properties and Methods to a Class

After naming the class it can be assigned properties and methods using the **properties** and **methods** keywords, respectively. These properties and methods can in turn be assigned attributes in the same way as class attributes to control characteristics such as visibility and accessibility. Once again, while properties and methods are inherited from the superclasses, their attributes are not. All the available property attributes are shown in Table 3.2 and all available method attributes are shown in Table 3.3. The handle classes and the **set** and **get** methods mentioned in these two tables will be discussed in depth later in this chapter.

Table 3.2 All available property attributes

| Attribute name | Description |
|---|---|
| AbortSet | For handle classes only<br>– **true**: MATLAB does not update the property value or call a **set** method if the new value is the same as the current value.<br>– **false**: The property is always updated (default). |
| Abstract | These properties cannot define **set** or **get** methods and cannot define initial values. Sealed classes cannot have abstract members<br>– **true**: The property is not used, but any subclass must overwrite this property. The new property must have **Abstract** set to **false.**<br>– **false**: The property does not need to be overwritten (default). |
| Access | Use **Access** to set **SetAccess** and **GetAccess** to the same value<br>– **public**: The property can be accessed from anywhere (default).<br>– **protected**: The property can be accessed by the defining class or its subclasses.<br>– **private**: The property can only be accessed by the defining class,<br>– If set equal to a cell array of class objects, those classes have **get** and **set** access to this property. |
| Constant | Subclasses inherit constant properties but cannot change them constant properties cannot be **Dependent**. The value of **SetAccess** is ignored for these properties<br>– **true**: The property value cannot be changed.<br>– **false**: The property value can be changed (default). |
| Dependent | These properties can have **set** methods, but those methods cannot set the value of these properties. When accessed by a **get** method, the returned value is ignored by the **isequal** function<br>– **true**: The property is not stored in the object. Instead, the property is calculated whenever it is accessed.<br>– **false**: The property is stored in the object (default). |
| GetAccess | All subclasses must specify the same values as the superclass for property **SetAccess** and **GetAccess** attributes. Possible values are the same as the **Access** attribute |
| GetObservable | – **true**: Listeners can be created for handle class properties. The listeners are called whenever the property values are read or queried.<br>– **false**: Listeners do not have access to this property (default). |
| Hidden | – **true**: The property is not visible in property lists or in results from the **get**, **set**, or **properties** functions.<br>– **false**: The property is visible (default). |
| NonCopyable | Only handle classes can set **NonCopyable** to **true**<br>– **true**: The property value is not copied when the defining object is copied.<br>– **false**: The property value is copied when the object is copied (default). |
| PartialMatchPriority | Can only be used with subclasses of **MATLAB.mixin.SetGet**<br>– Setting equal to a positive integer defines the relative priority of partial property name matches used in **get** and **set** methods. The default value is 1. |

(continued)

**Table 3.2** (continued)

| Attribute name | Description |
|---|---|
| SetAccess | All subclasses must specify the same values as the superclass for property **SetAccess** and **GetAccess** attributes. Possible values are the same as the **Access** attribute, with the addition of one more:<br>– **immutable**: The property can be set only in the constructor. |
| SetObservable | – **true**: Listeners can be created for handle class properties. The listeners are called whenever the property values are set.<br>– **false**: Listeners do not have access to this property (default). |
| Transient | – **true**: The property value is not saved when the object is saved to a file or sent from MATLAB to another program.<br>– **false**: The property value is saved when the object is saved (default). |

**Table 3.3** All available method attributes

| Attribute name | Description |
|---|---|
| Abstract | – **true**: The method is not used, but any subclass must overwrite this method. The subclass method does not need the same number of inputs and outputs.<br>– **false**: The method does not need to be overwritten (default). |
| Access | – **public**: The method can be accessed from anywhere (default).<br>– **protected**: The method can be accessed by the defining class or its subclasses.<br>– **private**: The method can only be accessed by the defining class.<br>– If set equal to a cell array of class objects, those classes have access to this method. |
| Hidden | – **true**: The method is not visible in method lists or in results from the **methods** or **methodsview** commands. **ismethod** does not return **true** for this method name.<br>– **false**: The method is visible (default). |
| Sealed | – **true**: The method cannot be redefined or overwritten in a subclass.<br>– **false**: The method can be overwritten (default). |
| Static | – **true**: The method does not require a class object as an input. The method is still called the same as others using dot notation.<br>– **false**: The method needs at least one input for a class object (default). |

```
classdef (attributeName = value, ...) ClassName < SuperclassName
        properties (attributeName1 = value1, attributeName2 = value2, ...)
                ...
        end
        methods (attributeName1 = value1, attributeName2 = value2, ...)
                ...
        end
end
```

When using the **properties** keyword to define new class properties, MATLAB is simply creating new variable objects parented to the class object. As such, the contents of **properties** are just variable names and numbers, not executable code.

A new property can be created named **m**, a new constant property could be created named **g** equal to 9.81, but a new property cannot be created named **W** equal to **m*g**. Variable definitions such as **W = m*g** are performed in methods, such as the constructor.

When defining new properties, the allowable size, data type, and any validators can be set in the same manner as they are set for argument blocks (Chap. 2). This is done by writing the dimensions in parentheses, the class handle, and a cell array of requirements (validators) separated by spaces after the variable name but before the equals sign (if an equal sign is included) as illustrated here:

```
variableName (rows, columns) class {validator1, validator2, ...} = value
```

These restrictions do not all need to be present, but they must always be provided in this order. Any values assigned to the property will be converted to the required dimensions and converted to the required class type, but MATLAB raises an error if either of these is not possible. If an assigned value violates any of the validators, an error is raised.

The following code creates a new class called **Link** with the properties mass (**m**) and weight (**W**), and a constant property **g** assigned the value 9.81. This **Link** class will be expanded throughout this chapter and used to represent one link or rigid member of a robot manipulator. The variable **m** must always be a 1x1 array of type double. The one validator function provided throws an error if **m** is not real. When attempting to create an instance of the class, however, an error is thrown because **m** and **g** do not exist yet, so **W** cannot be defined as a function of **m** or **g**. This could be corrected by defining **W** as an empty property and then setting it equal to **m*g** in the class's constructor. Since **g** is given the attribute **Constant**, its value cannot be changed later in any instances of **Link**.

```
classdef Link
        properties (Constant = true)
                g = 9.81;
        end
        properties
                m (1, 1) double {mustBeReal};
                W = m*g; % The class will instance correctly if this line is removed
        end
end
```

```
Link()
                Invalid default value for property 'W' in class 'Pendulum':
                        Unrecognized function or variable 'm'.
```

Unlike **properties**, defining methods using the **methods** keyword is very straightforward because it uses the same formatting as defining new functions. After the **methods** keyword, a new method is defined using the **function** keyword, followed by the outputs, method name, and inputs just like any other function. Where methods differ from functions is their inputs. Methods always require one input more than usual (unless the **Static** attribute is set to **true**, as described in Table 3.3) to hold an instance of the class object. MATLAB automatically assigns the class handle to the first input of the method, which can have any valid variable name. Argument blocks can be used at the start of class methods.

The following code alters the **Link** class to add two normal methods and one static method. The two normal methods require an input variable to hold the class instance, in this case called **self**. The static method does not require a class instance input, meaning it cannot be used to directly access class properties. The **printW** method does not return any variables, **setW** method returns the altered class instance, and the static **convertNtolb** method returns the altered input value just like a normal function without accepting a class instance as an input.

```
classdef Link
        properties (Constant = true)
                g = 9.81;
        end
        properties
                m (1, 1) double {mustBeReal};
                W
        end
        methods

                % Method that prints the property W
                function printW(self)
                        disp(self.W)
                end

                % Method that sets the value of W
                function self = setW(self)
                        self.W = self.m*self.g;
                end
        end
        methods (Static = true)

                % Method that converts numbers without any access to the
                % class's properties and methods
                function val = convertNtolb(val)
                        val = val * 0.2248089431;
                end
        end
end
```

As discussed previously, a class instance can be made by calling the name of the class like a function. The returned class object can then be assigned to a variable, as shown below. Notice how **g** is assigned a default value while **m** and **W** are assigned empty values matching their respective data types and sizes. The property **m** of instance **L1** is accessed and set to two by using dot notation like structure objects.

```
L1 = Link()
                    L1 = Link with properties:
                         g: 9.8100
                         m: 0
                         W: []
L1.m = 2
                    L1 = Link with properties:
                         g: 9.8100
                         m: 2
                         W: []
```

Similarly, the methods of **L1** can be called using dot notation and function call syntax. Methods can also be called like normal functions where the class instance is given as the first argument. The class instance is automatically given as an input when a non-static method is called. When properties are altered in a class's method, a new class instance is returned, but the existing class instance's properties are not updated. This can be remedied by setting the class variable equal to the new class instance or by using handle classes, which will be discussed in depth later on.

```
L1.setW() % Call with dot notation
                    ans = Link with properties:
                         g: 9.8100
                         m: 2
                         W: 19.6200
L1.printW()

L1 = setW(L1) % Call with instance as first argument
                    L1 = Link with properties:
                         g: 9.8100
                         m: 2
                         W: 19.6200
L1.printW()
                         19.6200
```

```
L1.convertNtolb( L1.W )
                         ans = 4.4108
L1.m = 2*i
          Error setting property 'm' of class 'Link'. Value must be real.
L1.m = [1, 2]
          Error setting property 'm' of class 'Link'. Value must be a scalar.
```

### 3.1.5 *Obtaining Class Information*

After defining a class, another programmer may wish to see what properties and methods a class has but may not have access to the source code. The **properties** and **methods** functions will return lists of all of a class's unhidden properties and methods, respectively, given the class instance as an input. Additionally, the **superclasses** function can be called with the same input to obtain a list of class handles for all the subclass's superclasses. Another helpful function that can be used to determine the class type of variable is **class**, which takes the variable as an input and returns the name of the variable's class type as a string.

## 3.2 Special Methods

### 3.2.1 *Constructor*

As mentioned previously, the constructor is a special method inside of a MATLAB class that shares the same name as the class and is called whenever the class is instanced. This method takes any number of inputs and returns a new class instance object, but the constructor does not take a class object as an input like other methods. The constructor effectively initializes the class instance by assigning input values to properties and should be able to handle cases where it is not given enough inputs. If no constructor method is defined for a class, MATLAB calls a default constructor which returns a class instance given no input arguments. To set a property in the constructor, use dot notation as if referencing the class instance from outside.

The following code is another version of the **Link** class, but this time **m** can be defined when the class is first instanced, and **W** is calculated without calling a separate method. The new class instance in this case is called **self**, and the properties of **self** are referenced using dot notation. Notice how this class's constructor works when provided with all or fewer of its inputs by using **nargin** and defining default values in the properties block.

```
classdef Link
    properties (Constant = true)
        g = 9.81;
    end
    properties
        m (1, 1) double {mustBeReal} = 0;
        W
    end
    methods
        function self = Link(m)

            % If the argument m is given, assign it to the property m
            if nargin == 1
                self.m = m;
            end

            % Define W when the class is initialized
            self.W = self.m*self.g;
        end
    end
end
```

```
L1 = Link()
            L1 = Link with properties:
                 g: 9.8100
                 m: 0
                 W: 0
L1 = Link(2)
            L1 = Link with properties:
                 g: 9.8100
                 m: 2
                 W: 19.6200
```

MATLAB will automatically call any superclass constructors, starting from the left and working right following the order in the class definition, but these constructors will be given no inputs. If the superclass constructor needs inputs, call the constructor with @ notation, where the function call is the class instance name followed by @ and then the superclass name. When directly calling a superclass constructor, call it before setting any class instance properties since the superclass constructor will return a new class instance object.

The following code is a subclass, with the above **Link** class as its superior. When the subclass is instanced, the constructor first assigns values to any inputs that were not given using an arguments block, then calls the superclass's constructor to initialize the class instance, and then sets its own properties using dot notation. The subclass's constructor returns the new class instance, which can be assigned to a variable as normal. Notice how the **RevoluteJoint** class gains all the properties of the **Link** class because it is subclassed to **Link**. If the superclass's constructor had not been called explicitly by the subclass constructor and given inputs, MATLAB would have automatically called the constructor with no inputs, so **m** and **W** would both remain zero.

```
classdef RevoluteJoint < Link
    properties
        r double % Gearbox Ratio
    end
    methods
        function self = RevoluteJoint(m, r)

            % This argument block sets the input m to zero if it is
            % not provided
            arguments
                m (1, 1) double {mustBeReal} = 0;
                r double = 1;
            end

            % Call the Link (superclass) constructor
            self = self@Link(m)

            % Set this class's properties after the superclass constructor
            self.r = r;
        end
    end
end
```

```
L1 = RevoluteJoint()
        L1 = RevoluteJoint with properties:
                r: 1
                g: 9.8100
                m: 0
                W: 0
L1 = RevoluteJoint(2)
        L1 = RevoluteJoint with properties:
                r: 1
                g: 9.8100
                m: 2
                W: 19.6200
L1 = RevoluteJoint(2, 0.5)
        L1 = RevoluteJoint with properties:
                r: 0.5
                g: 9.8100
                m: 2
                W: 19.6200
```

### 3.2.2 Deconstructor

Where a constructor method initializes a class, a deconstructor method cleans up a class instance for deletion (garbage collection). MATLAB automatically calls the deconstructor before destroying a class object, but the deconstructor can also be called by user-defined code. The deconstructor is also automatically called when an error occurs in the constructor, so deconstructor methods should be able to work even when not all properties are defined. This method must be named **delete** (all lowercase), take one input for the class instance object, and have no outputs. Additionally, the method cannot be static, abstract, or sealed; cannot use argument blocks to restrict the input; cannot raise any errors; cannot create new class objects; and cannot access properties or methods from subclasses. If the deconstructor does not follow all these rules, MATLAB will not automatically call the function when destroying a class object, which can prevent the destruction of the object. When destroying an object, MATLAB calls the subclass's delete method before the superclass', which is the opposite order of construction. It is important to note that the variable holding the class instance is not removed from the workspace when **delete** is called.

```
classdef Link
        properties (Constant = true)
                g = 9.81;
        end
        properties
                m (1, 1) double {mustBeReal} = 0;
                W
        end
        methods
                function self = Link(m)

                        % If the argument m is given, assign it to the property m
                        if nargin == 1
                                self.m = m;
                        end

                        % Define W when the class is initialized
                        self.W = self.m*self.g;
                end

                function delete(self)
                        disp('Link Object destroyed!')
                end
        end
end
```

```
L1 = Link();
delete(L1)
                                        Link Object destroyed!
```

### 3.2.3 Property Set/Get Methods

Aside from the constructor and deconstructor, MATLAB offers two additional method types that are called whenever a specified property's value is changed and whenever a specified property's value is read, named property **set** and **get** methods, respectively. To create the **set** and **get** methods, the function name must be either **set** or **get**, followed by a period and the name of the specific property. The **get** method takes the class instance as the input and must return the property as its only output. The **set** method takes the class instance and the property's new value as inputs and must return the modified class object if it is a value class (handle classes do not need to return anything). The differences between handle and value classes will be explained in more depth in the next section.

In the following example, the **Link** class is again defined with properties **m**, **g**, and **W**. Additionally, the class has a **get** function that is called whenever the value of **W** is requested and a **set** function that is called whenever the value of **m** is changed. As shown, the **set** function is not called when a default value is applied to **m** but when **m** is set to two or three, the **set** function is called, and the updated class instance is returned as shown by the command line code. Since the output is not suppressed by a semicolon, the value of **W** is requested to be printed in the terminal, so the **get** function is called and returns the value of **W**. Additionally, the **get** function for **W** is called when the property is explicitly queried using dot notation.

```
classdef Link
    properties (Constant = true)
        g = 9.81;
    end
    properties
        m (1, 1) double {mustBeReal} = 0;
        W
    end
    methods
        function self = Link(m)

            % If the argument m is given, assign it to the property m
            if nargin == 1
                self.m = m;
            end
        end

        % Function is called when W is queried. The value of W is not
        % saved, but rather calculated each time
        function W = get.W(self)
            W = self.m*self.g;
        end

        % Function is called when m is changed
        function self = set.m(self, val)
            self.m = val;
            disp('The mass is now ' + string(val) + ' kg.')
        end
    end
end
```

```
L1 = Link();

L1 = Link(2)
                    The mass is now 2 kg.

                  L1 = Link with properties:
                          g: 9.8100
                          m: 2
                          W: 19.6200
L1.m = 3
                    The mass is now 3 kg.

                  L1 = Link with properties:
                          g: 9.8100
                          m: 3
                          W: 29.4300
L1.W
                        ans = 29.4300
```

### 3.2.4 *Operator Methods*

Sometimes classes need to work with MATLAB operators such as **+** or * to be used as desired. Built-in data types such as **double** or **int64** have methods defined for all of MATLAB's operators so that MATLAB knows how to combine and compare instances of these classes when they are mixed into mathematical or Boolean expressions. Creating operator methods is as simple as defining methods with the specific names and inputs described in Table 3.4.

## 3.3 Class Types

### 3.3.1 *Value Classes*

MATLAB classes can be categorized into multiple types. The default class type and all of MATLAB's fundamental classes (such as **double**, **int32**, **struct**, and **cell**) are value classes, which return a class object that is associated only with the assigned variable. When this class object is passed to a modifying function, the function must return the modified version of the class object, otherwise the object will remain unaltered. If a new variable is assigned to an existing class object, the object is duplicated, and altering one copy of this class does not affect the other. Value class objects that are instanced from the same class are considered equal when compared using **isequal** or the **==** Boolean operator unless a property is changed to be different in one of the objects.

**Table 3.4** All available operator functions

| Operation | Method name and arguments |
|---|---|
| a + b | plus(a, b) |
| a – b | minus(a, b) |
| –a | uminus(a) |
| +a | uplus(a) |
| a .* b | times(a, b) |
| a * b | mtimes(a, b) |
| a ./ b | rdivide(a, b) |
| a .\ b | ldivide(a, b) |
| a / b | mrdivide(a, b) |
| a \ b | mldivide(a, b) |
| a.^ b | power(a, b) |
| a ^ b | mpower(a, b) |
| a < b | lt(a, b) |
| a > b | gt(a, b) |
| a <= b | le(a, b) |
| a >= b | ge(a, b) |
| a ~= b | ne(a, b) |
| a == b | eq(a, b) |
| a & b | and(a, b) |
| a \| b | or(a, b) |
| ~a | not(a, b) |
| a:d:b<br>a:b | colon(a, d, b)<br>colon(a, b) |
| a’ | ctranspose(a) |
| a.’ | transpose(a) |
| [a b] | horzcat(a, b,...) |
| [a; b] | vertcat(a, b,...) |
| a(s1, s2,..., sn) | subsref(a, s) |
| a(s1, s2,..., sn) = b | subsasgn(a, s, b) |
| b(a) | subsindex(a) |

In the code below, two copies of the built-in **double** class object are made (**a** and **b**). When one of the copies is changed (**b** is squared), the other class object instance remains unaltered. This is demonstrated by comparing **a** and **b** with the **==** Boolean operator before and after two changes. User-defined value classes (which are the default classes defined in section one) work the same way as all of MATLAB’s fundamental value classes, such as **double**.

```
a = double(2);
b = a;
a == b
                              ans = logical 1
b = b^2;
a == b
                              ans = logical 0
b = sqrt(b);
a == b
                              ans = logical 1
```

### 3.3.2 Handle Classes

Another kind of class is the handle class. Unlike value classes, handle classes return a connection to the same class object rather than a copy of the class object, which is known as the class's **handle**. This feature means multiple variables can reference the same object, and functions that alter the class object do not need to return a modified object. Handle classes are defined by subclassing the class to MATLAB's **handle** class (**handle** is one of the superclasses). MATLAB will consider two variables containing the same class object's handle equivalent, and when one handle's class is altered or deleted, all other handles associated with the same class object reflect that change.

```
classdef HandleClass < handle
        properties
                ...
        end
        methods
                ...
        end
end
```

### 3.3.3 Enumeration Classes

A third kind of class is the enumeration class, which is defined using an enumeration block. Enumerations allow the user to assign names to class objects that have certain property values, such as naming a servo class instance based on operating speed and voltage. The enumeration block is defined within the class block but outside of the properties and methods blocks using the **enumerations** keyword and closed using the **end** keyword. The contents of the block are a list of possible instance names separated by either commas or line breaks. The named class instance can be initiated with specific values by following the name with parentheses containing the specified constructor inputs. Unless the enumeration class is also made into a handle class by superclassing MATLAB's **handle** class, all properties in the enumeration class and superclasses must be immutable.

The following example defines an enumeration class named **Servo** that defines the servo's speed and operating voltage for three different servo types:

```
classdef Servo
    enumeration
        MG90S (60/0.1, 4.8)
        MG995 (60/0.2, 4.8)
        VTS08A (60/0.33, 4.8)
    end
    properties
        speed = 0
        V = 0
    end
    methods
        function self = Servo(speed, V)
            self.speed = speed; % deg/sec
            self.V = V; % V
        end
    end
end
```

A new instance of an enumeration class is created by calling the name of the class and the name associated with the desired instance values using dot notation. In this case, an instance of the **Servo** class is made for the MG90S servo type, and another is made for the MG995 servo type. As shown below, the properties for the different servo types are set in the class's constructor when the specific class enumeration is instanced.

```
s1 = Servo.MG90S;
s2 = Servo.MG995;
s1.speed
                                        s1 = 600
s2.speed
                                        s2 = 300
```

## 3.4 Encapsulation

### *3.4.1 Defining Encapsulation*

Encapsulation was a topic briefly touched upon in the first section of this chapter as a reason for using object-oriented programming, but exactly what it is, why it is useful, and how to use it will become clearer after reading through this section. Simply put, encapsulation is the method of breaking code into separate parts to avoid accidentally overwriting or altering saved data in large applications. A common and familiar form of encapsulation is a function, where inputs are received, variables are defined and altered, and then new values are returned. When a function defines new variables, these variables cannot overwrite existing saved data of the same name because the function does not have access to that existing data. Similarly, the rest of the code cannot access and change the variables in a function and can only save and alter the final output values.

### *3.4.2 Encapsulation in OOP*

Object-oriented programming is invaluable when it comes to the encapsulation of code. Where a function can complete a single operation without saving data, a class can complete multiple operations all while saving data in a way that does not overwrite variables in the workspace. Continuing with the robot example in section one, a program may need multiple variables for similar purposes (such as setting the masses and positions of two different links of a robot manipulator). As a code document gets larger and larger, simple mistakes like overwriting the mass of one of the links with the mass of a second can result in hours of debugging to finally find the source of the error. By instancing a link class twice rather than defining two separate mass variables or defining a list of mass values, the masses cannot easily overwrite each other because the two class instances have no access to each other's properties.

OOP encapsulation can also be useful when publishing applications that will be used by lots of people. By encapsulating properties and methods within classes, important information and data-changing functions can be made much more difficult to access. As an example, if an application runs the entry from a user input box without careful filtering, a user could potentially access confidential saved information, access areas of the program that they should not be able to by altering variables, or add modifications to the application that are unwanted.

### *3.4.3 Using OOP Encapsulation in MATLAB*

Most of the methods for using OOP encapsulation in MATLAB have already been included previously but have not been explicitly explained. The simplest encapsulation is breaking large, repeated sections of code into superclasses, subclasses, properties, and methods. This process can be expanded further by making use of the class, property, and method attributes shown in Tables 3.1 through **3.3**. More specifically, the **Hidden**, **Access**, and **Sealed** attributes are the most useful for encapsulating and protecting code. The **Hidden** attribute does not prevent external code from accessing the hidden class data, but it does make it harder for users to discover the classes, methods, and properties by preventing MATLAB from sharing their existence through functions like **help**, **properties**, **methods**, and **superclasses**. In contrast, the **Access** attribute can be used to restrict access to a property or method to only certain classes (**Access = 'protected'**), to only the class itself (**Access = 'private'**), or, for properties only, to be completely unalterable once set in the constructor (**SetAccess = 'immutable'**). Finally, the **Sealed** attribute does not prevent seeing or changing a property, method, or class, but it does prevent a superclass's method from being overwritten by a subclass's method with the same name.

## 3.5 Polymorphism

### *3.5.1 Defining Polymorphism*

Polymorphism is a useful feature of OOP that can be confusing at first glance. A simple definition of polymorphism is giving different classes a method with the same name to perform similar tasks but in different ways. The fact that MATLAB's **int** and **double** data types can both be multiplied using the same * operator despite being different classes is a quick example of polymorphism. A more in-depth example of polymorphism can be explained using the **Link** and **RevoluteJoint** classes from before. A robot can have links connected to two different joint types, prismatic (linear motion) and revolute (rotational motion) joints. These two types of connectors would have many similar properties and methods, such as needing to calculate the weights of the links they connect as shown in previous examples, but they would also have differing methods, such as how they move the links relative to themselves. By defining **PrismaticJoint** and **RevoluteJoint** classes, polymorphism allows both classes to have a method named **move** that updates their positions, making it easier to use the two link types interchangeably, despite the two **move** methods reaching their end results through different processes.

### *3.5.2 Using Polymorphism in MATLAB*

Polymorphism is very easy to use in MATLAB. To give two separate classes a method with the same name is as simple as writing two separate class definitions, each with a method of that name. This process can be taken a step further by overwriting methods inherited from a superclass so that all subclasses follow the same structure while carrying out their different operations. Overwriting a superclass's method is as simple as defining a new method in the subclass that has the same name.

The code below is another version of the generic **Link** class. This class has two additional properties called **pos** and **rot** to represent a link's position and rotation in two-dimensional space. This **Link** class also has one abstract method named **move**, meaning any subclass of **Link** must override the abstract method with a method of the same name.

```
classdef Link
    properties (Constant = true)
        g = 9.81;
    end
    properties
        m (1, 1) double {mustBeReal} = 0;
        W
        pos = [0, 0]
        rot = 0
    end
    methods
        function self = Link(m)
            if nargin == 1
                self.m = m;
            end
        end
        function W = get.W(self)
            W = self.m*self.g;
        end
    end

    % The abstract attribute requires all subclasses to override this
    % method with their own
    methods (Abstract = true)
        move
    end
end
```

The following two classes are both subclasses of the generic **Link** class but represent links joined with one of the two different joint types. Both of these classes inherit all five properties from **Link** and the **get** method that calculates the weights but override the **move** method so that **RevoluteJoint** only rotates and **PrismaticJoint** only translates.

```
classdef RevoluteJoint < Link
    methods
        function self = RevoluteJoint(m)

            % This argument block sets the input m to zero if it is
            % not provided
            arguments
                m (1, 1) double {mustBeReal} = 0;
            end

            % Call the link (superclass) constructor
            self = self@Link(m)
        end

        % The class must have this method because of the superclass's
        % abstract method
        function self = move(self)
            self.rot = self.rot + 1;
        end
    end
end
```

```
classdef PrismaticJoint < Link
    methods
        function self = PrismaticJoint(m)

            % This argument block sets the input m to zero if it is
            % not provided
            arguments
                m (1, 1) double {mustBeReal} = 0;
            end

            % Call the link (superclass) constructor
            self = self@Link(m)
        end

        % The class must have this method because of the superclass's
        % abstract method
        function self = move(self)
            self.pos = self.pos + [1, 0];
        end
    end
end
```

As demonstrated by the following code, polymorphism is useful in complex programs that use multiple different classes to carry out similar processes. The **RevoluteJoint** class and the **PrismaticJoint** class must be instanced differently, but can then be used the same way, allowing future users to easily understand and use all the different classes.

```
L1 = RevoluteJoint(2)
        L1 = RevoluteJoint with properties:
                 g: 9.8100
                 m: 2
                 W: 19.6200
               pos: [0 0]
               rot: 0
L2 = PrismaticJoint(3)
        L2 = PrismaticJoint with properties:
                 g: 9.8100
                 m: 3
                 W: 29.4300
               pos: [0 0]
               rot: 0
L1.move()
        L1 = RevoluteJoint with properties:
                 g: 9.8100
                 m: 2
                 W: 19.6200
               pos: [0 0]
               rot: 1
L2.move()
        L2 = PrismaticJoint with properties:
                 g: 9.8100
                 m: 3
                 W: 29.4300
               pos: [1 0]
               rot: 0
```

## 3.6 Events and Listeners

### *3.6.1 Introduction to Events and Listeners*

An event in programming is an action triggered when a defined change occurs, such as a property value changing, or a new class instance being created. When an event is triggered, any listeners watching that event will call their designated callback functions. The event-listener process is very similar to **if** statements, where triggering an event is similar to switching a variable from **false** to **true**. A listener calls a function when the altered variable is set to **true**. Events allow for programs to be decoupled, meaning the requirements to trigger an event and the actions taken once an event is triggered are independent. Essentially, listening classes could be added or removed without altering the event-triggering code, and the activating class can be modified without altering any of the listening classes. In MATLAB, only handle classes can define events, but any kind of class can define listeners.

### *3.6.2 Creating Events*

Events are initiated in event blocks, which are opened using the **events** keyword and closed using the **end** keyword and are located within the class definition similar to **properties** and **methods**. Additionally, the event block can be given attributes using the same format as the properties and methods blocks. The list of all three available event attributes is shown in Table 3.5.

The contents of event blocks are just the names of the events separated by line breaks or commas, similar to property blocks. Each event created using an event block can then be triggered using the **notify** function, which takes the class instance and the name of the event as inputs, where the event name is a string or char vector.

A third input could also be given if the user wants to provide custom event data. By default, the event data object contains the two properties **EventName** and **Source**, which store the name of the triggered event and handle of the triggering class, respectively. To add additional properties, create a subclass of **event.EventData** with a constructor that assigns any inputs to class properties. The third input to the **notify** function should be a handle to this new event data subclass which was given the desired inputs.

The following code creates a handle class named **Robot** with an event named **TurnServo**. The purpose of this class is to initialize each link in the robot manipulator and keep track of their instances so the links can be moved later. This **Robot** class has a cell array named **links** that holds instances of the **Link** class, which are created and added in the **createLink** method. When the robot class's **move** method is called, the **TurnServo** event is triggered by the robot class itself and given custom event data that includes a new **movementTime** property. Notice how the updated object is not returned from the **createLink** method since the class's handle was provided as an input rather than a copy of the object.

**Table 3.5** All available event attributes

| Attribute name | Description |
|---|---|
| Hidden | – **true**: The event does not appear in any event list such as the **events** function's output.<br>– **false**: The event appears in all event lists (default). |
| ListenAccess | – **public**: Listeners can access this event from anywhere (default).<br>– **protected**: Listeners can access this event from the event's class and its subclasses.<br>– **private**: Listeners can access this event from the event's class only.<br>– If set equal to a cell array of class objects, those classes have listen access to this event. |
| NotifyAccess | – **public**: Any code can trigger this event (default).<br>– **protected**: Only the event's class and subclasses can trigger this event.<br>– **private**: Only the event's class can trigger this event.<br>– If set equal to a cell array of class objects, those classes can trigger this event. |

```
classdef Robot < handle
    properties
        name string = ""
        links (1, :) cell
    end
    events
        TurnServo
    end
    methods
        function createLink(self, jointType, m)
            arguments
                self

                % Restricts type to be either "R" or "P"
                % and converts char objects to strings
                jointType string {mustBeMember(jointType, ["R", "P"])} ...
                    = "R"
                m (1, 1) double {mustBeReal} = 0;
            end

            % Add the joint type to the robot's name
            self.name = self.name + jointType;

            % Create a new link instance and add it to a list
            if jointType == "R"
                link = RevoluteJoint(m);
            elseif jointType == "P"
                link = PrismaticJoint(m);
            end
            self.links(end + 1) = {link};
        end

        % Trigger joint movement
        function move(self, time)
            notify(self, "TurnServo", CustomEventData(time))
        end
    end
end
```

The following code creates a new event data subclass named **CustomEventData** that adds the **movementTime** property:

```
classdef CustomEventData < event.EventData
    properties
        movementTime
    end
    methods
        function self = CustomEventData(time)
            self.movementTime = time;
        end
    end
end
```

### 3.6.3 Creating Listeners

Listeners can be added to a class anywhere, using either the **addlistener** or **listener** commands. Both of these functions take the same inputs, including a handle to the event's source class, the name of the event as a string, and a callback function handle (using @ notation). Additionally, both listener functions have four property events (Table 3.6) that can be used by providing the functions with the event's source handle, the property name as a string, the event name as a string, and a callback function handle. To use these property events, the class's property must have the **GetObservable** attribute set to **true**.

```
listener(SourceHandle, 'EventName', @CallbackName)
listener(SourceHandle, 'PropertyName', 'EventName', @CallbackName)
```

The **addlistener** function couples the listener to the source object so that when the source object is deleted, the listener is deleted as well. In contrast, the **listener** function does not couple the listener to the source, so MATLAB does not delete the listener with the source. Instead, the listener handle must be manually deleted later. Both listener-creating functions return the handle to the listener objects, which can be assigned to variables. The callback function for any listener must accept two inputs, including a variable for the source object's handle and a variable for the event data.

Listener objects can be both temporarily disabled and deleted. To disable and reenable a listener, set the listener's **Enabled** property to **false** and **true**, respectively. To delete a listener permanently, use the **delete** function with the listener's handle as its input.

```
listenerHandle.Enabled = false; % Disable the listener
listenerHandle.Enabled = true; % Reenable the listener
delete(listenerHandle); % Permanently delete the listener
```

The following code adjusts the **Robot** class from before so that a **TurnServo** listener is applied to each link. When the **TurnServo** event is triggered, this class calls the **move** method of every link object, whether it has a revolute or prismatic joint (making use of polymorphism). The event is triggered using the **CustomEventData** class to provide a **movementTime** variable to each link's **move** method.

**Table 3.6** All available event attributes

| Event name | Event trigger |
|---|---|
| **'PreSet'** | Immediately before the property value is set, before calling its **set** method |
| **'PostSet'** | Immediately after the property value is set |
| **'PreGet'** | Immediately before the property value is returned, before calling its **get** method |
| **'PostGet'** | Immediately after returning the property value to the query |

```
classdef Robot < handle
    properties
        name string = ""
        links (1, :) cell
        listeners (1, :) event.listener
    end
    events
        TurnServo
    end
    methods
        function createLink(self, jointType, m)
            arguments
                self

                % Restricts type to be either "R" or "P"
                % and converts char objects to strings
                jointType string {mustBeMember(jointType, ["R","P"])} ...

                    = "R"
                m (1, 1) double {mustBeReal} = 0;
            end

            % Add the joint type to the robot's name
            self.name = self.name + jointType;

            % Create a new link instance and add it to a list
            if jointType == "R"
                link = RevoluteJoint(m);
            elseif jointType == "P"
                link = PrismaticJoint(m);
            end
            self.links(end + 1) = {link};

            % Assign a listener to activate the move function
            self.listeners(end + 1) = addlistener(self, 'TurnServo', @link.move);
        end

        % Enable/Disable the specified listener
        function EnableServo(self, servoNum)
            self.listeners(servoNum).Enabled = true;
        end
        function DisableServo(self, servoNum)
            self.listeners(servoNum).Enabled = false;
        end

        % Trigger joint movement
        function move(self, time)
            notify(self, "TurnServo", CustomEventData(time))
        end
    end
end
```

The super and subclasses for the link objects were modified to work with the **Robot** event class. As shown below, the **Link** class was converted into a handle class so that modified class objects do not need to be returned after positions and rotations are updated in the **move** methods.

```
classdef Link < handle
    properties (Constant = true)
        g = 9.81;
    end
    properties
        m (1, 1) double {mustBeReal} = 0;
        W
        pos = [0, 0]
        rot = 0
    end
    methods
        function self = Link(m)
            if nargin == 1
                self.m = m;
            end
        end
        function W = get.W(self)
            W = self.m*self.g;
        end
    end

    % The abstract attribute requires all subclasses to override this
    % method with their own
    methods (Abstract = true)
        move
    end
end
```

Additionally, both the **RevoluteJoint** and the **PrismaticJoint** classes were modified so that their **move** methods accept the correct inputs for event callback functions, print out the event source's name, and update the **pos** and **rot** values using the custom event data.

```
classdef RevoluteJoint < Link
    properties
        speed = 10 % deg/sec
    end
    methods
        function self = RevoluteJoint(m)

            % This argument block sets the input m to zero if it is
            % not provided
            arguments
                m (1, 1) double {mustBeReal} = 0;
            end

            % Call the link (superclass) constructor
            self = self@Link(m)
        end

        % The class must have this method because of the superclass's
        % abstract method
        function move(self, eventSource, eventData)
            disp('Revolute joint of ' + eventSource.name + ' robot:')
            self.rot = self.rot + eventData.movementTime*self.speed;
            disp(self)
        end
    end
end
```

```
classdef PrismaticJoint < Link
    properties
        speed = 0.1 % m/sec
    end
    methods
        function self = PrismaticJoint(m)

            % This argument block sets the input m to zero if it is
            % not provided
            arguments
                m (1, 1) double {mustBeReal} = 0;
            end

            % Call the link (superclass) constructor
            self = self@Link(m)
        end

        % The class must have this method because of the superclass's
        % abstract method
        function move(self, eventSource, eventData)
            disp('Prismatic joint of ' + eventSource.name + ' robot:')
            self.pos = self.pos + [eventData.movementTime*self.speed, 0];
            disp(self)
        end
    end
end
```

The following code creates an RPR (describes the order of rotary and prismatic joints) robot using two revolute joint class instances and one prismatic joint instance. As shown by the command line output, a listener object is created and saved for each link.

```
R = Robot();
R.createLink("R", 1);
R.createLink("P", 2);
R.createLink("R", 3);
R
                     R = Robot with properties:
                         name: "RPR"
                        links: {1×3 cell}
                    listeners: [1×3 event.listener]
```

Next, the **move** method of the robot class is called, triggering the movement event and passing on a movement time of 2 seconds. The output proves that all three links' **move** methods are activated by the event and receive the 2 seconds of movement time.

```
R.move(2) % Move the robot for 2 seconds
                 Revolute joint of RPR robot:
                 RevoluteJoint with properties:
                         speed: 10
                          g: 9.8100
                            m: 3
                         W: 29.4300
                         pos: [0 0]
                          rot: 20

                 Prismatic joint of RPR robot:
                 PrismaticJoint with properties:
                        speed: 0.1000
                          g: 9.8100
                            m: 2
                         W: 19.6200
                       pos: [0.2000 0]
                           rot: 0

                 Revolute joint of RPR robot:
                 RevoluteJoint with properties:
                         speed: 10
                          g: 9.8100
                            m: 1
                          W: 9.8100
                         pos: [0 0]
                          rot: 20
```

The robot class's **EnableServo** and **DisableServo** methods can also be used to demonstrate temporary disabling of an event listener. Notice how the order that each callback function is called is the inverse order of when each listener was added or enabled (the first/oldest enabled listener is activated last).

```
R.DisableServo(1);
R.move(2)
        Revolute joint of RPR robot:
        RevoluteJoint with properties:
            speed: 10
            g: 9.8100
            m: 3
            W: 29.4300
            pos: [0 0]
            rot: 40

        Prismatic joint of RPR robot:
        PrismaticJoint with properties:
            speed: 0.1000
            g: 9.8100
            m: 2
            W: 19.6200
            pos: [0.4000 0]
            rot: 0
R.EnableServo(1);
R.move(1)
        Revolute joint of RPR robot:
        RevoluteJoint with properties:
            speed: 10
            g: 9.8100
            m: 1
            W: 9.8100
            pos: [0 0]
            rot: 30

        Revolute joint of RPR robot:
        RevoluteJoint with properties:
            speed: 10
            g: 9.8100
            m: 3
            W: 29.4300
            pos: [0 0]
            rot: 50

        Prismatic joint of RPR robot:
        PrismaticJoint with properties:
            speed: 0.1000
            g: 9.8100
            m: 2
            W: 19.6200
            pos: [0.5000 0]
            rot: 0
```

## Practice Problems

1. Create a new class named **fraction** that takes a numerator and denominator as separate inputs to the constructor. This new data type should be able to handle addition and subtraction with itself and all of MATLAB's other numeric data types. Additionally, the class should be able to convert to a **double**. The resulting fraction after all operations should be as simplified as possible, including no double negatives, and no decimals (all decimals with four or more decimal places can be rounded to avoid messing with infinite decimals). Improper fractions are okay. Make sure the class constructor can handle being given two, one, and even zero inputs.

   **Hint**: MATLAB's greatest common divisor function **gcd(a, b)** can be used to simplify fractions.
2. Create an enumeration class named **Battery** that sets properties for the nominal voltage and mass of at least three different batteries.
3. Given the **Vehicle** class shown below, define two new subclasses of this class named **Car** and **Motorcycle**. These subclasses should override the properties and methods of the superclass (choose your favorite car and motorcycle) so that the two subclasses can be used interchangeably despite representing different vehicles and having different property values and method outputs.

```
classdef Vehicle
        properties
                wheels (1, 1) int32 = 4
                cylinders (1, 1) int32 = 4
        end
        properties (Abstract)
                make string
                model string
                topSpeed double
        end
        methods
                function self = drive(self, distance)
                        disp("The _ drove " + string(distance) + " miles.")
                end
        end
end
```

4. Modify the **fraction** class from problem 1 to add a property called **value**. This value should be assigned a string showing a nice version of the fraction (E.g.: **"1/2"**). Assign a string to **value** using the different methods as follows:

   (a) Create a **get** method for **value** that returns a string whenever **value** is queried.
   (b) Use the **"PreGet"** event to assign a string to **value** before it is returned in a query. Remember that the **fraction** class must be transformed into a handle class to use events, so existing methods may need to be changed for the class to still work correctly.

5. Create a subclass of MATLAB's built-in **double** class called **whole**. This class should create a new data type to represent whole numbers (positive integers and zero) that maintains all of **double**'s operator methods. Essentially, the class should only accept real integers and zero as inputs.

# Chapter 4
# Mathematical Modeling in MATLAB

## 4.1 Symbolic Mathematics and Algebra

### *4.1.1 Symbolic Variables*

MATLAB's Symbolic Toolbox allows functions and expressions to be defined with unknown variable values to simplify and solve problems analytically. Symbolic numbers can have higher precision than other data types, but symbolic computations are slower and require more memory.

Symbolic variables, numbers, functions, and matrices can be defined with the **sym**, **syms,** and **str2sym** functions. The **syms** function creates symbolic variables and assigns them to MATLAB variables of the same name, the **sym** function returns a symbolic variable that can be assigned to any MATLAB variable, and the **str2sym** function converts a string object to a symbolic object. The **syms** function can be used to create multiple variables in one line, while the **sym** and **str2sym** functions can be used to create symbolic numbers as well as variables. Symbolic numbers reduce rounding error by saving fractions and irrational numbers as their symbolic representations rather than rounded decimals.

**syms x y** creates two symbolic variables **x** and **y** assigned to the MATLAB variables x and y, respectively.

**q1 = sym('x')** creates a symbolic variable x assigned to the MATLAB variable **q1**. This method can help avoid overwriting existing MATLAB variables, while still allowing the user to write equations with x as a symbol.

**c = sym('1/3')** creates a symbolic number representing 1 divided by 3. This helps to avoid rounding errors because MATLAB saves the number as a fraction containing two symbolic integers rather than a rounded decimal.

**q1 = str2sym('2*x - 1')** creates the symbolic expression **2x - 1** from a string and assigns it to the MATLAB variable **q1**.

Since symbolic computations can give multiple results, including complex or irrational numbers, symbolic variables can be defined with assumptions. The dimensions



M. Andalibi and S. A. Cherkauer, *Advanced MATLAB for Engineers*,
https://doi.org/10.1007/978-3-032-10677-3_4

of symbolic variables can also be defined to create matrices of symbolic variables. These symbolic matrices can be indexed the same way as numerical matrices.

**syms A [3, 1] real** creates a 3 by 1 symbolic column vector with the real number assumption assigned to the MATLAB variable **A**. This assumption forces MATLAB to use only the real solutions for the elements of **A**.

**B = sym('B', [2, 2], 'positive')** creates a 2 by 2 matrix with the positive number assumption assigned to the MATLAB variable **B**. This assumption forces MATLAB to use only the positive solutions for the elements of **B**.

Assumptions can be applied to existing symbolic variables as well. The **assume** function overwrites all existing assumptions assigned to the variable, while the **assumeAlso** function adds a new assumption to the list of existing assumptions. The **assumptions** function can be used to check all existing assumptions on a particular variable or on all variables.

**assume(x, 'integer')** replaces all existing assumptions assigned to the variable **x** with the integer assumption.

**assumeAlso(−2 < x & x < 2)** assigns two additional assumptions to the variable **x**.

**assume([x, y], 'clear')** removes all assumptions assigned to the variables **x** and **y**.

**assumptions** returns the assumptions on all variables.

**assumptions(x)** returns the assumptions on the variable **x**.

Table 4.1 shows possible assumptions that can be assigned to symbolic variables. The string expressions can be combined using a cell array, the expression assumptions can be combined using the **&** and | operators, and the string expressions can be combined with expression operators by placing the string assumptions in the **in(var, 'assumptionName')** function and then using the **&** and | operators. Table 4.2 shows how to implement some common assumption combinations.

**Table 4.1** Available symbolic assumptions

| Assumption | Details |
|---|---|
| x < a | Assumes that **x** is less than **a**. **x** and **a** can be symbolic expressions. Similar assumptions can be made with any other Boolean comparison operators, including = =, ~ =, < =, > =, and > |
| 'real' | Assumes that the variable is real |
| 'positive' | Assumes that the variable is positive |
| 'integer' | Assumes that the variable is an integer |
| 'rational' | Assumes that the variable is rational |
| 'clear' | Removes all assumptions on the variable. Cannot be used with **assumeAlso** |

**Table 4.2** Symbolic assumption examples

| Assumption | Syntax |
|---|---|
| Positive integer | assume(x, {'positive', 'integer'}) |
| Less than −2 or greater than 3 | assume(x < −2 \| x > 3) |
| An integer between 2 and 10 | assume(in(x, 'integer') & x > 2 & x < 10) |
| Not an integer | assume(~in(x, 'integer')) |
| Even | assume(x/2, 'integer') |
| Odd | assume((x – 1)/2, 'integer') |
| A multiple of pi | assume(x/pi, 'integer') |

### *4.1.2 Symbolic Functions*

Symbolic variables can be used to create symbolic functions which can be given numerical arguments similar to anonymous functions.

Symbolic function handles can be defined ahead of time using the **syms** function, or they can be created later using either the **symfun** function or normal variable assignment syntax with existing symbolic variables.

1. syms f(x, y)
   **f(x, y) = x^2 + y^2;**
   The first line creates a symfun object assigned to the MATLAB variable f that is a function of two input variables x and y. The variables x and y do not need to be defined separately. The second line defines the function f.
2. **syms f x y**
   **f(x, y) = x^2 + y^2;**
   The first line creates symbolic variables assigned to the MATLAB variables f, x, and y, respectively. The second line defines the variable f as a function of x and y, turning f into a symfun object.
3. **syms x y**
   **f = symfun( x^2 + y^2, [x, y] );**
   The first line creates symbolic variables assigned to the MATLAB variables x and y, respectively. The second defines the variable f as a function of x and y. In this method, the variable f does not need to be defined as symbolic before the equation is assigned.

Calling these symbolic functions later with numbers as inputs returns the function evaluated as symbolic objects, as shown below:

```
f(1, 1)
                    ans = 2
class(ans)
                    ans = 'sym'
```

Symbolic function objects can be added to matrices, but every symbolic function object in the matrix must be functions of the same variables. That is, they each have the same symbolic inputs. Adding symbolic functions to a matrix converts the entire matrix to a single symbolic function, which cannot be indexed normally. Attempting to index a symbolic function matrix with values in parentheses will treat the values as input arguments to the function. The symbolic function matrix will be evaluated at the input values and then returns the output as a symbolic matrix that can be indexed normally.

```
syms x(t) y(t)
A = [x, y];
A(1)
                    ans = [x(1), y(1)]
```

To index the unevaluated function matrix, the matrix can be given its symbolic variable arguments as inputs. This is shown below, continuing from the code above:

```
B = A(t);
B(1)
                    ans = x(t)
```

Indexing function matrices can be done in one line instead of two by initially defining the matrix with each symbolic function element given symbolic inputs.

```
A = [x(t), y(t)];
A(1)
                    ans = x(t)
```

### *4.1.3 Symbolic Toolbox Mathematical Functions*

Once symbolic variables, expressions, and functions have been defined, the symbolic toolbox allows them to be processed with various mathematical functions analytically. The symbolic toolbox can perform algebra, trigonometry, calculus, and linear algebra, and can then evaluate the solutions with numerical values. The symbolic toolbox has numerous functions, so this section will only cover the most used and most important functions.

Aside from the covered symbolic functions, symbolic functions, expressions, and equations can be used with typical algebraic and matrix functions such as **sin**, **acos**, **exp**, **transpose**, and **eig** to name a few.

**solve Function:** The **solve** function takes a symbolic equation and the variable to solve for as inputs. It then returns the solutions of the equation as symbolic objects, even if they are just numbers. These solutions are restricted based on the assumptions assigned to the variables. The input equation is defined using Boolean operators such as ==, ~=, =>, =<, <, and >. The **solve** function can also be used to find the solutions to a system of equations by inputting a vector of equations and a vector of desired variables. The **solve** function returns a structure with fields for each variable.

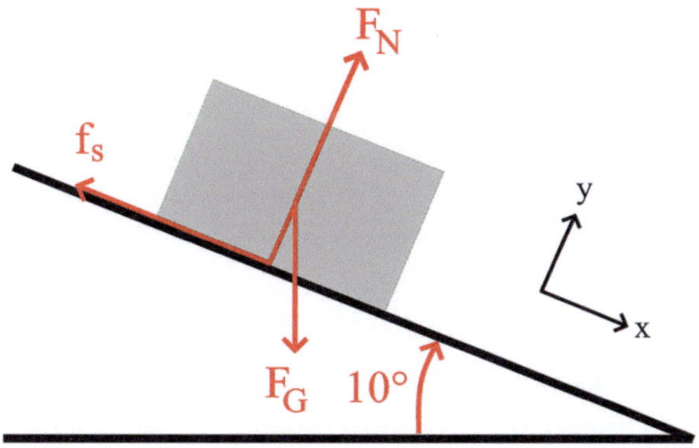


**Fig. 4.1** Stationary block

In the simple example shown in Fig. 4.1, a 25 kg block is in static equilibrium on a 10° slope just before sliding. Using the rotated frame shown, solving for the maximum static friction force can be done with a single equation:

```
syms fs
fs = solve( 25*9.81*sind(10) - fs == 0, fs )
                                fs = 5993617755764847/140737488355328
```

The returned value for **fs** is a symbolic fraction rather than a rounded decimal. This is confirmed by checking the class type of the answer, which shows that the returned fraction is a **sym** object. This **sym** object can be converted to a decimal by rounding it with **round**, converting it to a different data type such as a double, or using the **vpa** function which will be discussed later.

```
class(fs)
                                ans = 'sym'
double(fs)
                                fs = 42.5872
```

To continue on with the stationary block problem in Fig. 4.1, solving for the static friction coefficient of the maximum static friction force requires two separate equations (one for the coefficient, **us**, and one for the normal force, **Fn**). MATLAB can solve both of these equations simultaneously by combining the equations into a vector:

```
syms us Fn
eqns(1) = 25*9.81*sind(10) - us*Fn == 0;
eqns(2) = Fn - 25*9.81*cosd(10) == 0;
solve( eqns, [us; Fn] )
                                ans = struct with fields:
                        us: 5993617755764847/33991495412007076
                           Fn: 8497873853001769/35184372088832
double(ans.us)
                                ans = 0.1763
```

When solving functions with multiple solutions such as the Pythagorean Theorem, apply assumptions to the dependent variables to get only the desired solutions.

```
syms a b c
solve( a^2 + b^2 == c^2, c )
                                    ans = -(a^2 + b^2)^(1/2)
                                           (a^2 + b^2)^(1/2)
assume( c, 'positive' )
solve( a^2 + b^2 == c^2, c )
                                    ans = (a^2 + b^2)^(1/2)
```

**diff Function:** The **diff** function returns the derivative of a symbolic expression with respect to the specified variable. The default **diff** function takes the first derivative, but the order can be specified. The returned value is always symbolic, even if it is just a number.

If a particle is moving along the path **x** over the course of time, **t**:

```
syms t
x(t) = t^2 + sin(t);
```

Then that particle's velocity (**v**) is the first derivative of **x** with respect to **t**:

```
v = diff( x, t )
                                    v(t) = 2*t + cos(t)
```

And that particle's acceleration (**a**) is the second derivative of **x** with respect to **t** or the first derivative of **v** with respect to **t**:

```
a = diff( x, t, 2 )
                                    a(t) = 2 - sin(t)
a = diff( v, t )
                                    a(t) = 2 - sin(t)
```

**int Function**: The **int** function returns the integral of a symbolic expression with respect to the specified variable. The default **int** function finds the indefinite integral solution, but the limits of integration can be specified. The returned value is always symbolic, even if it is just a number.

If the same particle as before still has an acceleration of **2 - sin(t)**, then its velocity can be found using an indefinite integral with respect to **t**. Note that MATLAB does not add any unknown constant to the output of indefinite integrals:

```
v = int( a, t )
                                    v(t) = 2*t + cos(t)
```

In most practical cases, it is better to use a definite integral rather than an indefinite integral to avoid any unknown error. The change in velocity of the particle from time **t = 0** to time **t = 2** can be found using a definite integral:

```
v = int( a, [0, 2] )
                                    v(t) = cos(2) + 3
```

**limit Function**: The **limit** function returns the limit of a symbolic expression with respect to the specified variable approaching a specified value. The direction the limit approaches from can be specified. The returned value is always symbolic, even if it is just a number.

If the same particle continues moving along its path, **x**, endlessly, its final position can be shown by the limit of its position function as time, **t**, approaches positive infinity:

```
limit( x, t, inf )
                                        ans = Inf
```

Some functions may approach different values when the limit is checked from different directions due to discontinuities or asymptotes. An example would be calculating the slope of a ramp when given its inclination angle. Limit of **tan(theta)** as **theta** approaches **pi/2** from the left (negative) side:

```
syms theta
limit( tan(theta), theta, pi/2, 'left' )
                                        ans = Inf
```

Limit of **tan(theta)** as **theta** approaches **pi/2** from the right (positive) side:

```
limit( tan(theta), theta, pi/2, 'right' )
                                        ans = -Inf
```

**taylor Function:** The **taylor** function returns the Taylor series of a symbolic expression as a function of the specified variables. The order of the Taylor series is five and the expansion point is zero by default, but they can both be changed. The independent variables of the series can be given as a single variable or a vector. While Taylor functions are less accurate than evaluating an expression symbolically, MATLAB can process and evaluate the polynomial much more efficiently.

Fifth-order Taylor series of **sin(x) + y** with two independent variables:

```
taylor( sin(x) + y, [x, y] )
                                ans = x^5/120 - x^3/6 + x + y
```

Fifth-order Taylor series of **log(x)** with an expansion point of 2:

```
taylor( log(x), x, 2 )
        ans = x/2 + log(2) - (x - 2)^2/8 + (x - 2)^3/24 - (x - 2)^4/64 + (x - 2)^5/160 – 1
```

Third-order Taylor series of **log(x)**. The number following the **'Order'** keyword is the order when MATLAB stops the algorithm, which is one greater than the order of the series.

```
taylor( log(x), x, 2, 'Order', 4 )
                        ans = x/2 + log(2) - (x - 2)^2/8 + (x - 2)^3/24 – 1
```

**fourier Function:** The **fourier** function returns the Fourier transform of a symbolic expression. By default, the Fourier transform is in terms of the variable **w**, but this can be changed. The independent variables can be specified as well, but one independent variable is chosen by MATLAB by default.

The Fourier transform with the transform variable **w** and the independent variable **x**:

```
syms a x
fourier( exp(-a^2 - x^2) )
                          ans = pi^(1/2)*exp(- a^2 - w^2/4)
```

The Fourier transform with the transform variable **b** and the independent variable **x**:

```
syms a x b
fourier( exp(-a^2 - x^2), b )
                          ans = pi^(1/2)*exp(- a^2 - b^2/4)
```

The Fourier transform with the transform variable **b** and the independent variable **a**:

```
syms a x b
fourier( exp(-a^2 - x^2), a, b )
                          ans = pi^(1/2)*exp(- b^2/4 - x^2)
```

**ifourier Function**: The **ifourier** function returns the inverse Fourier transform of a symbolic expression. By default, **w** is assumed to be the independent variable and **x** is assumed to be the transformation variable, but MATLAB automatically selects an independent variable if **w** is not in the expression and an independent variable is not specified.

The inverse Fourier transform with the transform variable **x** and the independent variable **w**. The number pi is turned into a symbolic number to get an exact analytic solution rather than an approximate solution by avoiding rounding error:

```
syms a w
ifourier( sym(pi)^(1/2)*exp(- a^2 - w^2/4) )
                          ans = exp(- a^2 - x^2)
```

The inverse Fourier transform with the transform variable **b** and independent variable **w**:

```
syms a w b
ifourier( sym(pi)^(1/2)*exp(- a^2 - w^2/4), b )
                          ans = exp(- a^2 - b^2)
```

The inverse Fourier transform with the transform variable **b** and independent variable **a**:

```
syms a w b
ifourier( sym(pi)^(1/2)*exp(- a^2 - w^2/4), a, b )
                          ans = exp(- b^2/4 - w^2/4)/2
```

**laplace Function**: The **laplace** function returns the Laplace transform of a symbolic expression. By default, the Laplace transform is in terms of the variable **s**, but this can be changed. The independent variables can be specified as well, but one independent variable is chosen by MATLAB by default.

The Laplace transform with the transform variable **s** and the independent variable **x**:

```
syms a x
laplace( exp(-a*x) )
                                ans = 1/(a + s)
```

The Laplace transform with the transform variable **b** and the independent variable **x**:

```
syms a x b
laplace( exp(-a*x), b )
                                ans = 1/(a + b)
```

The Laplace transform with the transform variable **b** and the independent variable **a**:

```
syms a x b
laplace( exp(-a*x), a, b )
                                ans = 1/(b + x)
```

**ilaplace Function**: The **ilaplace** function returns the inverse Laplace transform of a symbolic expression. **s** is assumed to be the independent variable, and **t** is assumed to be the transformation variable by default, but MATLAB automatically selects an independent variable if s is not in the expression and an independent variable is not specified.

The inverse Laplace transform with the transform variable **t** and independent variable **s**:

```
syms a s
ilaplace( 1/(a + s) )
                                ans = exp(-a*t)
```

The inverse Laplace transform with the transform variable **b** and independent variable **s**:

```
syms a s b
ilaplace( 1/(a + s), b )
                                ans = exp(-a*b)
```

The inverse Laplace transform with the transform variable **b** and independent variable **a**:

```
syms a s b
ilaplace( 1/(a + s), a, b )
                                ans = exp(-b*s)
```

**Fig. 4.2** Two-DOF robot

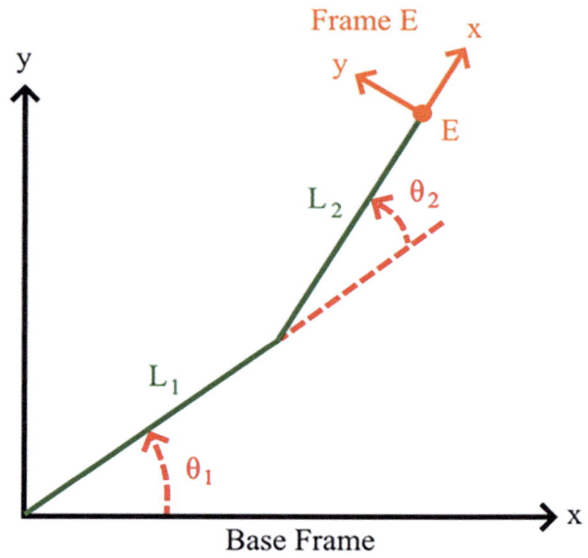


**jacobian Function**: The **jacobian** function returns the Jacobian matrix of a symbolic function with respect to a vector of symbolic variables. The function and variables can be row or column vectors and can be 1 × 1.

The following example finds the Jacobian matrix for the 2-degree-of-freedom (DOF) robotic arm shown in Fig. 4.2, where the two independent joint variables changing with time are $\mathbf{\Theta_1}$ and $\mathbf{\Theta_2}$ and the two arm lengths are $\mathbf{L_1}$ and $\mathbf{L_2}$. The Jacobian matrix of the robot describes the rotational and linear velocity of the end effector (point E) for given joint velocities. This relationship is represented by Eq. 4.1, where the end effector velocity vector, known as twist ( $\vec{\xi}$ ), is equal to the Jacobian matrix (J) times the vector of joint velocities ( $\vec{\dot{q}}$ ).

$$\vec{\xi} = J\vec{\dot{q}} \tag{4.1}$$

```
syms L1 L2 theta1(t) theta2(t) % Define the joint variables as functions of time (t)
eqn(1) = L1*cos(theta1) + L2*cos(theta1 + theta2); % x position of E from base
eqn(2) = L1*sin(theta1) + L2*sin(theta1 + theta2); % y position of E from base
eqn(3) = theta1 + theta2; % Angle of axis E relative to the base frame
jacobian( eqn, [theta1, theta2] )
                                        ans =
    [ - L1*sin(theta1(t)) - L2*sin(theta1(t) + theta2(t)), - L2*sin(theta1(t) + theta2(t))]
    [  L2*cos(theta1(t) + theta2(t)) + L1*cos(theta1(t)),  L2*cos(theta1(t) + theta2(t))]
    [                                                  1,                               1]
```

**gradient Function:** The **gradient** function returns the gradient of a symbolic function with respect to a vector of symbolic variables. The variables can be a row or column vector and can be $1 \times 1$, but the function must be $1 \times 1$. The output of the **gradient** function is the transpose of the **jacobian** function.

Computes the gradient of the function **x^2 + y^2 + z^2** with respect to **x**, **y**, and **z**:

```
syms x y z
gradient( x^2 + y^2 + z^2, [x, y, z] )
                                        ans = 2*x
                                              2*y
                                              2*z
```

Compare the above output with the transpose of **jacobian** given the same inputs:

```
transpose( jacobian( x^2 + y^2 + z^2, [x, y, z] ) )
                                        ans = 2*x
                                              2*y
                                              2*z
```

### 4.1.4 Symbolic Toolbox Helper Functions

Some functions in the Symbolic Toolbox are not necessary for the symbolic analytical computation, but make the symbolic expressions nicer to look at, easier to use, and better optimized for certain use cases. Again, not every possible function is covered, but only those more commonly used.

**simplify Function:** The **simplify** function algebraically simplifies a symbolic expression, returning an expression that is usually clearer and easier to read. By default, the **simplify** function completes one iteration of simplification, but it can be forced to perform more using the **'Steps'** keyword. Additionally, MATLAB does not always fully simplify logarithms and powers. The **'IgnoreAnalyticConstraints'** keyword forces the **simplify** function to apply more simplification rules. For an expression where multiple different simplification methods are possible, the **'All'** keyword forces MATLAB to display all different simplifications. Finally, the **'Criterion'** keyword applies another simplifying rule that prioritizes real numbers over imaginary, which can lead to simpler results.

```
syms x
simplify( [0.5*sin(x)^2 + (1/2)*cos(x)^2; exp(log(x))] )
                                        ans = 1/2
                                              x
```

Apply the **'Steps'** keyword to run **simplify** once, 20 times, and 31 times:

```
simplify( (exp(i*x) - exp(-i*x))/((exp(i*x) + exp(-i*x))*i) )
                    ans = -(exp(x*2i)*1i - 1i)/(exp(x*2i) + 1)
simplify( (exp(i*x) - exp(-i*x))/((exp(i*x) + exp(-i*x))*i), 'Steps', 20 )
                    ans = (exp(-x*1i)*1i)/cos(x) - 1i
simplify( (exp(i*x) - exp(-i*x))/((exp(i*x) + exp(-i*x))*i) , 'Steps', 31 )
                    ans = tan(x)
```

Apply the **'IgnoreAnalyticConstraints'** keyword:

```
simplify( sqrt(x^2) )
                    ans = (x^2)^(1/2)
simplify( sqrt(x^2), 'IgnoreAnalyticConstraints', true )
                    ans = x
```

Apply the **'All'** keyword, which shows two possible simplifications:

```
simplify( sin(x)*cos(x) + cos(x)*sin(x), 'All', true )
                    ans = sin(2*x)
                    2*cos(x)*sin(x)
```

Apply the **'Criterion'** keyword:

```
simplify( sym(i)^(2*i + 1), 'Steps', 10 )
                    ans = (-1)^(1/2 + 1i)
simplify( sym(i)^(2*i + 1), 'Steps', 10, 'Criterion', 'preferReal' )
                    ans = exp(-pi)*1i
```

**expand Function**: Sometimes, making an expression easier to read involves multiplying and distributing to remove sets of parentheses. The **expand** function multiplies out all parentheses in the input expression and simplifies the inputs to functions. The expansion of function inputs can be suppressed. Logarithms can also be expanded, but they require the additional keyword **'IgnoreAnalyticConstraints'** to be set to true.

```
syms x y
expand( [sin(x + y); (x - y + 5)^2; log(x/y)] )
                    ans = cos(x)*sin(y) + cos(y)*sin(x)
                    x^2 - 2*x*y + 10*x + y^2 - 10*y + 25
                    log(x/y)
```

Suppress the function input expansion, so only the parentheses are multiplied out:

```
expand( [sin(x + y); (x - y + 5)^2; log(x/y)], 'ArithmeticOnly', true )
                    ans = sin(x + y)
                    x^2 - 2*x*y + 10*x + y^2 - 10*y + 25
                    log(x/y)
```

Enable the expansion of the logarithm function:

```
expand( [sin(x + y); (x - y + 5)^2; log(x/y)], 'IgnoreAnalyticConstraints', true )
                    ans = cos(x)*sin(y) + cos(y)*sin(x)
                    x^2 - 2*x*y + 10*x + y^2 - 10*y + 25
                    log(x) - log(y)
```

**subs Functions:** Symbolic functions allow symbolic expressions to be evaluated using numeric values, but numbers can only be substituted in for the lowest level function inputs. If **f** is a function of **x**, but **x** is a function of **t**, **f(1)** will be evaluated when **t = 1** and not when **x = 1**. To circumvent this problem, the **subs** function can be used to substitute new numbers, variables, or expressions in for existing numbers, variables, or expressions. The **subs** function returns the expression containing the new values when given the original expression, what symbolic objects to replace, and with what to replace the symbolic objects.

Evaluate the following function when **x = 2** and **y = 3** rather than a value of **t**:

```
syms x(t) y(t)
f = [x^2, x^y];
subs( f, [x(t), y(t)], [2, 3] )
                                        ans(t) = [4, 8]
```

Make the returned value easier to read by substituting in **xdot** for **diff(x, t)**:

```
syms x(t) xdot(t)
f = diff(x, t) + 5*x;
subs( f, diff(x(t), t), xdot )
                                        ans(t) = xdot(t) + 5*x(t)
```

**odeFunction:** Using the **subs** function to evaluate symbolic functions can be quite slow and computationally heavy. To speed up code, the **odeFunction** function can be used to convert a symbolic function to a MATLAB function handle, which can handle numeric inputs and computation much more quickly. Once converted, the function cannot be processed further by Symbolic Toolbox functions. The **odeFunction** accepts the symbolic function to be converted, a vector of input arguments, and any other parameters. Additionally, the generated function handle could be saved as a function file, overwriting any file with the same name provided. The symbolic function must only have one input argument at the lowest level, meaning every input argument specified for the **odeFunction** must be functions of the same, single variable.

Create a function handle for **x^2 + y^2**, where **x** and **y** are the input arguments. **x** and **y** are only functions of **t**. Notice that **odeFunction** always places the shared independent variable as the first input argument (in this case, the variable **t**). Additionally, the specified higher level input variables (**x** and **y**) are always combined into a column vector represented by the variable **in2**:

```
syms x(t) y(t)
odeFunction( x^2 + y^2, [x, y] )
                                        ans = @(t,in2)in2(1,:).^2+in2(2,:).^2
ans(0, [2; 2])
                                        ans = 8
```

Create a function handle for **x + y + g**, where **g** is a constant parameter. The value of **g** can be set using an anonymous function. The **args** variable used in the anonymous function can have any name and is used to represent the **x** and **y** input column vectors:

```
syms x(t) y(t) g
f = odeFunction( x + y + g, [x, y], g )
                          f = @(t,in2,param2)param2+in2(1,:)+in2(2,:)
f = @( t, args ) f( t, args, 9.81 )
                                f = @(t,args)f(t,args,9.81)
```

Save the function generated as a function file:

```
syms x(t) y(t) g
odeFunction( x + y + g, [x, y], g, 'File', 'Filename', 'Comments', 'Hello World');
```

This code saved a file under the name **Filename.m**, with contents shown below:

```
function expr = Filename(t,in2,param4)
%Filename
%    EXPR = Filename(T,IN2,PARAM4)

%    This function was generated by the Symbolic Math Toolbox version 9.0.
%    23-May-2024 17:33:42

%Hello World
x = in2(1,:);
y = in2(2,:);
expr = param4+x+y;
```

**vpa Function:** Symbolic numbers, while precise, can make reading code outputs unnecessarily difficult. Rounding helps, but the **round** function does not work when there are symbolic variables present. The **vpa** function allows rounding even in expressions containing unknown symbolic variables. The inputs to **vpa** are a symbolic expression and the number of significant digits to leave.

Round the symbolic number to four significant digits while leaving the symbolic variables untouched:

```
syms x
vpa( sym('20000/71773')*x, 4 )
                                    ans = 0.2787*x
```

### 4.1.5 Symbolic Units

MATLAB's symbolic toolbox can also be used to keep track of units throughout complex computations.

To define units, an instance of the **symunit** object must be made first:

```
u = symunit;
```

Then, any Metric or Imperial unit can be applied to a variable or number by multiplying by the corresponding **symunit** structure field. In the case that the number is a zero, the unit can be applied instead using a cell array.

Define some physics constants with their corresponding units:

```
T = {0, u.Celsius};
g = 9.81 * u.m/((u.s)^2);
m = 10 * u.slug;
L = 3 * u.cm;
```

MATLAB does not automatically combine different units of the same measurement, but calling **simplify** forces MATLAB to simplify the units as well:

```
g*m*L
                    ans = (2943/10)*((1*[cm]*[m]*[slug])/[s]^2)
simplify(g*m*L)
                    ans = 29430*((1*[cm]^2*[slug])/[s]^2)
```

Now all the distance units are combined into centimeters, but the units are not standard. The **unitConvert** function can help by either individually converting units, or by converting everything to a common standard at one time. The unit standards that MATLAB supports are **'CGS'**, **'EMU'**, **'ESU'**, **'GU'**, **'SI'**, and **'US'**.

Convert everything to SI units. The returned number is an example of MATLAB using fractions to avoid rounding cutoff errors in decimals:

```
unitConvert( g*m*L, 'SI' )
        ans = (8727410809141101/2032000000000000)*((1*[kg]*[m]^2)/[s]^2)
```

Convert the **cm*slug/s^2** to Newtons:

```
unitConvert( g*m*L, u.N )
        ans = (8727410809141101/2032000000000000)*1*[N]*[m]
```

Finally, units can be removed with the **separateUnits** function. This function returns both the number and the associated units as two separate symbolic objects:

```
[num, unit] = separateUnits( g )
                    num = 981/100
                    unit = 1*([m]/[s]^2)
```

### 4.1.6 Symbolic Plotting

Plotting symbolic functions works very similar to normal MATLAB plotting, except the function does not need to be evaluated with a list of numeric values beforehand. The **fplot** function takes a singular symbolic function, or two functions for the x- and y-axes as inputs and creates a plot figure as normal. The plotting interval can be specified as well. All other post-processing of the figures can be done using the same methods as other MATLAB plot functions.

An example of plotting a symbolic function is shown below and in Fig. 4.3:

```
syms f(x)
f(x) = tan(x);
fplot( f )
title('Symbolic plot of tan(x)')
```

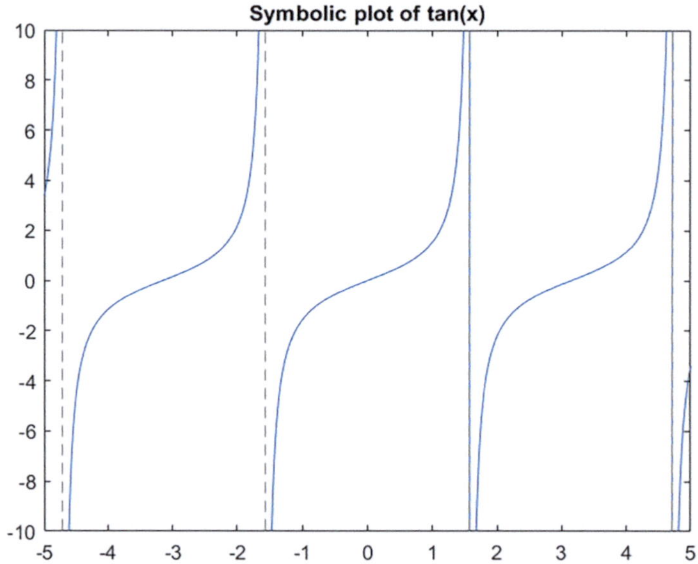


**Fig. 4.3** Plot of the symbolic function f(x) = tan(x)

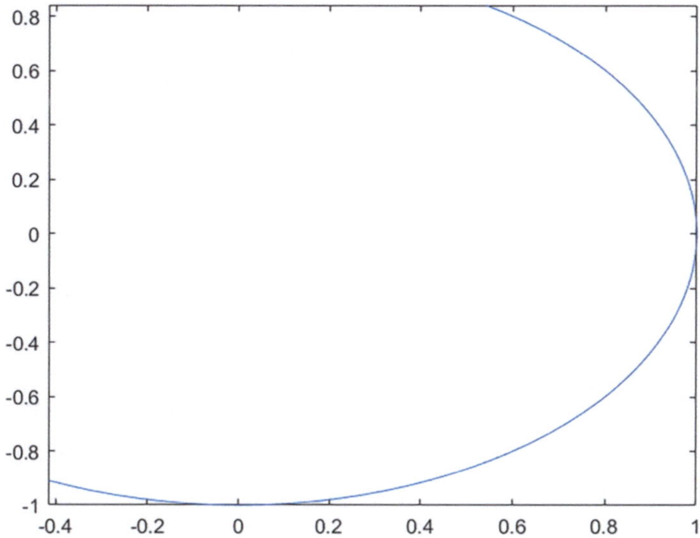


**Fig. 4.4** Plot of a symbolic parametric function

An example of plotting a symbolic function for the x- and y-axes (parametric function) with a specified range of $\mathbf{t = -2}$ to $\mathbf{t = 1}$ is shown below and in Fig. 4.4:

```
syms x(t) y(t)
x(t) = cos(t);
y(t) = sin(t);
fplot( x, y, [-2, 1] )
```

## 4.2 Solvers for ODEs and PDEs

### *4.2.1 Introduction to MATLAB ODE Solvers with Ode45*

Ordinary differential equations **(ODEs)** can be defined using MATLAB's symbolic toolbox using the methods described in Sect. 4.1, but ODEs generally cannot be solved analytically. Luckily, MATLAB has a variety of numerical ODE solvers that can offer solutions with different tradeoffs between accuracy and efficiency. All of MATLAB's numerical solvers are limited to solving first-order differential equations, so differential equations of higher orders must be reduced either manually or by using the **reduceDifferentialOrder** function. This function takes the symbolic differential equation and independent variable as inputs and returns a column vector of first-order differential equations that represent the original and a column vector of the state variables.

Any robot's dynamics can be represented by Eq. 4.2, where **M** is the robot's mass matrix, **C** is the robot's Christoffel symbols of the First Kind matrix, **B** is the robot's friction matrix, **g** is the gravity affecting each of the robot's joints, **Q** is the torques and forces applied to each joint of the robot, and **q** is the joint position vector.

$$M\ddot{\vec{q}} + C\dot{\vec{q}} + B\vec{q} + \vec{g} = \vec{Q} \tag{4.2}$$

This second-order differential equation will be used to represent the same 2-DOF robot shown in Fig. 4.2 for some examples in this section, where the matrices and vectors are defined below. How these matrices and vectors are obtained is not important for this chapter. In this example, the elements of the vector **q**, **q1** and **q2**, represent the joint variables $\mathbf{\Theta_1}$ and $\mathbf{\Theta_2}$, respectively. In addition to what is given in the figure, assume the robot experiences no friction, its joints have no gearboxes, and the masses of the two links are **m1** and **m2**, respectively.

```
syms L1 L2 m1 m2 q1(t) q2(t) g T1 T2
q = [q1; q2];
M = [m1*L1^2/3 + m2*(L1^2 + L2^2/3 + L1*L2*cos(q2)), ...
      m2*(L2^2/3 + L1*L2*cos(q2)/2);
      m2*(L2^2/3 + L1*L2*cos(q2)/2), ...
      m2*L2^2/3];
C = [-L1*L2*m2*sin(q2)*diff(q2, t)/2, -L1*L2*m2*sin(q2)*(diff(q1, t) + diff(q2, t))/2;
      L1*L2*m2*sin(q2)*diff(q1, t)/2, 0];
B = [0, 0; 0, 0];
gVec = [-g*m2*(L1*cos(q1) + L2*cos(q1 + q2)/2) - L1*g*m1*cos(q1)/2;
         -L2*g*m2*cos(q1 + q2)/2];
Q = [T1; T2];

eqns = M*diff(q, t, 2) + C*diff(q, t) + B*q + gVec == Q;
```

The first step in solving any ODE in MATLAB is simplifying higher order problems such as this second-order equation into first-order differential equations. One way to reduce orders is using the **reduceDifferentialOrder** function, which takes vectors of the equations and variables as inputs. Notice that the variables required by the function are **q1** and **q2**, not the variable **t**. The variables **Dq1t** and **Dq2t** represent the first time derivative of **q1** and **q2**, respectively, and higher derivative orders would be represented using more t's at the end of the variable name. The returned equations are all equal to zero. In the robot example, **L1**, **L2**, **m1**, **m2**, and **g** are all treated as constant parameters with respect to **q1**, **q2**, and **t**.

```
[eqns, vars] = reduceDifferentialOrder( eqns, q )
eqns =
diff(Dq1t(t), t)*(m2*(L1^2 + cos(q2(t))*L1*L2 + L2^2/3) + (L1^2*m1)/3) - T1 +
m2*diff(Dq2t(t), t)*(L2^2/3 + (L1*cos(q2(t))*L2)/2) - g*m2*((L2*cos(q1(t) + q2(t)))/2 +
L1*cos(q1(t))) - (L1*g*m1*cos(q1(t)))/2 - (L1*L2*m2*sin(q2(t))*Dq1t(t)*Dq2t(t))/2 -
(L1*L2*m2*sin(q2(t))*Dq2t(t)*(Dq1t(t) + Dq2t(t)))/2

m2*diff(Dq1t(t), t)*(L2^2/3 + (L1*cos(q2(t))*L2)/2) - T2 + (L2^2*m2*diff(Dq2t(t), t))/3 -
(L2*g*m2*cos(q1(t) + q2(t)))/2 + (L1*L2*m2*sin(q2(t))*Dq1t(t)^2)/2

Dq1t(t) - diff(q1(t), t)

Dq2t(t) - diff(q2(t), t)

vars =
q1(t)
q2(t)
Dq1t(t)
Dq2t(t)
```

ODEs can be broken into two categories, stiff and non-stiff. An ODE is considered stiff if some numerical methods become unstable while solving, when given certain initial conditions. If the step size is small enough, non-stiff ODE solvers can be used to solve stiff ODEs, otherwise more specialized stiff ODE solvers must be implemented. While stiff solvers are computationally heavy, they can converge to the stiff solutions in fewer iterations than non-stiff solvers with extremely small steps, making them potentially more efficient for solving stiff ODEs.

Since non-stiff ODEs are quite common, and non-stiff numerical methods can be used to solve many stiff problems as well, non-stiff ODE solvers are a good choice to try first when tackling ODEs. Of all of MATLAB's non-stiff ODE solvers, **ode45** has a good balance between efficiency and accuracy, and uses the robust Runge-Kutta Dormand-Prince algorithm.

As inputs, the **ode45** function takes an ODE function handle, a range for time, and the initial conditions. Additionally, an optional input in the form of an **odeset** object can be used to control aspects of the solving algorithm including setting the absolute and relative error tolerances. The number of parameters that can be set using **odeset** is too extensive to cover in this section and are typically not required for solving most ODEs, so only a few settings are covered in this chapter.

The ODE function can be created multiple ways, but it needs to follow a general format where it takes a derivative variable (usually time represented by **t**), a column vector of the independent variables, and any extra constants as inputs and returns a column vector of the first time derivatives of the independent variables.

One method of generating an ODE function handle is by simplifying a differential equation using **reduceDifferentialOrder**, and then converting the equations into the form **M*diff(vars, t) = F** using the **massMatrixForm** function. The mass matrix form can be solved for the vector of first time derivatives of the independent variables, and these expressions can then be converted into an ODE function using the **odeFunction** command described previously. The **odeFunction** command returns the function handle in the exact form required by all of MATLAB's numerical ODE solvers.

```
[M, F] = massMatrixForm( eqns, vars )
M =
[ 0,  0, m2*(L1^2 + cos(q2(t))*L1*L2 + L2^2/3) + (L1^2*m1)/3, m2*(L2^2/3 +
(L1*cos(q2(t))*L2)/2)]
[ 0,  0,  m2*(L2^2/3 + (L1*cos(q2(t))*L2)/2),  (L2^2*m2)/3]
[-1,  0,                                    0,            0]
[ 0, -1,                                    0,            0]

F =
T1 + g*m2*((L2*cos(q1(t) + q2(t)))/2 + L1*cos(q1(t))) + (L1*g*m1*cos(q1(t)))/2 +
(L1*L2*m2*sin(q2(t))*Dq1t(t)*Dq2t(t))/2 + (L1*L2*m2*sin(q2(t))*Dq2t(t)*(Dq1t(t) +
Dq2t(t)))/2
T2 + (L2*g*m2*cos(q1(t) + q2(t)))/2 - (L1*L2*m2*sin(q2(t))*Dq1t(t)^2)/2
-Dq1t(t)
-Dq2t(t)
```

```
fHandle = odeFunction( M\F, vars, L1, L2, m1, m2, g , T1, T2 )
fHandle =  function_handle with values:
@(t,in2,param1,param2,param3,param4,param5,param6,param7)[...]
```

A second method of generating an ODE function handle is by manually reducing the differential equation to first order, and then writing the reduced form into a function. Both methods produce the exact same results when solved.

```
function varDots = fHandle( t, vars, L1, L2, m1, m2, g, T1, T2 )
    % vars is of the form [q1; q2; q1dot; q2dot]
    % varDots is of the form [q1dot; q2dot; q1ddot; q2ddot]

    % Define all the matrices with the new numeric values
    q = vars(1:2);
    qdot = vars(3:4);
    M = [m1*L1^2/3 + m2*(L1^2 + L2^2/3 + L1*L2*cos(q(2))), ...
         m2*(L2^2/3 + L1*L2*cos(q(2))/2);
         m2*(L2^2/3 + L1*L2*cos(q(2))/2), ...
         m2*L2^2/3];
    C = [-L1*L2*m2*sin(q(2))*qdot(2)/2, -L1*L2*m2*sin(q(2))*(qdot(1) + qdot(2))/2;
         L1*L2*m2*sin(q(2))*qdot(1)/2, 0];
    B = [0, 0; 0, 0];
    gVec = [-g*m2*(L1*cos(q(1)) + L2*cos(q(1) + q(2))/2) - L1*g*m1*cos(q(1))/2;
            -L2*g*m2*cos(q(1) + q(2))/2];
    Q = [T1; T2];

    % q1ddot and q2ddot are found by solving the equations at each
    % iterative value for the vars
    qddot = M\(Q - C*qdot - B*q - gVec);

    varDots = [qdot; qddot];
end
```

Next, the **odeset** function can be used to define additional requirements for the numerical solver, such as setting the relative error tolerance to $10^{-8}$ and the absolute error tolerance to $10^{-10}$:

```
options = odeset( 'RelTol', 1e-8, 'AbsTol', 1e-10 );
```

Finally, the **ode45** function can be called to solve the desired differential equation. In the following example, **ode45** is given a time range of 0 to 10 s; initial conditions of **q1 = 0.5236** rad, **q2 = 0.2618** rad, **q1dot = 0** rad/s, and **q2dot = 0** rad/s; the constant parameters set to **L1 = 1** m, **L2 = 2** m, **m1 = 1.5** kg, **m2 = 3** kg, **g = 9.81** m/s^2, **T1 = 1** Nm, and **T2 = 2** Nm; and additional options defined using **odeset** (as shown above). The function **ode45** returns a column vector of all the time points and a matrix of the values of **vars** at every time point. If assigned to only one variable, it returns all contents as a structure object, and if it is assigned to no variables, then MATLAB automatically plots the results (Fig. 4.5).

```
fHandle = @(t, vars) fHandle(t, vars, 1, 2, 1.5, 3, 9.81, 1, 2);
[t, vars] = ode45( fHandle, [0, 10], [0.5236; 0.2618; 0; 0], options )
                    t = [3797x1 double]
                    vars = [3797x4 double]
```

```
struct = ode45( fHandle, [0, 10], [0.5236; 0.2618; 0; 0], options )
                    struct = struct with fields:
                    solver: 'ode45'
                    extdata: [1×1 struct]
                    x: [0 1.6824e-05 1.0094e-04 5.2154e-04 ... ]
                    y: [4×950 double]
                    stats: [1×1 struct]
                    idata: [1×1 struct]
```

```
ode45( fHandle, [0, 10], [0.5236; 0.2618; 0; 0], options )
```

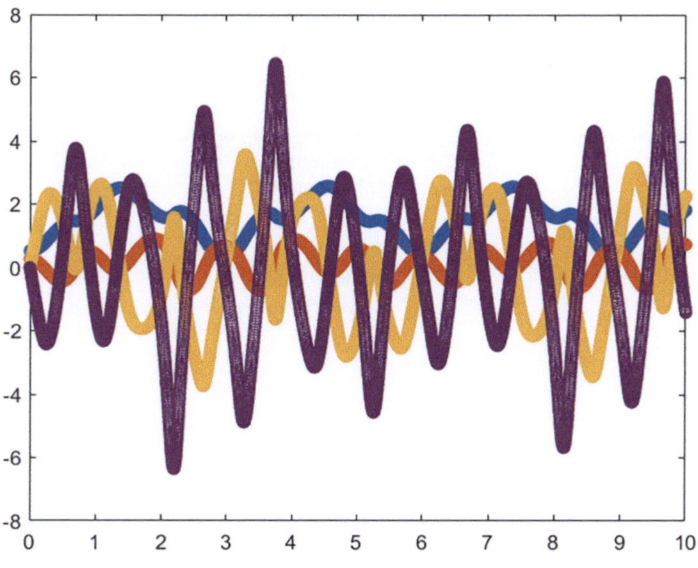


**Fig. 4.5** Plot of q1, q2, q1dot, and q2dot versus time from the solution of the function fHandle. q1 is blue, q2 is orange, q1dot is yellow, and q2dot is purple

### 4.2.2 Variable-Step ODE Solvers

The **ode45** function as well as many other ODE functions that will be discussed later are variable-step and variable-order (VSVO) solvers. VSVO solvers adjust the step size throughout the calculation rather than using a constant step size to ensure that per-step error remains at a given relative and absolute tolerance. Variable step-size ODE solvers are used when there is a large variation in the size of the derivative. They take small time steps when the variables vary rapidly and can take larger time steps when the variables vary slowly. Computing the step size adds to the computational overhead at each step, but can reduce the total number of steps, and hence simulation time. One common way to adjust the step size, **h**, in adaptive solvers is using Eq. 4.3, where $\varepsilon$ is a given relative error constant and **r** is the absolute value of the difference between the calculated **y** value (solution) for a step size **h** and for a step size of **h/2** divided by the **h** value (Eq. 4.4).

$$h_{new} = 0.9\frac{\varepsilon}{r}h_{old} \tag{4.3}$$

$$r = \frac{\left|y_{h,\,n+1} - y_{h/2,\,n+1}\right|}{h} \tag{4.4}$$

Variable step-size ODE solvers are not appropriate for deterministic real-time applications because the computational overhead of taking a time step varies over the course of an application. You can use fixed step size, deterministic ODE solvers for real-time applications, since these ODE solvers do not estimate the local per-step error and maintain a fixed step size throughout a calculation. VSVO solvers can be forced to use a constant step size, or any other step pattern, by providing a vector of time values rather than the start and end times. In the following example, rather than letting MATLAB automatically choose a step size, the step size was specified as 0.5 s (Fig. 4.6).

```
[t, vars] = ode45( fHandle, 0:0.5:10, [0.5236; 0.2618; 0; 0], options )
                              t = [21x1 double]
                              vars = [21x4 double]
```

```
ode45( fHandle, 0:0.5:10, [0.5236; 0.2618; 0; 0], options )
```

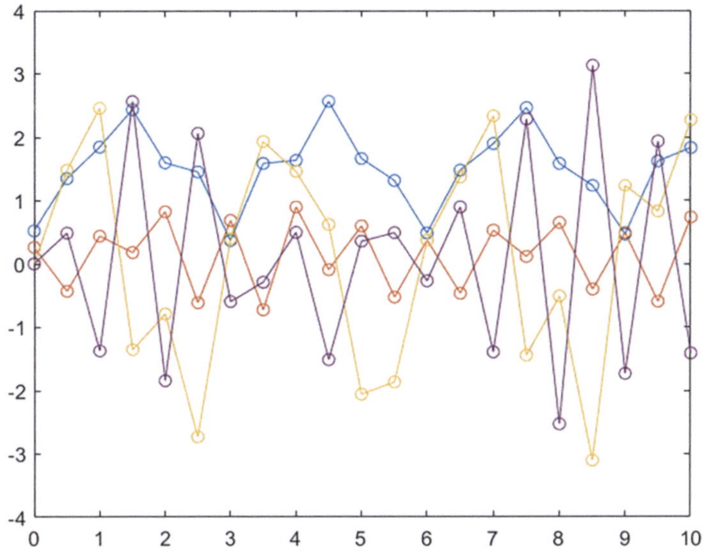


**Fig. 4.6** Plot of q1, q2, q1dot, and q2dot versus time with a 0.5 s time step from the solution of the function fHandle. q1 is blue, q2 is orange, q1dot is yellow, and q2dot is purple

In MATLAB's **odeset** command, there is a **'Refine'** option that specifies a factor by which the number of output points should increase in each step. While most solvers have a default refine value of 1, **ode45** uses a default value of 4 to compensate for its tendency to take large steps.

```
options = odeset('Refine', 5)
```

The **odeset** command also has options for initial step size and maximum step size:

```
options = odeset('InitialStep', 1e-3, 'MaxStep', 1e-2)
```

Furthermore, it has an option to show failed and successful steps:

```
options = odeset('Stats', 'on')
```

### *4.2.3 Solving Non-stiff ODEs Without Ode45*

While it is often a good start, **ode45** is not always the best solution for solving non-stiff ODEs. The **ode45** function is a good balance between efficiency and accuracy, meaning there are solvers that are faster but less accurate, and solvers that are slower but more accurate. ODE solvers other than **ode45** include **ode23**, **ode113**, **ode78**, and **ode89**. The **ode23** function uses the Bogacki-Shampine Runge–Kutta method, which

**Table 4.3** Reference table for MATLAB's non-stiff ODE solvers

| Solver | Accuracy | When to use |
|---|---|---|
| ode45 | Medium | Always works to try first. If it runs slow or fails, try another |
| ode23 | Low | More efficient than **ode45** for problems with large error tolerance or if a problem is moderately stiff |
| ode113 | Low to high | More efficient than **ode45** for problems with small error tolerance or when the ODE function is intensive to calculate |
| ode78 | High | More efficient than **ode45** if the solution is smooth and has a high accuracy requirement |
| ode89 | High | More efficient than **ode78** when the solution is extremely smooth, the solution interval is very large, or when error tolerances are extremely small |

is much faster than **ode45**, but is less robust and accurate. Both **ode45** and **ode23** are single step solvers, meaning they only need the solution immediately preceding the current time point. In contrast, **ode113** is a VSVO differential equation solver. It uses the Adams–Bashforth-Moulton method to solve equations of orders 1 to 13, but often requires the solutions for multiple previous time points. The **ode113** function can be faster and more accurate than **ode45** when the ODE is a smooth function. Similarly, **ode78** and **ode89** can be faster and more accurate than **ode45** when solving smooth functions by performing the Runge–Kutta seventh- and eighth-order continuous extensions, respectively. All four of these functions take the same inputs and can be used in the same ways as **ode45**. Table 4.3 can be used as a reference table for when to use each of the non-stiff ODE solvers.

### *4.2.4 Solving Stiff ODEs More Efficiently*

One reason that the **ode45** function may run inefficiently or fail to solve is if the differential equation is stiff. An ODE is stiff, as defined previously, when a solver such as **ode45** becomes unstable when solving at certain initial conditions. If a problem is suspected of being stiff, specialized stiff ODE solvers can be used to get solutions in a shorter amount of time and with fewer iterations. MATLAB's built-in stiff solvers include **ode15s**, **ode23s**, **ode23t**, and **ode23tb**. Once again, all four of these functions use the same inputs and formats as **ode45**.

The **ode15s** function is a VSVO solver that has the best accuracy of the four stiff ODE solvers. Alternatively, **ode23s**, **ode23t**, and **ode23tb** could all potentially be more efficient than **ode15** for problems that have less strict tolerances. The **ode23s** function is a single step solver that uses a modified Rosenbrock formula, which can make this function more efficient than **ode15s** when solving stiff problems with rapidly changing solutions. **ode23s** requires the Jacobian of the differential equation, which can significantly slow down the solving algorithm unless the Jacobian is provided ahead of time using **odeset**.

The following code solves the robot equations from before using **ode23s**. The time range is much larger to show the difference in **ode23s** run time.

```
tic
[t, vars] = ode23s( fHandle, [0, 1000], [0.5236; 0.2618; 0; 0] );
toc
                                    Elapsed time is 5.108239 seconds.
```

The following code solves the robot equations from before using **ode23s** again, this time providing the Jacobian. The Jacobian must be provided as a function handle if it is variable, similar to the ODEs themselves. The **vars** variable used in the **jacobian** and **odeFunction** functions is the output from the **reduceDifferentialOrder** function from before, and the **M** and **F** are the outputs from the **massMatrixForm** function.

```
J = jacobian(M\F, vars);
JHandle = odeFunction( J, vars, L1, L2, m1, m2, g, T1, T2 );
options = odeset( 'Jacobian', @(t, vars) JHandle(t, vars, 1, 2, 1.5, 3, 9.81, 1, 2) );

tic
[t, vars] = ode23s( fHandle, [0, 1000], [0.5236; 0.2618; 0; 0], options );
toc
                                    Elapsed time is 2.162663 seconds.
```

While the time difference is not large in this simple example, providing the Jacobian in more complex stiff problems could greatly increase the efficiency of **ode23s**, potentially making it even faster than **ode15s**. If the equations have a variable mass matrix, **ode23s** cannot be used. The **ode23t** function uses the trapezoidal rule, which can make it more efficient than **ode23s** when the problem is only moderately stiff, or when the solution should not contain any numerical damping. Finally, the **ode23tb** function uses the TR-BDF2 method, which could solve stiff ODEs more efficiently than **ode15s** if the solution does not require tight tolerances (result is imprecise). Table 4.4 can be used as a reference table for when to use each of the stiff ODE solvers.

**Table 4.4** Reference table for MATLAB's stiff ODE solvers

| Solver | Accuracy | When to use |
|---|---|---|
| ode15s | Low to medium | Try when **ode45** fails or is inefficient and the problem is stiff. Also use to solve differential algebraic equations (DAE) |
| ode23s | Low | More efficient than **ode15s** for problems with large error tolerance. Provide the Jacobian for best efficiency |
| ode23t | Low | Use if the problem is only moderately stiff to get a solution without numerical damping. Can be used to solve DAEs |
| ode23tb | Low | Possibly more efficient than **ode15s** for problems with large error tolerance |

### 4.2.5 Solving Implicit ODEs with MATLAB

There is one last numerical ODE solver provided by MATLAB. This function, **ode15i**, is a VSVO solver designed to solve implicit differential equations. Unlike all previously discussed ODE solvers, **ode15i** takes two separate initial conditions, one for the independent variable and one for the first time derivative of the independent variable. Additionally, the ODE function handle must be in the form of **f(t, y, y') = 0**. Similar to **ode23s**, **ode15i** solves the Jacobian every step, so using **odeset** to define the Jacobian ahead of time can improve performance.

The code below solves an implicit ODE using **ode15i**. The solution plot output by **ode15i** is shown in Fig. 4.7.

```
syms x(t) xdot(t)
f(t) = xdot*x + x^2 + t*x;
fHandle = odeFunction( f, [x, xdot] );
ode15i( @(t, x, xdot) fHandle(t, [x; xdot]), [0, 5], 0, 1 );
```

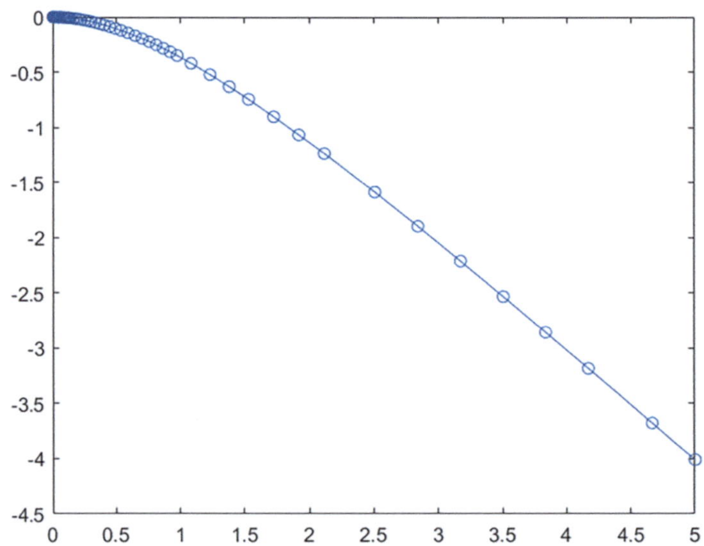


**Fig. 4.7** Solution of an implicit ODE using ode15i

### 4.2.6 Solving PDEs with MATLAB

As with ordinary differential equations, MATLAB can solve **partial differential equations (PDEs)**, where PDEs are equations that depend on multiple different independent variables. Unlike with ODEs, MATLAB only has one PDE solver, called **pdepe**. MATLAB divides PDEs into two categories, those with time derivatives are considered parabolic while those without time derivatives are considered elliptic. For **pdepe** to solve the PDE system, at least one of the provided equations must be parabolic. Additionally, **pdepe** can only solve one-dimensional problems, meaning the systems of equations can only have three variables **x**, **u**, and **t**, where.

1. **x** is the independent special variable (such as distance in one direction).
2. **t** is the independent time variable.
3. **u** is the dependent variable being differentiated with respect to **x** and **t**. **u** is a column vector with a size matching the number of equations in the system.

To define the inputs for the **pdepe** solver, the PDE equations must be put into the same format as Eq. 4.5, where **c**, **f**, and **s** are functions of **x**, **t**, **u**, and **du/dx**.

$$c\frac{du}{dt} = x^{-m}\frac{d}{dx}(x^m f) + s \tag{4.5}$$

In this equation, $\mathbf{a} \leq \mathbf{x} \leq \mathbf{b}$ and $\mathbf{t_0} \leq \mathbf{t} \leq \mathbf{t_f}$, where **a**, **b**, $\mathbf{t_0}$, and $\mathbf{t_f}$ are finite values. The variable **m**, known as the symmetry constant, can be 0, 1, or 2 whether the equation is symmetrical across a plane, axis, or point (planar, cylindrical, spherical symmetries), respectively. If **m** is greater than zero, then the lower limit of **x** (the variable **a**) must be greater than or equal to zero. The variable **f**, known as the flux term, must depend on the term **du/dx**. The variable **c** is a diagonal matrix, where the diagonal elements can only be greater than or equal to zero.

The **pdepe** function takes the variable **m**, a function representing the PDEs, a function representing the initial conditions, a function representing the boundary conditions, a range of **x** values, and a range of time values and outputs the solution to the PDE system over the given mesh. The PDE function should accept **x**, **t**, **u**, and **du/dx** as inputs and return the **c**, **f**, and **s** matrices.

An example of a one-dimensional PDE commonly used in mechanical engineering is the wave equation, shown in Eq. 4.6, where **u** is the displacement of the wave from its equilibrium position at a certain distance **x** at a time **t**, and **c** is the speed of the wave propagation (Fig. 4.8).

$$\frac{d^2u}{dt^2} = c^2\frac{d^2u}{dx^2} \tag{4.6}$$

The propagation speed can be calculated using Eq. 4.7, where **T** is the tension in the string or cable and **μ** is the linear mass density of the string (mass divided by the length).

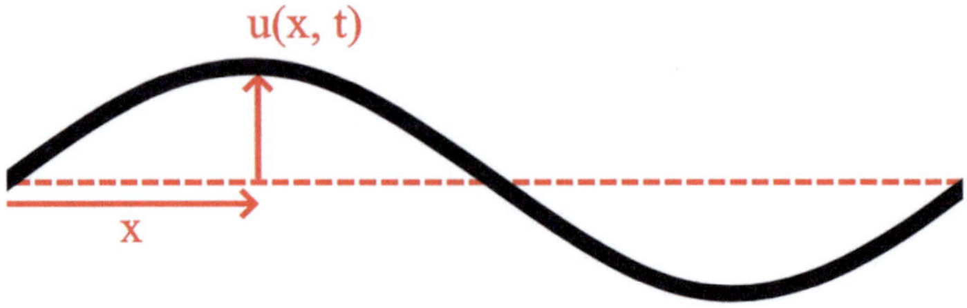


**Fig. 4.8** Wave propagation reference

$$c = \sqrt{\frac{T}{\mu}} \tag{4.7}$$

If the two ends of a string with length **L** are fixed in place, the boundary conditions of the problem are defined by Eqs. 4.8 and 4.9.

$$u(0, t) = 0 \tag{4.8}$$

$$u(L, t) = 0 \tag{4.9}$$

Additionally, if the string is raised in the middle initially so that it forms an upside-down parabola, the initial condition can be defined by Eq. 4.10.

$$u(x, 0) = -x(x - L) = Lx - x^2 \tag{4.10}$$

For this example, assume that **L = 10** m, **T = 2** Nm, and **mu = 0.3**. The first step to solve this PDE is to rewrite the system into the form described by Eq. 4.5 (Eq. 4.11). If the system has more than one equation, each of the **u**, **f**, and **s** terms becomes a column vector where each row represents one equation.

$$[1]\frac{d}{dt}\left[\frac{du}{dt}\right] = \frac{d}{dx}\left[c^2\frac{du}{dx}\right] + [0] \tag{4.11}$$

In this example, the $\mathbf{x^m}$ term is not present, meaning **m** is zero. Additionally, the equations for **c**, **f**, and **s** are easily pulled out and placed into a function:

```
function [c, f, s] = pdeFunc(x, t, u, dudx)
        % x, u, and dudx are column vectors each containing one value

        c = 1; % The function only requires the diagonal values of c
        f = 2/0.3 .* dudx; % This is the c^2 * dudx term
        s = 0;
end
```

Next, a function for the initial conditions needs to be written. This function returns the value of **u** as a column vector given values of **x**:

```
function u = icFunc(x)
        u = 10*x - x^2;
end
```

Writing the function for the boundary conditions is a little more complicated. The boundary conditions all need to be combined and written into the form shown in Eq. 4.12, where **f** represents the flux found using the PDE system, **p** is a column vector, and **q** is a diagonal matrix:

$$p(x, t, u) + q(x, t)*f\left(x, t, u, \frac{du}{dx}\right) = 0 \tag{4.12}$$

In this example, the boundary conditions when **x = a = 0** would be written in the form shown by Eq. 4.13:

$$[u] + [0]\left[c^2\frac{du}{dx}\right] = 0 \tag{4.13}$$

And the boundary conditions when **x = b = L** would be written in the form shown by Eq. 4.14:

$$[u] + [0]\left[c^2\frac{du}{dx}\right] = 0 \tag{4.14}$$

Using these two boundary condition equations, the vector **p** and the diagonal of **q** can be defined and then written into a MATLAB function as follows:

```
function [pa, qa, pb, qb] = bcFunc(xa, ua, xb, ub, t)
        pa = ua; % The a refers to the first BC
        qa = 0;
        pb = ub; % The b refers to the second BC
        qb = 0;
end
```

Now that all the input functions have been defined, the system of PDEs can be solved. The ranges for **x** and **t** must be provided as full vectors rather than just start and end points. The **pdepe** function does not automatically graph the solution like the ODE solvers. Instead, it outputs one matrix containing meshes of values (if there are multiple u's, the meshes are separated by the third index). The solution of the code below is shown in Fig. 4.9.

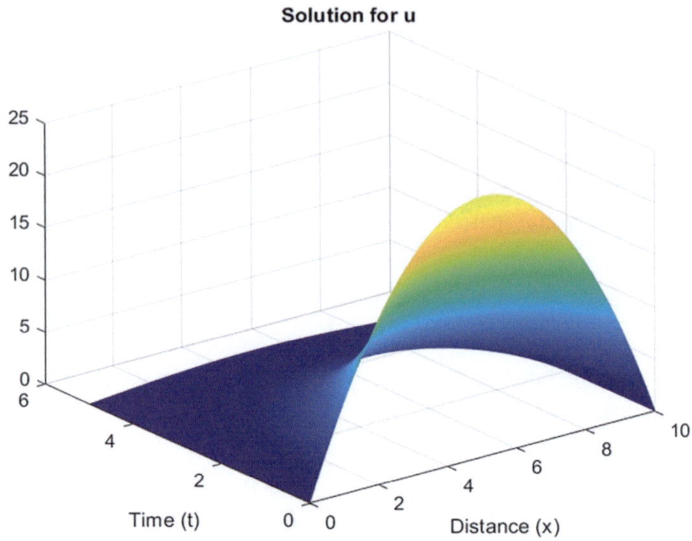


**Fig. 4.9** Solution to the wave propagation PDE using MATLAB's pdepe function

```
x = 0:0.01:10;
t = 0:0.01:5;
u = pdepe( 0, @pdeFunc, @icFunc, @bcFunc, x, t );

surf( x, t, u )
title( 'Solution for u' )
xlabel( 'Distance (x)' )
ylabel( 'Time (t)' )
shading interp
```

## 4.3 Ode Events

### *4.3.1 Introduction to ODE Events*

Event functions take an expression in terms of the state-space (dummy) variables used to define the first-order governing ODEs and detect an event (raise a flag) when that expression crosses zero. The integration of the ode solver momentarily stops when an event happens, allowing a user-defined function called the **event response function** to be run.

### 4.3.2 *Creating ODE Events*

To illustrate this topic, consider the motion of a projectile without the air drag. The projectile is launched from an initial location with an initial velocity and launch angle and falls back down with an acceleration in the y-direction of **-g**. Each time that the projectile hits the ground, it bounces back with a modified y-component of velocity given by Eq. 4.15, where **e** is the coefficient of restitution in the range [0 1]. The x-component of velocity does not change due to impact if the ground is horizontal.

$$\left(V_y\right)_{\text{after impact}} = -e \times \left(V_y\right)_{\text{before impact}} \tag{4.15}$$

An anonymous function for governing the ODE is shown below. This function matches the form returned by **odeFunction**, where the input is a column vector containing the two position coordinates **x** and **y** and their first derivatives **xdot** and **ydot**, and the output is the first and second derivatives of **x** and **y**.

```
zdot = @(t, z) [z(3); z(4); 0; -9.81];    % z(1) = x, z(2) = y, z(3) = xdot, z(4) = ydot
```

In this simulation, the projectile is supposed to bounce once it reaches the ground. Without any event applied, the projectile's y value, **z(2)**, will change from positive to negative when passing through the ground. An ODE event is triggered when a given function changes signs (crosses zero), therefore the ODE event that detects when the projectile hits the ground can be triggered when **z(2)** changes sign, as shown by the anonymous function below:

```
bounceEvent = @(t, z) z(2);
```

Next, an event response function must be defined which determines what must be done should an event occur. The function below sets the y value to zero (when y value crosses zero, it might go from a positive to a negative value and never exactly become zero), changes the y-component of velocity according to Eq. 4.15, and does not stop the integration operation permanently (**stop = false**).

```
function [stop, z] = bounceResponse(t, z)
        stop = false;
        e = 0.6;
        z(2) = 0;
        z(4) = -e*z(4);
end
```

Finally, the ode event can be defined using the command **odeEvent**:

```
E = odeEvent(EventFcn = bounceEvent, Direction = "descending", ...
             Response = "callback", CallbackFcn = @bounceResponse);
```

Note that the anonymous function **bounceEvent** is defined as the event function, and the user-defined function **bounceResponse** is defined as the callback function.

**Direction = "descending"** means y values are crossing zero by going from positive to negative values. It can be set to "**ascending**" or "**both**" as well.

Furthermore, the type of response to an event can be set using a callback function defined by **Response = "callback"**, "**proceed**", or "**stop**".

### 4.3.3 *Using ODE Events*

With the ODE event defined, the ODE can be solved by creating an **ode** object and then calling the **solve** command. The most important inputs to **ode** are the ode function, initial values, and the ode event. The **ode** function automatically selects a solver, but the user can choose the solver by setting the **Solver** property equal to a string containing the name of one of the previously discussed ODE solvers. The automatic selection of **ode** can be sped up slightly by setting **Solver** equal to "**stiff**" or "**nonstiff**" instead. The **ode** object has many other properties that can be set and shares many properties with **odeset**, but these properties are too numerous to cover in depth in this chapter. Once an **ode** object is created, it can be solved by passing the object to the **solve** command along with the solution time span, as shown below:

```
x0 = 0;
y0 = 0;
v0 = 50;
alpha = 60;
z0 = [x0 y0 v0*cosd(alpha) v0*sind(alpha)];

F = ode(ODEFcn = zdot, InitialValue = z0, EventDefinition = E);
S = solve(F, 0, 20)
```

The output of the **solve** function is a class object that includes the following properties:

- **Time** (all time steps of the solution).
- **Solutions** (all state-space variables in the form of an N × 4 matrix, where N is the number of time steps).
- **EventTime** (all the time points when the provided event function crossed zero in the specified direction).
- **EventSolution** (state-space variables at the time of events).
- **EventIndex** (index of each triggered event, in case multiple events are given).

```
S = ODEResults with properties:
Time: [0 1.1602e-06 2.3204e-06 3.4806e-06 … ] (1×181 double)
Solution: [4×181 double]
EventTime: [8.8280 14.1248 17.3029 19.2097]
EventSolution: [4×4 double]
EventIndex: [1 1 1 1]
```

The following code is used to plot the projectile trajectory (y versus x) and annotate it with impact times (**EventTime**). Results of the code can be seen in Fig. 4.10.

```
% Plot the trajectory
plot(S.Solution(1, :), S.Solution(2, :), 'LineWidth', 4);
hold on;
xlabel('x'); ylabel('y');
title('projectile motion with no air drag bouncing back from the ground')

% Annotate the impact times
for ii = 1:length(S.EventSolution)
        plot(S.EventSolution(1,ii), 0, 'ro', 'MarkerSize', 6, 'MarkerFaceColor', 'r');
        text(S.EventSolution(1,ii), 5, sprintf('t=%.2fs', S.EventTime(ii)), ...
            'VerticalAlignment', 'bottom', 'HorizontalAlignment', 'right');
end
```

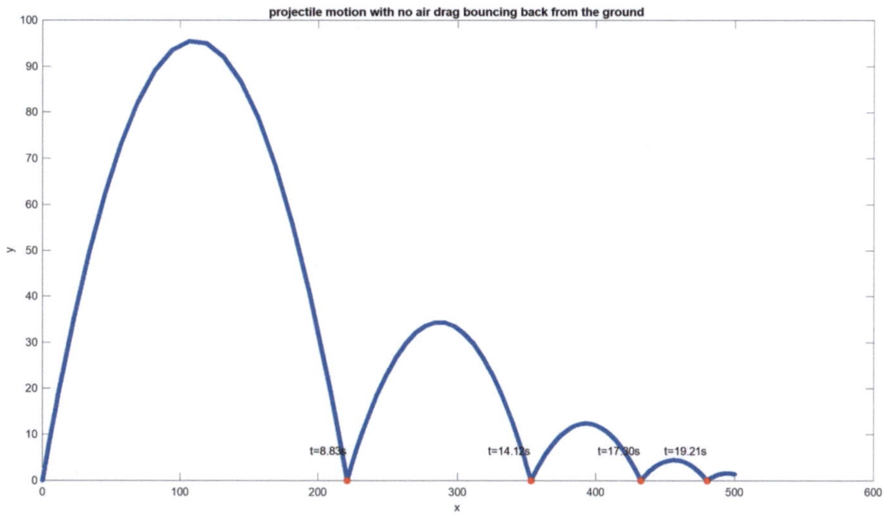


**Fig. 4.10** Using ode events to simulate a bouncing projectile

It must be emphasized that the root-finding mechanism employed by the ODE solver with the event function has these limitations (Fig. 4.11):

1. If a terminal event occurs during the first step of the integration, then the solver registers the event as nonterminal and continues integrating.
2. If more than one terminal event occurs during the first step, then only the first event registers and the solver continues integrating.
3. Zeros are determined by sign crossings between steps. Therefore, zeros of functions with an even number of crossings between steps can be missed.

**Practice Problems**

(1) The matrix **C** is a direction cosine matrix, meaning the sum of the squares of elements of any row or column is equal to one. Using this knowledge, define **C** symbolically and then use MATLAB's **solve** method to find **x** and **y**. Use assumptions to assume both **x** and **y** are positive. Round the answers to four decimal places.

$$C = \begin{bmatrix} x & 0.6124 & -0.6124 \\ 0.0000 & -0.7071 & -0.7071 \\ -0.8660 & y & -0.3536 \end{bmatrix}$$

(2) If a particle is moving at the speed **v = 3*sin(t)*cos(3*t) + 1.5*exp(−4*t)**, find the functions of the particle's position and acceleration. Assume there is no integration error. Print out the initial position and acceleration values if the initial time is **t = 0**.

(3) Substitute **cos(a)** for **ca**, **cos(b)** for **cb**, **sin(a)** for **sa**, **sin(b)** for **sb**, **cos(a + b)** for **cab**, and **sin(a + b)** for **sab** in the code below using the **subs** function. Then use the **simplify** command to get the expression into its simplest form. This expression's simplest form is a single trigonometric function.

```
syms ca cb sa sb cab sab
ex = ((-sa*sb + cb*ca)*cab + (sb*ca + sa*cb)*sab - ca) / sa;
```

(4) If a 5 lbf particle is falling at 74 m per minute 16 inches above the ground, calculate the particle's total energy relative to the ground. Use symbolic units and give the answer in Joules.

(5) Define a parametric function that returns the x- and y-coordinates as a column vector for an ellipse when given a time value. The ellipse should have a minor axis of 2 in the y-direction, should have a major axis of 5 in the x-direction, and should be centered on the point (−2, 3). Plot the ellipse and the parametric function in separate subplots from **t = 0** to **t = 7**. Make sure to title the figure, label the axes, and add a legend to the parametric function plot.

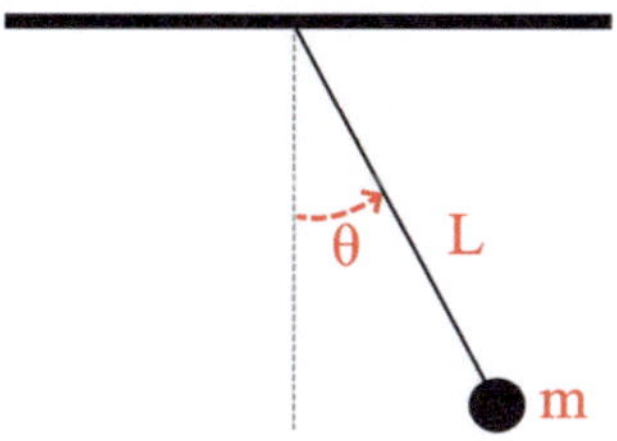


**Fig. 4.11** Simple pendulum of mass m and length L

(6) A simple pendulum of mass **m = 2** kg and length **L** = **1** m is shown in Fig. 4.11. Find the ODE for this system and then use **ode45** to solve the problem. Plot the position and velocity of the pendulum (**Θ**) versus time from **t = 0** to **t = 6** in separate subplots. Assume the pendulum starts from rest at 60°.

(7) The one-dimensional heat equation (Eq. 4.16) is a PDE that can be solved using **pdepe**, where **u** is the temperature of the material and **x** is the distance along the material (**0** < **x** < **3** m). Assume the initial temperature of the material is 110 °C, the left side of the material is also at 110 °C, and the right side has been dropped to 25 °C. Solve this PDE using **pdepe** and plot the results for 5 s.

$$\frac{du}{dt} = \frac{d^2u}{dx^2} \tag{4.16}$$

(8) Using the ODE from Problem 6, add an ODE event that bounces the pendulum back with 75% of its velocity whenever it reaches **Θ = 0**. Plot the position and velocity of the pendulum (**Θ**) versus time from **t = 0** to **t = 6** in separate subplots.

# Chapter 5
# Data Visualization in MATLAB

## 5.1 2D Plots

There are several 2D plot commands in MATLAB.

**Line plots**: command **plot(x,y,linespec)** plots vector **y** on the vertical axis versus vector **x** on the horizontal axis using the specifications for line, color, and marker in the **linespec**.

The commands below will first define the **x** and **y** vectors for the function **y = sin(x)** over the range $[0, 2\pi]$ using circle markers connected by red solid lines, and then add labels for **x** and **y** axes. Next, a title is added, the axes limits are determined, and finally the grid is turned on as can be seen in Fig. 5.1.

```
x=linspace(0,2*pi,100);
y=sin(x);
plot(x,y,'r-o');
xlabel('x');
ylabel('y');
title('y=sin(x)');
axis([0 2*pi -1 1]);
grid on;
```

Different marker shapes, different line styles, and different default colors can be seen in Fig. 5.2.

If a color other than the ones in Fig. 5.2 is desired, it can be specified using its RGB values. For instance, for the color gray, values of 0.5 for Red, Green, and Blue can be used.

```
plot(x,y,'color', [0.5 0.5 0.5]);
```

To turn the grid off, we can use:



M. Andalibi and S. A. Cherkauer, *Advanced MATLAB for Engineers*,
https://doi.org/10.1007/978-3-032-10677-3_5

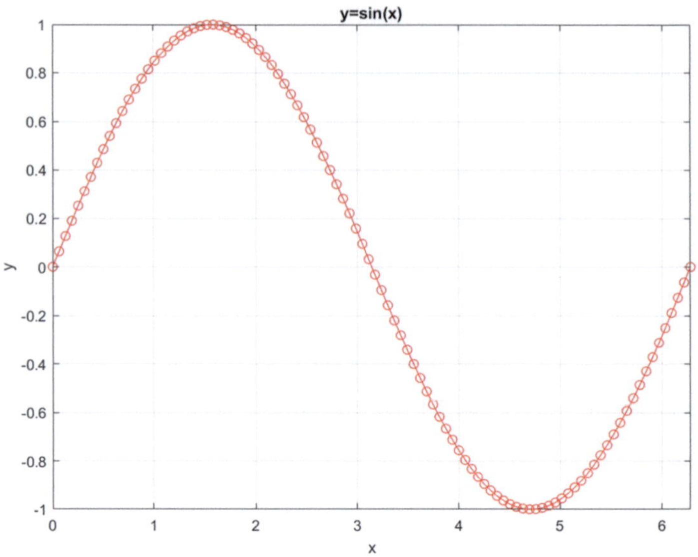


**Fig. 5.1** Plot of the function y = sin(x) using circle markers connected by solid red lines

**Marker Specifiers**

| Specifier | Marker Type |
|---|---|
| '+' | Plus sign |
| 'o' | Circle |
| '*' | Asterisk |
| '.' | Point |
| 'x' | Cross |
| 's' or 'square' | Square |
| 'd' or 'diamond' | Diamond |
| '^' | Upward-pointing triangle |
| 'v' | Downward-pointing triangle |
| '>' | Right-pointing triangle |
| '<' | Left-pointing triangle |
| 'p' or 'pentagram' | Five-pointed star (pentagram) |
| 'h' or 'hexagram' | Six-pointed star (hexagram) |
| 'none' | No marker (default) |

**Line Style Specifiers**

| Specifier | Line Style |
|---|---|
| '-' | Solid line (default) |
| '--' | Dashed line |
| ':' | Dotted line |
| '-.' | Dash-dot line |
| 'none' | No line |

**Color Specifiers**

| Specifier | Color |
|---|---|
| 'r' | Red |
| 'g' | Green |
| 'b' | Blue |
| 'c' | Cyan |
| 'm' | Magenta |
| 'y' | Yellow |
| 'k' | Black |
| 'w' | White |

**Fig. 5.2** Markers, line styles, and default colors for the plots in MATLAB

```
grid off;
```

We can plot several pairs of (x, y) data on the same plot either by using them all in the same plot command or using the command **hold on** and several plot commands as shown below:

```
x=linspace(0,2*pi,100);
y=sin(x);
z=cos(x);
plot(x,y,'r-o',x,z,'b:^');
xlabel('x');
legend('y=sin(x)','z=cos(x)','Location','best');
```

Or:

```
x=linspace(0,2*pi,100);
y=sin(x);
z=cos(x);
plot(x,y,'r-o');
hold on;
plot(x,z,'b:^');
xlabel('x');
legend('y=sin(x)','z=cos(x)','Location','best');
hold off;
```

These commands will produce the plot in Fig. 5.3, where the **legend** command is used to provide a legend for both curves at the location determined "best" (least occluding) by MATLAB.

We can access the handle of the current figure (the window containing all graphs) or the current axis (the sub window) using the command **gcf** or **gca** and use them for modifying the figure properties. For example, the default name for a figure is "**Figure #**", where # is the number of this figure, but this name can be changed as well as the color, position, and size of the figure using the code shown below and as can be observed in Fig. 5.3. Also, the values of the x-axis (or y-axis) ticks can be modified by changing the property of **gca** called **'XTickLabel'**.

```
fgHdl= gcf
set(fgHdl, 'Name', 'This is my figure');
set(fgHdl, 'Color', [1 1 1]);
set(fgHdl, 'Position', [300 100 500 500]);

axHdl = gca;
set(axHdl,'XTick',0:pi/2:2*pi)
set(gca,'XTickLabel',{'0','pi/2','pi','3*pi/2','2*pi'});
```

The direction of an axis can also be reversed (values increasing from right to left instead of left to right):

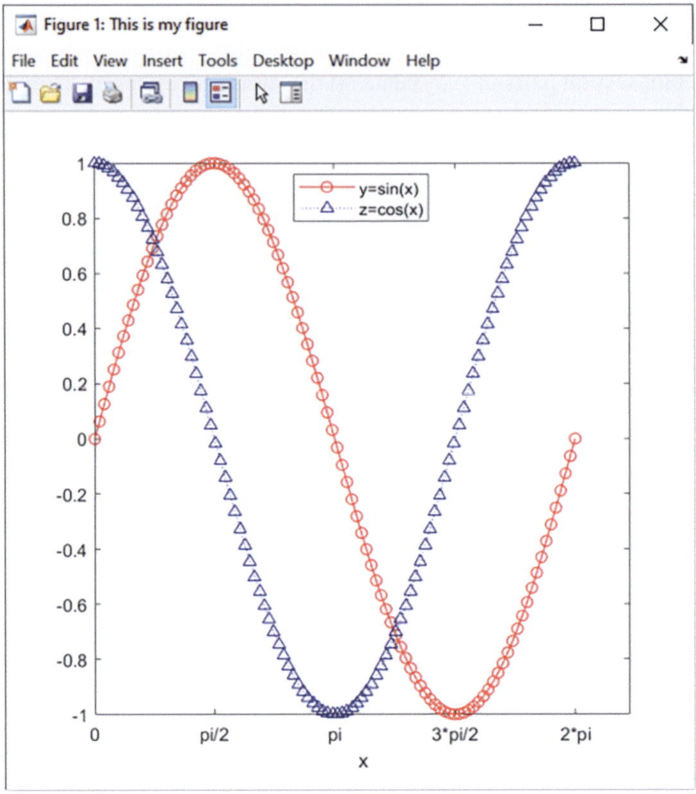


**Fig. 5.3** Plots of the functions y = sin(x) and z = cos(x) in the same figure using different line styles and a legend; figure and axes properties have been modified

```
set(gca,'XDir', 'reverse')
```

Other options for changing the color of the edges and faces, size, and thickness of the markers and lines are:

```
x=linspace(0,2*pi,100);
y2=exp(-x).*sin(pi*x);
plot(x,y2,'-ro','LineWidth', 2, 'MarkerSize', 6, 'MarkerEdgeColor', 'k', ...
'MarkerFaceColor', 'r');
xlabel('x');
ylabel('y');
title('y_{2}=e^{-x}sin(\pi x)');
```

Here, in the title command, Greek letter $\pi$ is inserted using **\pi** and subscripts and superscripts are added using _{} and ^{}. Results for this code can be seen in Fig. 5.4, while character sequences for inserting Greek and Mathematical symbols are shown in Fig. 5.5.

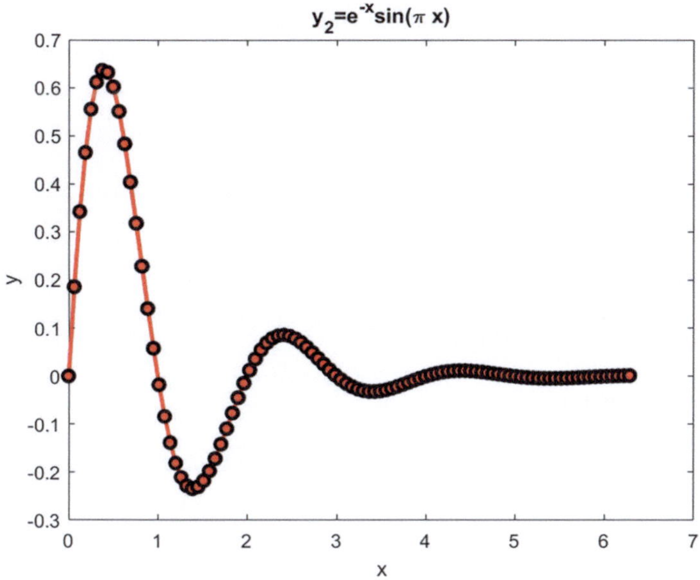


**Fig. 5.4** Modifying line, size, and thickness of markers and lines; adding Greek letters, subscripts and superscripts in the figure title

| Character Sequence | Symbol | Character Sequence | Symbol | Character Sequence | Symbol |
|---|---|---|---|---|---|
| \alpha | $\alpha$ | | | \int | $\int$ |
| \beta | $\beta$ | | | \cong | $\cong$ |
| \gamma | $\gamma$ | \Gamma | $\Gamma$ | \sim | $\sim$ |
| \delta | $\delta$ | \Delta | $\Delta$ | \infty | $\infty$ |
| \epsilon | $\varepsilon$ | | | \pm | $\pm$ |
| \eta | $\eta$ | | | \leq | $\leq$ |
| \theta | $\theta$ | | | \geq | $\geq$ |
| \lambda | $\lambda$ | \Lambda | $\Lambda$ | \neq | $\neq$ |
| \mu | $\mu$ | | | \propto | $\propto$ |
| \nu | $\nu$ | | | \div | $\div$ |
| \pi | $\pi$ | \Pi | $\Pi$ | \circ | $\circ$ |
| \phi | $\phi$ | | | \leftrightarrow | $\leftrightarrow$ |
| \rho | $\rho$ | | | \leftarrow | $\leftarrow$ |
| \sigma | $\sigma$ | \Sigma | $\Sigma$ | \rightarrow | $\rightarrow$ |
| \tau | $\tau$ | | | \uparrow | $\uparrow$ |
| \omega | $\omega$ | \Omega | $\Omega$ | \downarrow | $\downarrow$ |

**Fig. 5.5** Character sequence for Greek and Mathematical symbols

Although text and shapes/annotations, such as a textbox, arrow, ellipse, etc. can be inserted in a figure using the Insert tab, the code below demonstrates how to add them at the command window:

```
x=linspace(0,2*pi,100);
y2=exp(-x).*sin(pi*x);
plot(x,y2,'-ro','LineWidth', 2, 'MarkerSize', 6, 'MarkerEdgeColor', 'k', ...
'MarkerFaceColor', 'r');
xlabel('x'); ylabel('y');
title('y_{2}=e^{-x}sin(\pi x)');
text(3,0.4,'damped harmonic');
annotation('arrow',[0.6,0.45],[0.62,0.45]);
axHdl = gca;
set(axHdl,'FontSize',14)
```

Note that the syntax for the text command is **text(x,y,'string')** and the syntax for the annotation command is **annotation('linetype',[$x_1$ $x_2$],[$y_1$ $y_2$])**. Results can be seen in Fig. 5.6.

Two other important items to pay attention to in Fig. 5.6 is the font size increase from default to 14, which is very useful when MATLAB figures are inserted in report documents **so that the axes ticks are sufficiently visible**; and also the option for "**data tips**" in the plot window, which allows the user to find the coordinates of any marker point on the graph. If more than one data tip point is needed, you need to keep the "**alt**" key down and then press on the "data tip" button again.

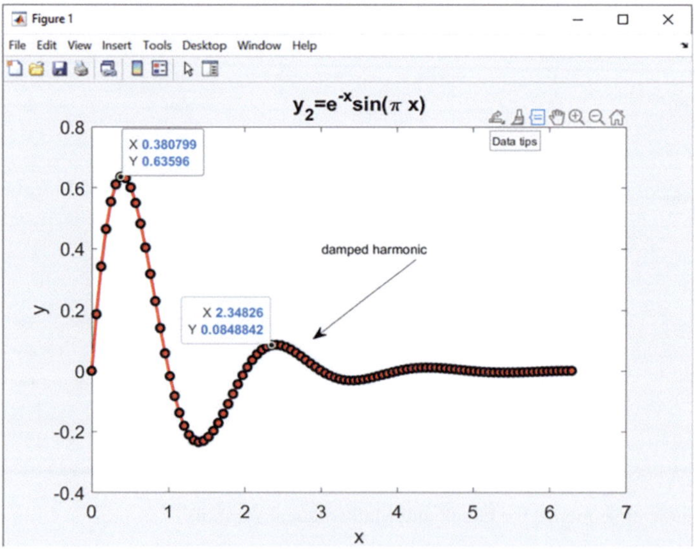


**Fig. 5.6** Adding text and annotation to a plot; Using data tips

The next option for the **plot** command is creating subplots, which are separate axes within the same figure for different graphs. An example can be seen in Fig. 5.7 using the codes below for 4 different plotting functions:

```
figure(1);
x=0.1:0.1:10;
y=exp(x);
subplot(2,2,1);
plot(x,y);xlabel('x');ylabel('y');title('y=e^{x}: plot command');
subplot(2,2,2);
semilogx(x,y);xlabel('log(x)');ylabel('y');title('y=e^{x}: semilogx command');
subplot(2,2,3);
semilogy(x,y);xlabel('x');ylabel('log(y)');title('y=e^{x}: semilogy command');
subplot(2,2,4);
loglog(x,y);xlabel('log(x)');ylabel('log(y)');title('y=e^{x}: loglog command');
```

Note that the syntax **subplot(m,n,ii)** is to create an $m \times n$ matrix of axes and plot on the ii$^{th}$ axis, which is counted from 1 up in a **row-by-row manner**. Also, note that each subplot has its own axes labels and titles.

The new plotting commands **semilogx**, **semilogy**, and **loglog** are used when one or both of the axes have to be in logarithmic scale, which is particularly applicable to functions with rapid growth, such as exponential or power functions. If you check the top left corner subplot, where the standard plot command is used, you might believe

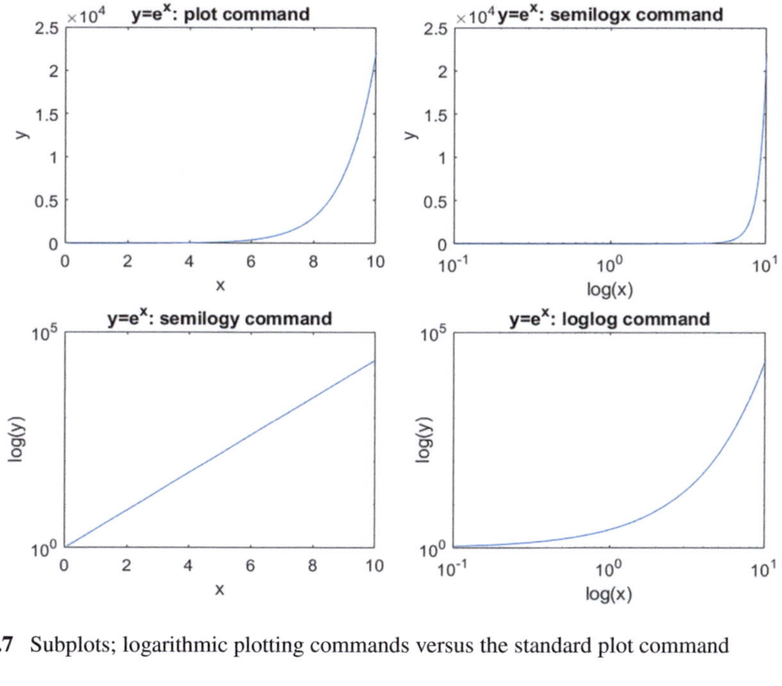


**Fig. 5.7** Subplots; logarithmic plotting commands versus the standard plot command

that the values of **y** are almost zero until **x = 4**, but this is not true. The reason is that $e^4$ is much smaller than $e^{10}$ and in a limited-size graph, it will show as zero. So, in such cases, typically the y axis or both axes are rendered in a logarithmic scale. Command **semilogx** is useful when **x** is a power function of **y**.

Also note that if a figure is given a specific number, such as **figure(1)**, and the user has several figures open and is plotting on all of them at the same time in a specific order, each figure can be made the current figure by calling its number with the **figure** command. The plotting will only be done on the current figure.

Two other advanced commands for line plots are **fimplicit** and **yyaxis**.

Command **fimplicit(f,axeslimits,linespec)** plots the implicit function **f(x,y) = 0**, which is defined using anonymous functions over the range specified by **axes-limits**. Once again, **linespec** is a placeholder for plot options as can be seen in Fig. 5.8 using the code below:

```
f = @(x,y) x.^2 + y.^3 +5*y - 3;
fimplicit(f,[-3 0 -2 2],'LineWidth',2)
xlabel('x');ylabel('y');
title('Plot of the implicit function x^{2}+y^{3}+5y-3=0')
```

Command **yyaxis** is an auxiliary tool for the **plot** command, which is used to plot two functions with different y-axis scaling, where the plot will not be rendered properly using only the plot command. Figure 5.9 shows the result of using the code below for plotting two different functions on two different y axes but on the same x

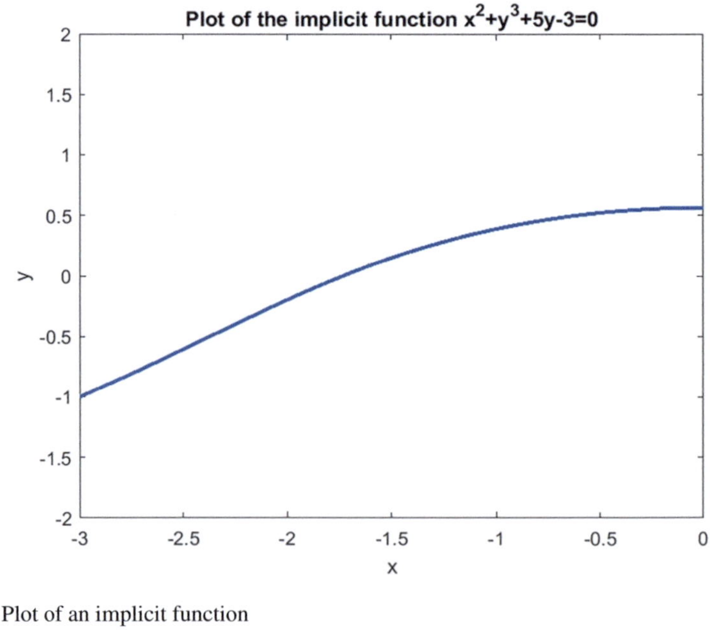


**Fig. 5.8** Plot of an implicit function

axis. Note that each axis has to be made current using **yyaxis left** and **yyaxis right** before using the **plot** command:

```
x=linspace(0,2*pi,100);
y1=sin(x);
y2=x.^2-2*x+3;
yyaxis left
plot(x,y1,'LineWidth',2)
yyaxis right
plot(x,y2,'LineWidth',2);
xlabel('x');title('yy axis demo');
legend('y_{1}=sin(x)','y_{2}=x^{2}-2x+3','Location','northeastoutside');
```

Also note that the legend can be placed outside the axes if they are cluttered.

(II) **Bar plots, Stairs and Stem plots, Pie plot:** Commands **bar(x,y)** and **barh(x,y)** can be used to create bar plots as can be seen in the top subplots of Fig. 5.10. Commands **stairs(x,y)** and **stem(x,y)** can be used to create stair and stem plots as can be seen in the bottom subplots of Fig. 5.10.

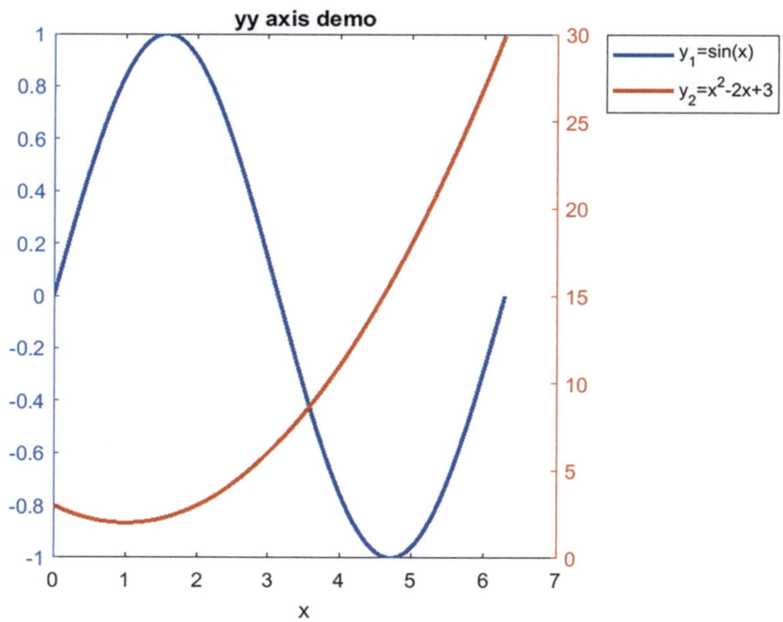


**Fig. 5.9** Using yyaxis for plotting two different y axes

```
x=linspace(0,2*pi,40);
y=sin(x);
subplot(2,2,1);
bar(x,y);
xlabel('x');ylabel('y');title('bar plot');
subplot(2,2,2);
barh(x,y);
xlabel('x');ylabel('y');title('horizontal bar plot');
subplot(2,2,3);
stairs(x,y);
xlabel('x');ylabel('y');title('stairs plot');
subplot(2,2,4);
stem(x,y);
xlabel('x');ylabel('y');title('stem plot');
```

Pie plots can be created using the command **pie(x, explode, labels)** as can be seen in Fig. 5.11 with the options of **explode** for the land areas of different continents:

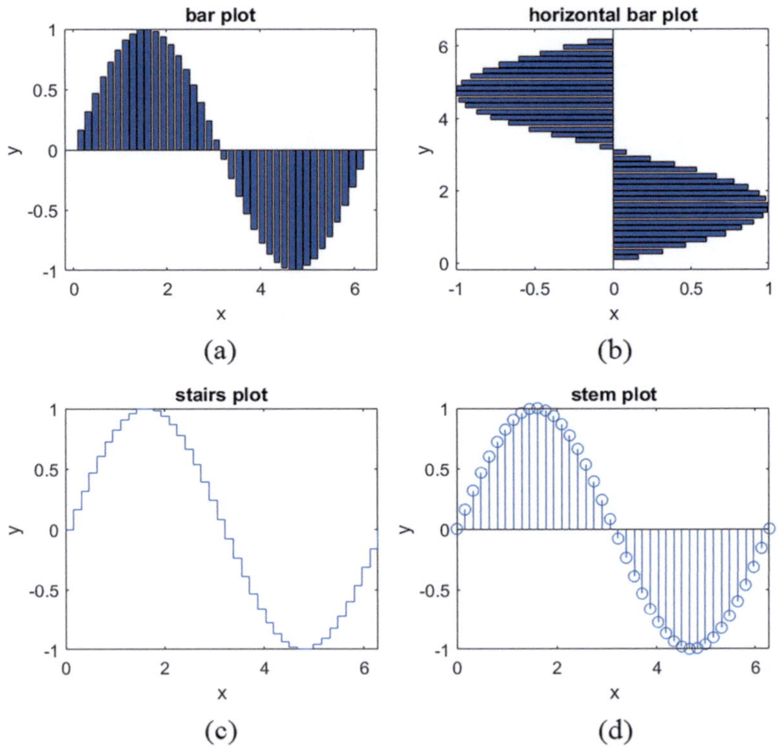


**Fig. 5.10** Bar plot (**a**), Horizontal bar plot (**b**), Stairs plot (**c**), Stem plot (**d**)

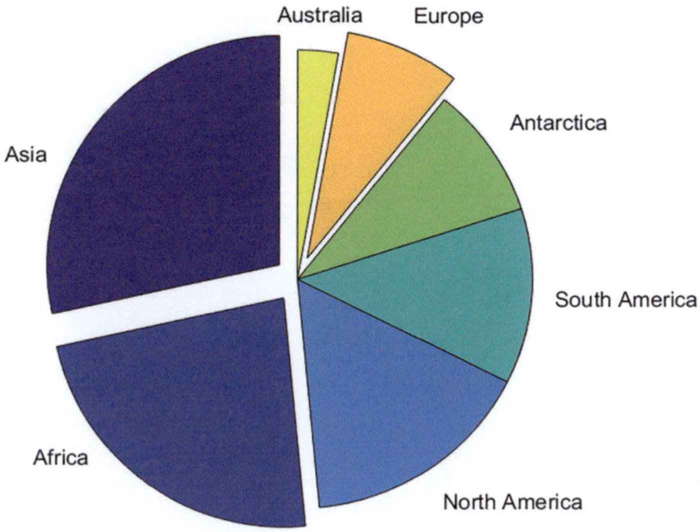


**Fig. 5.11** Pie plot for land areas of different continents with labels added and Asia, Europe, and Africa exploded

```
land_areas = [30.37, 24.71, 17.21, 13.21, 9.98, 8.5, 2.97];
continents = {'Asia', 'Africa', 'North America', 'South America', 'Antarctica', 'Europe',
'Australia'};
explode = [1, 1, 0, 0, 0, 1, 0];
pie(land_areas, explode, continents);
```

(III) **Scatter plots:** Commands **scatter(x,y,sz,c,'filled','mkr')** can be used to create scatter plots with the desired marker color, marker size, marker fill, and marker shape as can be seen in Fig. 5.12. Marker colors and sizes can be variable as can be seen in this figure.

```
x = linspace(0,3*pi,200);
y = cos(x) + rand(1,200);
sz = linspace(1,100,200);
c = linspace(1,10,length(x));
scatter(x,y,sz,c,'filled','d');
xlabel('x');ylabel('y');
title('Scatter Plot with varying colors and marker sizes');
```

There is also a command called **gscatter** which creates a scatter plot for grouped data. The code below shows a scatter plot for 3 different groups of cars made in 1970, 1976, and 1982 plotted using different marker shapes and colors, where the horizontal axis is the engine displacement, and the vertical axis is the horsepower as can be seen in Fig. 5.13.

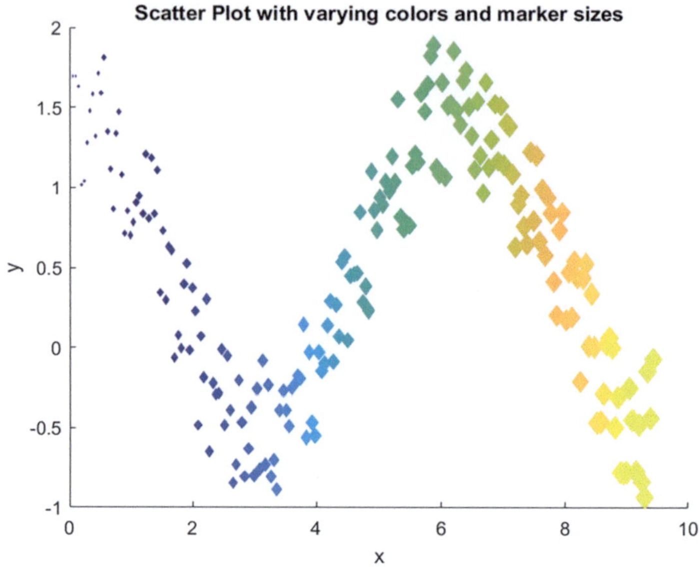


**Fig. 5.12** - Scatter plot for with size, shape, and color modifications for marker and filled

```
load carsmall;
gscatter(Displacement,Horsepower,Model_Year,'rbk','o*^',8,'on','Displacement','Horsepo
wer')
legend('location','best')
```

(IV) **Histograms:** Histograms are a type of bar plot that group data into bins. We can specify the number of bins or the ranges of the bins and normalize the frequencies to make it a Probability Density Function (PDF) as can be seen in the code below.

Results can be seen in Fig. 5.14, where two normal distributions with $\mu_1 = 2, \sigma_1 = 5$ and $\mu_2 = 3, \sigma_2 = 4$ are plotted in the same axis.

Note that an option **'FaceAlpha'** is used for the first histogram, which controls the opacity of the plot bars. Values for this parameter change from 0 to 1, where 1 indicates fully opaque.

```
x1 = 2*randn(5000,1) + 5;
x2 = 4*randn(5000,1) + 3;
nbins=30;
histogram(x1,nbins,'Normalization','pdf','FaceColor','g','FaceAlpha',0.3);
hold on;
histogram(x2,nbins,'Normalization','pdf','FaceColor','r','FaceAlpha',0.3);
title('Histogram of Random Data from a Normal Distribution');
xlabel('Value');ylabel('Frequency');
legend(' \mu=5, \sigma=2', '\mu=3, \sigma=4');
```

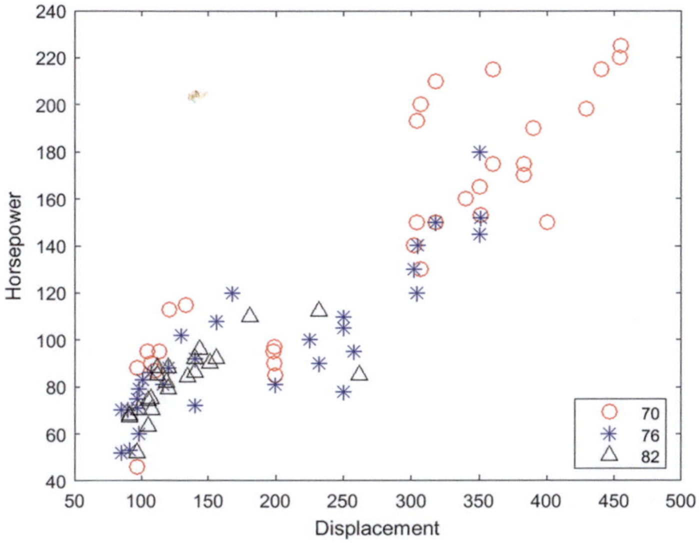


**Fig. 5.13** Grouped Scatter plot for three different groups of cars

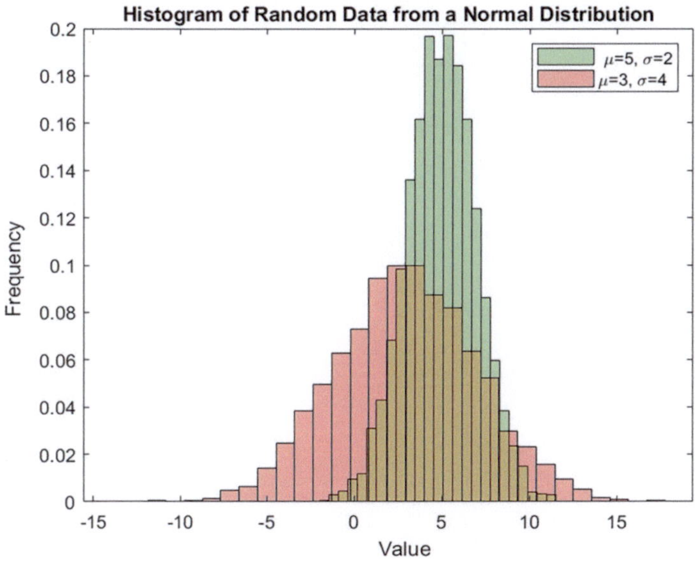


**Fig. 5.14** Histogram plots for two normal distributions with different parameters and using opacity control

(V) **Quiver Plots:** Command **quiver(X,Y,U,V)** plots arrows with directional components **U** and **V** at the Cartesian coordinates specified by **X** and **Y**, i.e., the vector field defined by (U,V) at points (X,Y). The result of using this command to plot a rotating vector field can be seen in Fig. 5.15.

```
[x, y] = meshgrid(-2:0.2:2, -2:0.2:2);
u = -y;
v = x;
quiver(x, y, u, v);
title('Quiver Plot of a rotating fluid');
xlabel('x');ylabel('y');
```

(VI) **Contour Plots:** Commands **contour** and **contourf** are introduced in this section. Command **contour(X,Y,Z,n)** plots the contour plot for the 3D surface **Z = f(X,Y)** using **n** contour lines as can be seen in Fig. 5.16a, while command **contourf(X,Y,Z)** plots the **filled** contour plot for the 3D surface **Z(X,Y)** as shown in Fig. 5.1b. You can show the contour levels on the contour lines using the option "**ShowText", true.** Command **meshgrid** will be explained in Sect. 5.2.

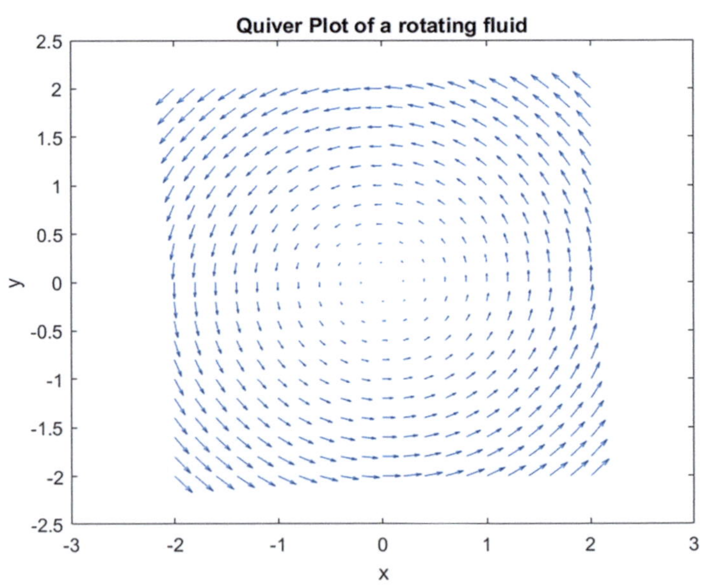


**Fig. 5.15** Quiver plot of a rotating vector field

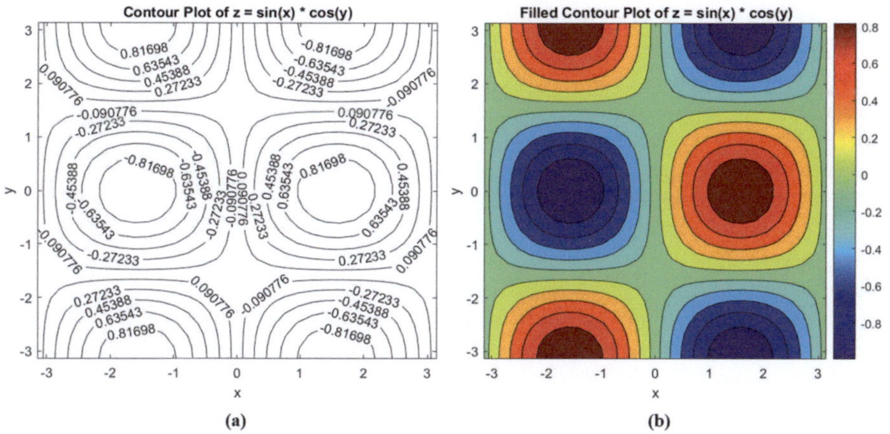


**Fig. 5.16** Contour plot (**a**); Filled contour plot (**b**)

```
x = linspace(-pi, pi, 30);
y = linspace(-pi, pi, 30);
[X, Y] = meshgrid(x, y);
Z = sin(X) .* cos(Y);
subplot(1,2,1);
contour(X, Y, Z, 10, 'LineColor', 'k',"ShowText",true);
title('Contour Plot of z = sin(x) * cos(y)');
xlabel('x');ylabel('y');
subplot(1,2,2);
contourf(X, Y, Z, 10, 'LineColor', 'k');
colormap(jet);colorbar;
title('Filled Contour Plot of z = sin(x) * cos(y)');
xlabel('x');ylabel('y');
```

(VII) **Area Plots:** The commands below can be used to fill a 2D region defined by vertices in (x,y):

**area(x,y):** plots the values in **y** against **x**.

**fill(x,y,c):** plots filled polygonal regions as patches with vertices at the (**x,y**) locations.

**patch(X,Y,C):** creates one or more patches of colored polygons (color vector is **C**) by specifying **X** and **Y** as the coordinates for each vertex.

The commands below will create the area under a damped harmonic function, a red star, and a simple house using the commands **area**, **fill**, and **patch**, respectively as shown in Fig. 5.17.

```
subplot(1,3,1);
x = linspace(0, 10, 100);
y = exp(-x/5) .* cos(x);
area(x,y);
title('area plot: y=exp(-x/5) * cos(x)');
xlabel('x');ylabel('y');
subplot(1,3,2);
angles = (0:4) * 2 * pi / 5 - pi / 2;
outer_radius = 3;
inner_radius = 3 / 2;
x_star = reshape([outer_radius * cos(angles); inner_radius * cos(angles + pi/5)], 1, []);
y_star = reshape([outer_radius * sin(angles); inner_radius * sin(angles + pi/5)], 1, []);
fill(x_star, y_star,'r');
title('fill plot: red star');
xlabel('x');ylabel('y');
axis square;
subplot(1,3,3);
x_base = [1 1 4 4];
y_base = [1 3 3 1];
x_roof = [0.5 2.5 4.5];
y_roof = [3 5 3];
patch(x_base, y_base, 'b', 'EdgeColor', 'k', 'LineWidth', 2);
patch(x_roof, y_roof, 'r', 'EdgeColor', 'k', 'LineWidth', 2);
title('patch plot: a simple house');
xlabel('x');ylabel('y');
axis([0 5 0 5]);
```

Note that the command **axis square** makes the aspect ratio of the axis 1:1. **This is useful when drawing circles, since non-square axes can render a circle an ellipse**.

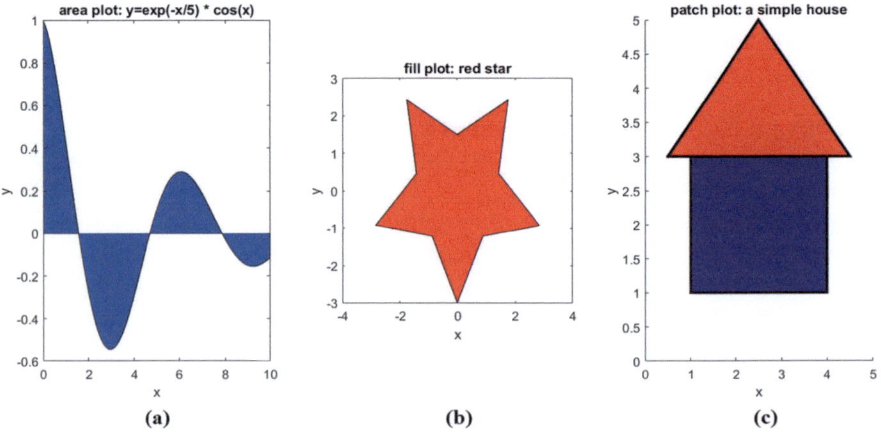


**Fig. 5.17** Area plot (**a**); Fill plot (**b**); Patch plot (**c**)

(VIII) **Heatmap Plots:** Command **heatmap(tbl,xvar,yvar)** creates a heatmap by aggregating the variables in the table **tbl**. Specify **xvar** and **yvar** as the table variables that you want to visualize.

The code below creates a heatmap plot using the average temperatures in different months of the year and modifies the properties of the plot handle variable **h** (plot handles will be discussed in more detailed in Chap. 7). Result can be seen in Fig. 5.18.

```
months = {'Jan', 'Feb', 'Mar', 'Apr', 'May', 'Jun', 'Jul', 'Aug', 'Sep', 'Oct', 'Nov', 'Dec'};
average_temperature = [30, 32, 45, 55, 65, 75, 80, 78, 70, 60, 50, 35];
h = heatmap(months, {'Avg Temp (°F)'}, average_temperature);
h.Title = 'Average Temperature and Precipitation by Month';
h.XLabel = 'Month';
h.YLabel = 'Variable';
colormap(parula);
colorbar;
h.ColorLimits = [min(average_temperature(:)), max(average_temperature(:))];
```

Two new commands are used in the code above, one is **colorbar** that adds a color bar to the side of the plot, and the **colormap(colormapname)** command which picks a specific colormap for the plot. Figure 5.19 shows the predefined colormaps in MATLAB.

(IX) **Polar Plots:** Command **polarplot(theta,rho)** plots the values of radial distances defined by **rho** versus angles defined by **theta**. The code below plots a flower using the function $r = \cos(4\theta)$ as can be seen in Fig. 5.20.

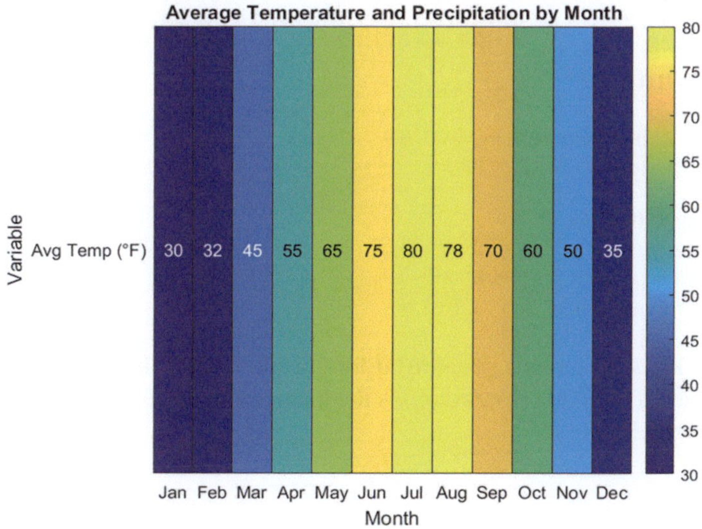


**Fig. 5.18** Heatmap plot for average temperature versus month

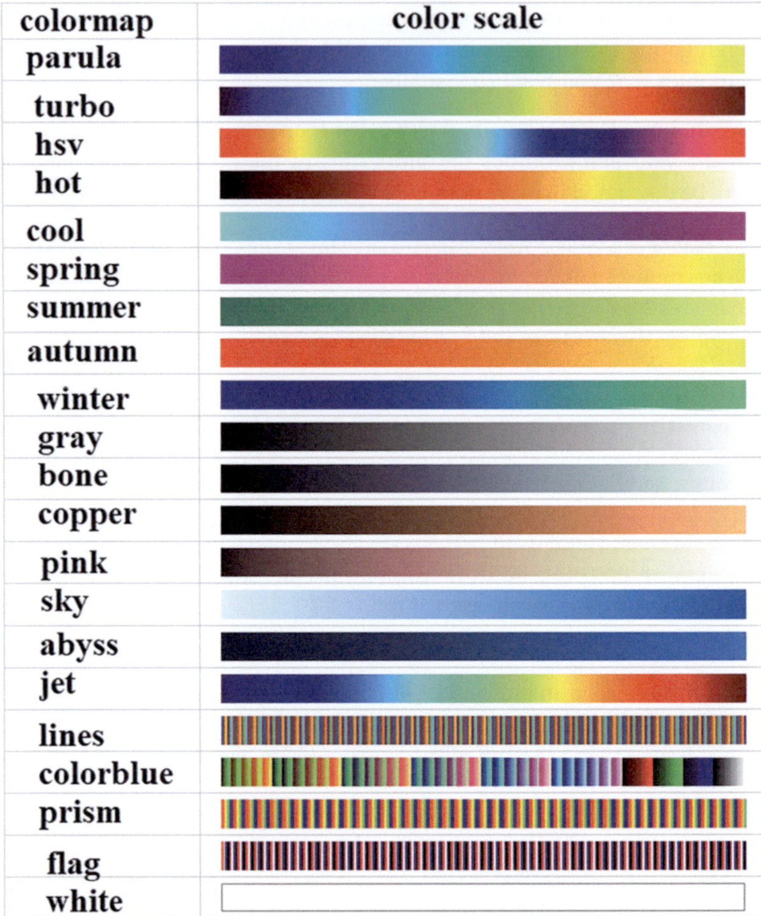


**Fig. 5.19** Predefined colormaps in MATLAB

```
theta = linspace(0, 2*pi, 1000);
r = cos(4*theta);
polarplot(theta, r, 'r', 'LineWidth', 2);
title('Flower Polar Plot');
```

(X) **Geo Plots:** Command **geoplot(lat,lon,linspec)** plots a line in geographic coordinates. The code below connects the major cities in California as can be seen in Fig. 5.21.

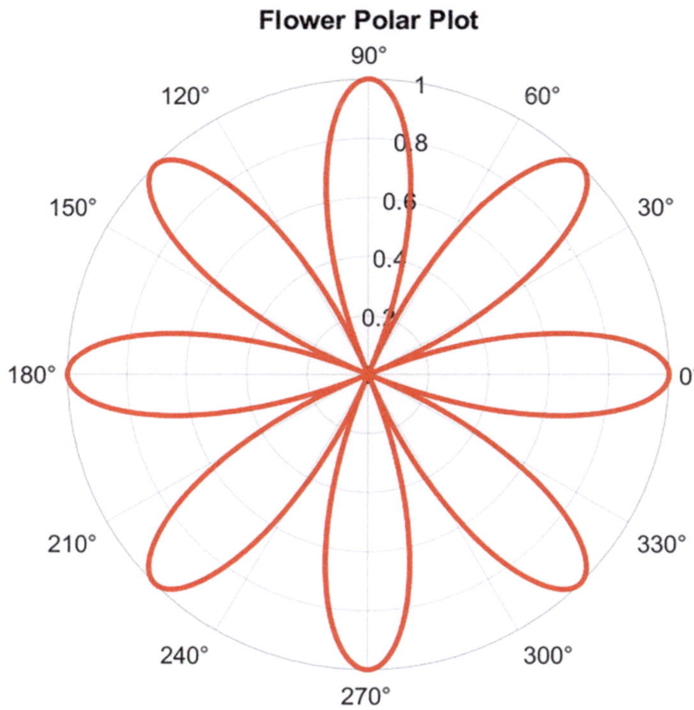


**Fig. 5.20** Polar plot of a flower

```
geobasemap('streets');
latitudes = [34.0522, 36.7783, 37.7749, 32.7157];
longitudes = [-118.2437, -119.4179, -122.4194, -117.1611];
cities = {'Los Angeles', 'Fresno', 'San Francisco', 'San Diego'};
geoplot(latitudes, longitudes, '-o', 'LineWidth', 2, 'MarkerSize', 5, 'MarkerEdgeColor', 'b',
'MarkerFaceColor', 'r');
for ii = 1:length(cities)
        text(longitudes(ii), latitudes(i), cities{ii}, 'VerticalAlignment', 'bottom',
        'HorizontalAlignment', 'right', 'Color', 'k');
end
geolimits([32 38], [-124 -114]);
title('Path Connecting Cities in California');
```

The command **geobasemap('option')** sets the basemap property for the current geographic axes or geographic bubble chart to the value specified by **'option'**. For example, the option **'streets'** is a general-purpose road map that emphasizes accurate, legible styling of roads and transit networks.

The command **geolimits** sets the latitude and longitude for the geoplot.

**Fig. 5.21** Geoplot of major cities in California connected

## 5.2 3D Plots

In this section, we look at the 3D plot commands in MATLAB for which a 3D function, surface, or matrix is needed.

One of the first steps that is typically needed in plotting 3D functions is to create 2D variables, such as **x** and **y**, over which the height variable (let's say **z**) can be defined. For instance, let's consider plotting the 3D function $z = x^2 + y^2$ over a range of [−2 2] for both **x** and **y**. While the first two lines of code below will create equally spaced vectors **x** and **y** in the range [−2 2], these values only cover the grid points on the x and y axes and not inside the region, so we need to use the command **meshgrid** to construct all required grid points for the 3D function as can be seen in the third line. Using command **whos** will reveal the difference in dimension between (x, y) and (X, Y):

```
x=-2:0.1:2;
y=-2:0.1:2;
[X,Y]=meshgrid(x,y);
whos
                    Name    Size        Bytes    Class
                    X       41x41       13448    double
                    Y       41x41       13448    double
                    x       1x41        328      double
                    y       1x41        328      double
```

Note that instead of all 3 lines of code above, we could have used the single line of code below:

```
[X,Y]=meshgrid(-2:0.1:2);
```

Now, we can define the function **Z = f(X,Y)**:

```
Z=X.^2+Y.^2;
```

Then, we can plot **Z** using different 3D function plotter commands discussed below.

I. **Mesh Plot:** it will create a mesh from the 3D surface, i.e., only surface curves are shaded while the insides remain empty. Using the code shown below, we can generate the mesh plot for **Z**, where the result is shown in Fig. 5.22. Command **view(az,el)** sets the azimuth and elevation angles of the 3D view for the plot.

```
[X,Y]=meshgrid(-2:0.1:2);

mesh(X,Y,Z);title('mesh plot for Z=X^{2}+Y^{2}');
xlabel('x');ylabel('y');zlabel('z');
view(-62,3);
```

II. **Surf Plot:** surf plots can display the same function using different shading options, as can be seen in Fig. 5.23. In part (a) of this figure, edges of surface

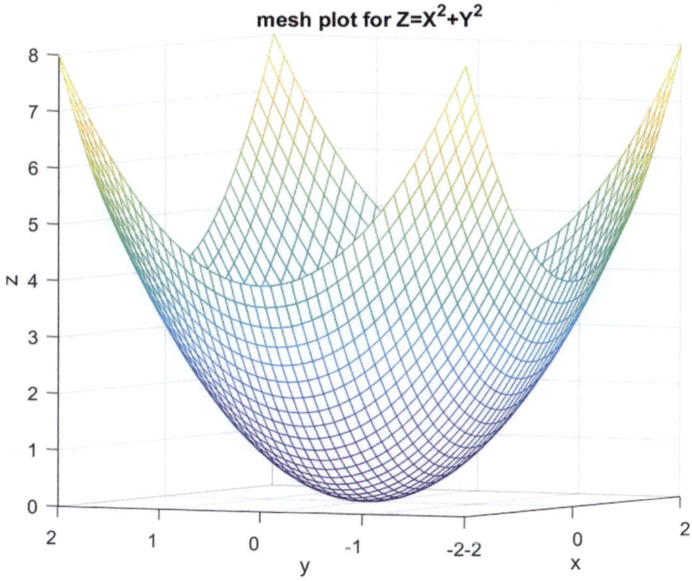


**Fig. 5.22** Mesh plot for function $Z = X^2 + Y^2$

curves are shaded with black (faceted shading which is the default shading option), whereas in part (c), they are not shaded (flat shading), and in part (b), the surface curves are eliminated by using an interpolated shading. Note that in surf plots, the insides of the surface curves are also shaded/filled as opposed to the mesh plots.

```
subplot(1,3,1)
surf(X,Y,Z);title('surf plot with faceted shading');
xlabel('x');ylabel('y');zlabel('z');
shading faceted;
subplot(1,3,2)
surf(X,Y,Z);title('surf plot with interpolated shading');
xlabel('x');ylabel('y');zlabel('z');
shading interp;
subplot(1,3,3)
surf(X,Y,Z);title('surf plot with flat shading');
xlabel('x');ylabel('y');zlabel('z');
shading flat;
```

Note that the option **'FaceAlpha'** can be used for surf plots to alter the surface opacity.

III. **Surface Plot:** a command similar to the surf command called **surface(x,y,z)** exists in MATLAB to create a primitive, three-dimensional surface plot. The function plots the values in matrix **Z** as heights above a grid in the *x*–*y* plane

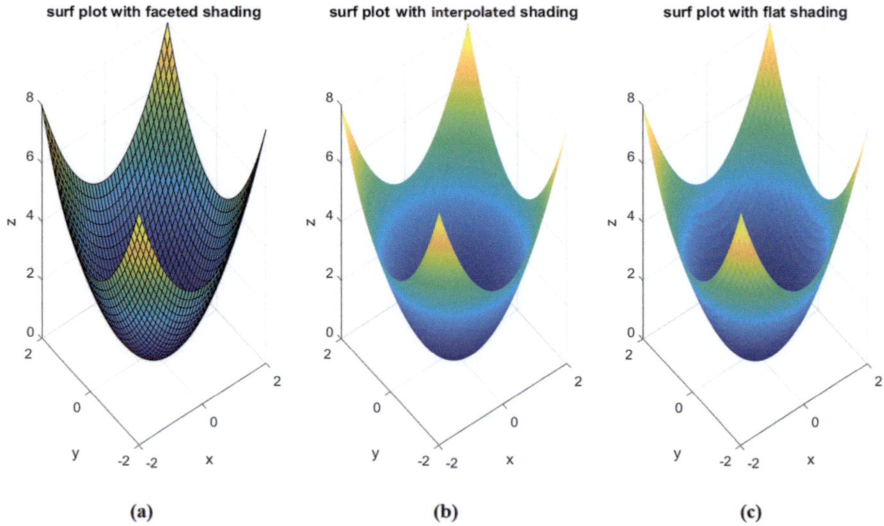


**Fig. 5.23** Surf plots for function Z = $X^2 + Y^2$: **a** faceted shading; **b** interpolated shading; **c** flat shading

defined by **X** and **Y**. The color of the surface varies according to the heights specified by Z. This command allows for an image to be overlaid on the surface, which makes it more powerful for rendering.

In the code below, a complex 3D function that resembles a non-flat land is created as can be seen in Fig. 5.24a. An image of a flat grassland is then laid over the surface as shown in Fig. 5.24b using the code below:

```
img=imread('grassland.jpg');
xx=linspace(-1.8,2.2,50);
yy=linspace(-2.15,2.85,50);
[x,y]=meshgrid(xx,yy);
z=-0.2*exp(-(x-0.25).^2-(y-0.25).^2)+0.3*exp(-3*(x-1.2).^2-y.^2)-0.5*exp(-(x-2.6).^2-y.^2)-
0.5*exp(-(x-2.4).^2-(y-2.5).^2)-exp(-(x+3).^2-3*(y-1.85).^2)+0.17*((2*(1-x).^2).*exp(-(x.^2)-
(y+1).^2)-7*(x/5-x.^3-y.^5).*exp(-x.^2-y.^2)-(1/3)*exp(-(x+1).^2-y.^2))-0.95*exp(-(x-
2.6).^2-(y-2.8).^2);
subplot(1,2,1);
surface(x,y,z);
xlabel('x');ylabel('y');zlabel('z');title('surface plot');
view(126,16);
subplot(1,2,2);
surface(x,y,z,img,'FaceColor','texturemap','EdgeColor','none','CDataMapping','direct');
xlabel('x');ylabel('y');zlabel('z');title('surface plot with image overlay');
view(126,16);
```

IV. **Contour3 Plot:** this command creates the contour curves but displays the results in 3D. The contour levels can be shown on the curves as seen in Fig. 5.25 for the same function used in Fig. 5.24 using the code below:

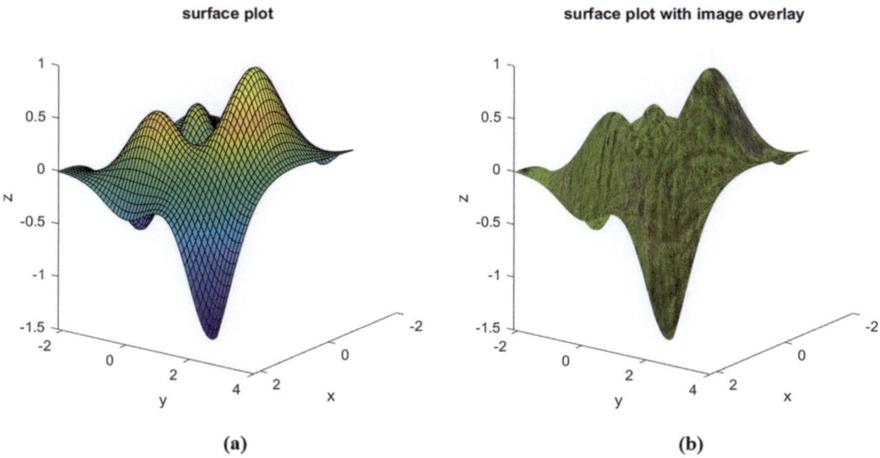


**Fig. 5.24** A Surface plot; **b** Surface plot with image overlay

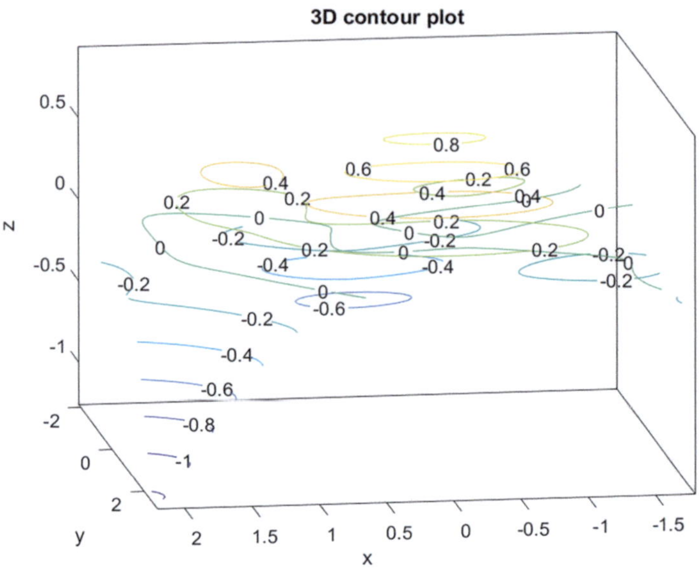


**Fig. 5.25** Contour3 plot for the function in Fig. 5.24

```
contour3(x,y,z,'ShowText','on');
xlabel('x');ylabel('y');zlabel('z');title('3D contour plot');
grid off; view(150,75);
```

V. **3D Scatter Plot:** command **scatter3(X,Y,Z,S,C)** displays scattered circles at locations **(X,Y,Z)** with sizes and colors defined by **S** and **C**. The following code generates points on the surface of a sphere that has an $n \times n$ surface (n = 50 here) with a radius of one, then distorts the axes to make it an ellipsoid surface, and centers it off origin. Next, it shows the scatter plot of the circles on the ellipsoid surface with marker size and color altered continuously as can be seen in Fig. 5.26. Note that this command needs the **X**, **Y**, and **Z** to be vectorized.

```
[X,Y,Z] = sphere(50);
x=4*X(:)+1;
y=2*Y(:)-2;
z=Z(:)+3;
S=linspace(5,15,length(x));
C=linspace(1,20,length(x));
scatter3(x,y,z,S,C)
xlabel('x');ylabel('y');zlabel('z');
```

VI. **Parametric 3D curves:** command **plot3(X,Y,Z)** is useful in drawing parametric 3D curves, such as a helix. The code below generates a tapered helix with a pitch of 1 and a base diameter of 2 as can be seen in Fig. 5.27.

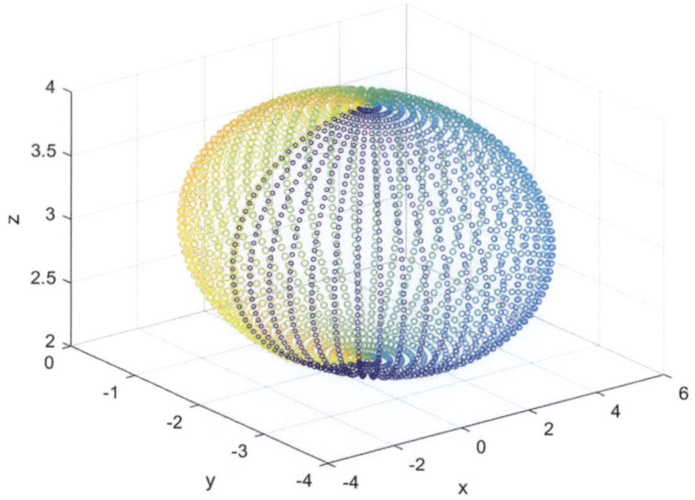


**Fig. 5.26** Scatter3 plot for an ellipsoid surface

```
t = 0:pi/50:10*pi;
p=1;
d=2;
st = 0.5*d*sin(t).*exp(t/50);
ct = 0.5*d*cos(t).*exp(t/50);
plot3(st,ct,p*t,'LineWidth',3);
xlabel('x');ylabel('y');zlabel('z');
```

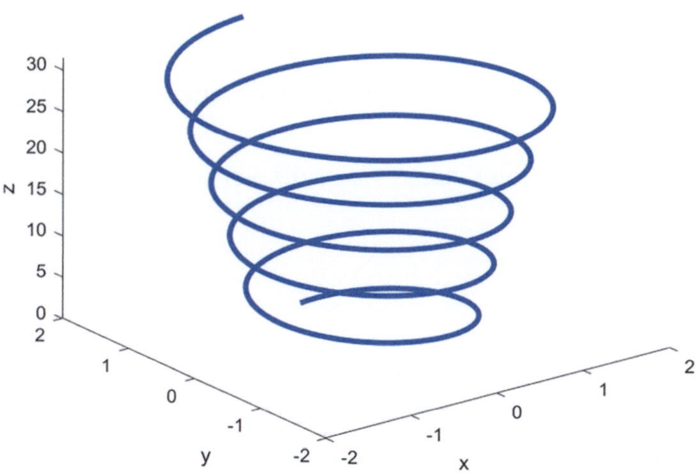


**Fig. 5.27** Plot3 for a tapered ellipsoid

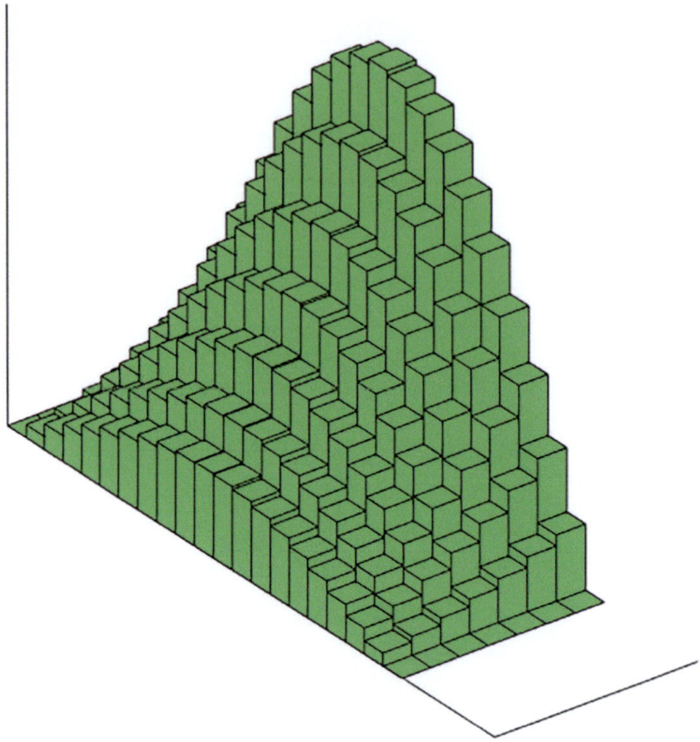

**Fig. 5.28** Bar3 plot example

VII. **3D Bar Plot:** command **bar3(Y,Z,S,C)** can be used to create 3D bar plots for **(Y,Z)** values using bars with width and color specified by **S** and **C**. The code below plots an example of a 3D bar plot that is shown in Fig. 5.28.

Note that the ticks for all 3 axes are turned off by setting the ticks vector to **[]**.

```
y = 0:pi/20:pi;
z = [sin(y')/5 sin(y')/3.5 sin(y')/2.5 sin(y')/1.9 sin(y')/1.5 sin(y')/1.2 sin(y')];
bar3(y,z,1,'g');
set(gca,'XTick',[ ]);
set(gca,'YTick',[ ]);
set(gca,'ZTick',[ ]);
```

VIII. **3D Quiver Plot:** command **quiver3(X,Y,Z,U,V,W)** will plot the 3D vector field given by **(U,V,W)** at points given by **(X,Y,Z)**. The code below plots a 3D surface, then extracts the normal to the surface patches using the **surfnorm** command and plots this normal vector field on the surface as can be seen in Fig. 5.29.

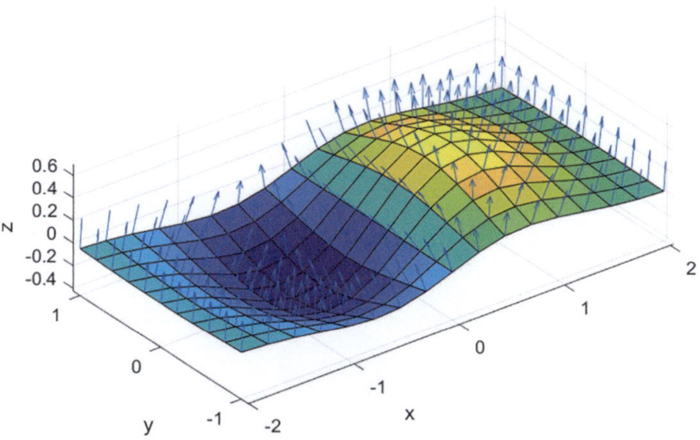


**Fig. 5.29** Quiver3 plot for Normals to a surface

```
[X,Y] = meshgrid(-2:0.25:2,-1:0.2:1);
Z = X.*exp(-X.^2 - Y.^2);
[U,V,W] = surfnorm(X,Y,Z);
quiver3(X,Y,Z,U,V,W);
hold on;
surf(X,Y,Z);
axis equal;
xlabel('x');ylabel('y');zlabel('z');
```

IX. **3D Implicit Function Plot:** command **fimplicit3(f,interval,linespec)** will plot the 3D implicit function **f(x,y,z) = 0** over the range **interval** using the **linespec** as can be seen in Fig. 5.30 using the code below:

```
f = @(x, y, z) x.^2 + y.^2 + z.^2 - sin(5*x).*cos(5*y).*cos(5*z);
fimplicit3(f, [-2, 2, -2, 2, -2, 2])
xlabel('x');ylabel('y');zlabel('z');
title('3D Implicit Function: x^2 + y^2 + z^2 - sin(5x)cos(5y)cos(5z) = 0')
```

X. **3D Volume Plots:** command **fill3(X,Y,Z,C)** and **patch(X,Y,Z,C)** can be used to created shaded 3D polygons/patches. The code below plots a simple house with a door and two windows using the **patch** command as can be seen in Fig. 5.31.

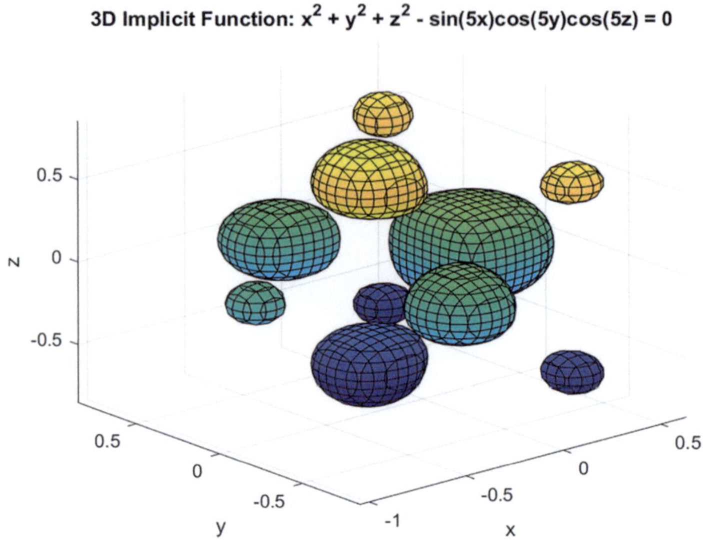


**Fig. 5.30** Implicit3 plot example

```
xf=[0 0 0 0];yf=[0 10 10 0];zf=[0 0 10 10];
xre=[-10 -10 -10 -10];yre=[0 10 10 0];zre=[0 0 10 10];
xb=[-10 0 0 -10];yb=[0 0 10 10];zb=[0 0 0 0];
xl=[0 0 -10 -10];yl=[0 0 0 0];zl=[0 10 10 0];
xrg=[0 0 -10 -10];yrg=[10 10 10 10];zrg=[0 10 10 0];
xft=[0 0 -5];yft=[0 10 5];zft=[10 10 15];
xrt=[-10 -10 -5];yrt=[0 10 5];zrt=[10 10 15];
xlt=[-10 0 -5];ylt=[0 0 5];zlt=[10 10 15];
xrgt=[-10 0 -5];yrgt=[10 10 5];zrgt=[10 10 15];
xd=[0 0 0 0];yd=[4 6 6 4];zd=[0 0 7 7];
xw1=[-9 -6 -6 -9];yw1=[0 0 0 0];zw1=[4 4 8 8];
xw2=[-4 -1 -1 -4];yw2=[0 0 0 0];zw2=[4 4 8 8];
patch(xf,yf,zf,'r');hold on;
```

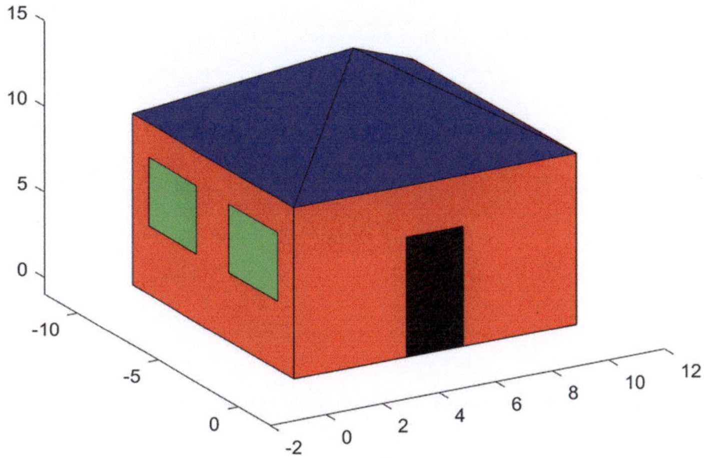


**Fig. 5.31** Patch command used to render a simple 3D house

```
patch(xre,yre,zre,'r');
patch(xb,yb,zb,'r');
patch(xl,yl,zl,'r');
patch(xrg,yrg,zrg,'r');
patch(xft,yft,zft,'b');
patch(xrt,yrt,zrt,'b');
patch(xlt,ylt,zlt,'b');
patch(xrgt,yrgt,zrgt,'b');
patch(xd,yd,zd,'k');
patch(xw1,yw1,zw1,'g');
patch(xw2,yw2,zw2,'g');
axis([-12 2 -2 12 -1 15]);
view(60,29);
```

The following code plots a twisted torus using the **fill3** command as can be seen in Fig. 5.32.

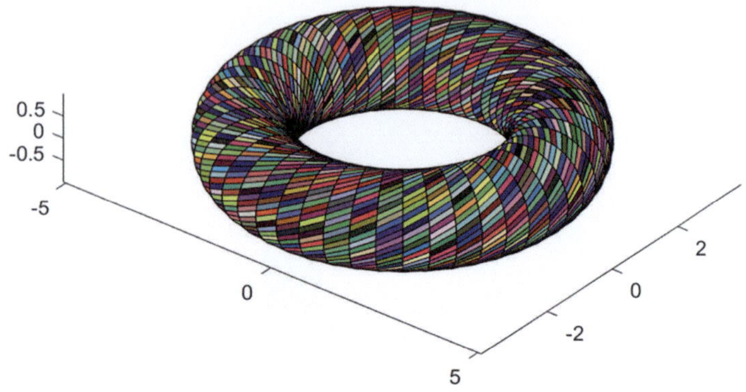


**Fig. 5.32** Fill3 command used to render a twisted torus with random colors

```
theta = linspace(0, 2*pi, 50);
phi = linspace(0, 2*pi, 50);
[Theta, Phi] = meshgrid(theta, phi);
X = (3 + 1*cos(Theta + 2*Phi)).*cos(Phi);
Y = (3 + 1*cos(Theta + 2*Phi)).*sin(Phi);
Z = 1*sin(Theta + 2*Phi);
hold on;
for ii = 1:length(theta)-1
        for j = 1:length(phi)-1
                fill3([X(ii,j), X(ii+1,j), X(ii+1,j+1), X(ii,j+1)],[Y(ii,j), Y(ii+1,j), Y(ii+1,j+1),
                Y(ii,j+1)], ...
                [Z(ii,j), Z(ii+1,j), Z(ii+1,j+1), Z(ii,j+1)],rand(1,3));
        end
end
axis equal;view(38,32);
```

## 5.3 Image and Signal Visualization

In this section, first we will look at the commands for image and video visualization, then we look at commands for signal visualization (other than images).

To read, see, and write images, we can utilize the commands below:

I. **Reading an image into a matrix:** command **imread** is the first command to use in MATLAB for image visualization. When reading an image in MATLAB, you want to use the semicolon at the end; otherwise, the entire matrix will be displayed on the screen. If the file you want to read is not in the current folder, you need to provide the full address of the file as the input argument. As can be seen, even a simple RGB image that is not HD will have lots of information.

```
img=imread('EEA logo ChatGPT.png');
Whos
```

| Name | Size | Bytes | Class |
|---|---|---|---|
| img | 838x790x3 | 1986060 | uint8 |

II. **Displaying the image:** command **imshow** is typically used for this purpose. Figure 5.33 displays the result of showing the above matrix/image.

While the command **imshow** is specifically designed for displaying grayscale or RGB images at full scale/range of values, there is another more general-purpose command called **imagesc,** which can be used to display any matrix as an image with scaled colors and any colormap. For instance, the code below is used to visualize a $100 \times 100$ two-dimensional matrix of random numbers taken from a normal distribution and displays them using a specific scale/range and a "**bone**" colormap as shown in Fig. 5.34.

**Fig. 5.33** Using imread and imshow to display an image in MATLAB

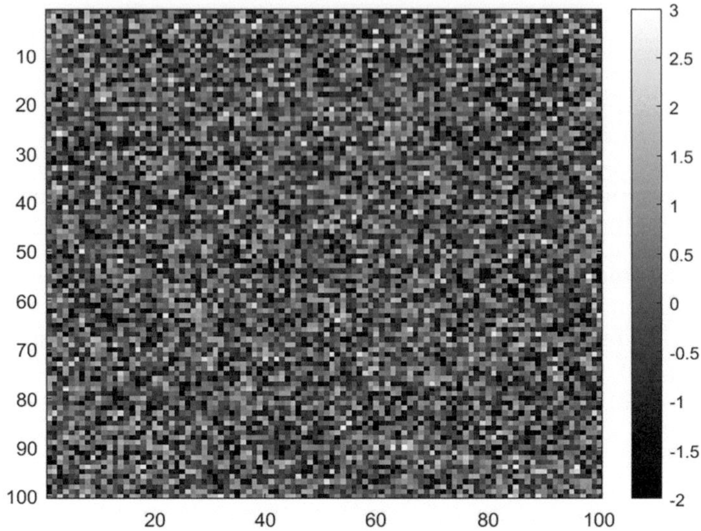


**Fig. 5.34** Using imagesc to display a matrix of random number as an image with a specific color scale and colormap

```
data = randn(100, 100);
clims = [-2 3];
imagesc(data,clims);
colormap bone;
colorbar;
```

III. **Displaying the image information:** while not directly used for visualization, command **iminfo** can extract the information for a specific image as shown in Fig. 5.35.

IV. **Interactive Image Display Tool:** command **imtool** can display images with interactive tools, such as cropping, distance measurement, and pixel region as shown in Fig. 5.36.

V. **Displaying a pair of images:** command **imshowpair** creates a composite image from two images to compare their differences. For instance, the commands below read an RGB image, create a grayscale version of it, and then shows the side-by-side as a montage shown in Fig. 5.37:

```
img=imread('Lenna.png');
imggray = rgb2gray(img);
imshowpair(img,imggray,'montage')
```

VI. **Writing an image:** while not intended for displaying an image, command **imwrite** is used to write a matrix as an image with a specific extension to the

```
info = struct with fields:

Filename: 'C:\Users\andalibm\Downloads\EEA logo ChatGPT.png'
FileModDate: '23-May-2024 09:18:08'
FileSize: 799365
Format: 'png'
FormatVersion: []
Width: 790
Height: 838
BitDepth: 24
ColorType: 'truecolor'
FormatSignature: [137 80 78 71 13 10 26 10]
Colormap: []
Histogram: []
InterlaceType: 'none'
Transparency: 'alpha'
SimpleTransparencyData: []
BackgroundColor: []
RenderingIntent: 'perceptual'
Chromaticities: [0.3127 0.3290 0.6400 0.3300 0.3000 0.6000 0.1500 0.0600]
Gamma: 0.4546
XResolution: 5669
YResolution: 5669
ResolutionUnit: 'meter'
XOffset: []
YOffset: []
OffsetUnit: []
SignificantBits: []
ImageModTime: []
Title: []
Author: []
Description: []
Copyright: []
CreationTime: []
Software: []
Disclaimer: []
Warning: []
Source: []
Comment: []
OtherText: []
```

**Fig. 5.35** Using iminfo to extract the information for a specific image

local drive. For instance, the code below reads and converts an RGB image into a grayscale matrix, then performs **histogram equalization** (an operation used in image processing to enhance image contrast) on it and assign it to a new matrix. Next, it shows both images using **imshowpair**, grabs a snapshot/frame **F** from the open window using the command **getframe** (which will be

**Fig. 5.36** Interactive image display using imtool

**Fig. 5.37** Showing pairs of images using imshowpair

**Fig. 5.38** Using imwrite to save a matrix as an image to local drive (here the image is a montage of a grayscale image and a histogram-equalized version of it)

discussed later in chapter 7), saves the frame into an image **X**, and then writes **X** to a new image on the local drive that can be seen in Fig. 5.38.

```
img=rgb2gray(imread('cameraman.jpg'));
imgmod = histeq(img);
obj=imshowpair(img,imgmod,'montage')
F = getframe(gcf);
[X, Map] = frame2im(F);
imwrite(X,'cameraman_histeq.png')
```

While we cannot properly render the tools for video display, the example below demonstrates useful commands for reading, playing, and writing videos. The snapshot of the original and written/modified videos can be seen in Fig. 5.39.

```
% creates an object ot read a video from a file
video = VideoReader('viptraffic.avi');

% creates a videowriter object for an output file and opens it
videoWriter = VideoWriter('output.avi');
open(videoWriter);

% reads the video frame by frame until the end and shows the frames
while hasFrame(video)
        frame = readFrame(video);
        imshow(frame);
        %write the grayscale frame to the output video
        writeVideo(videoWriter, rgb2gray(frame));
end
```

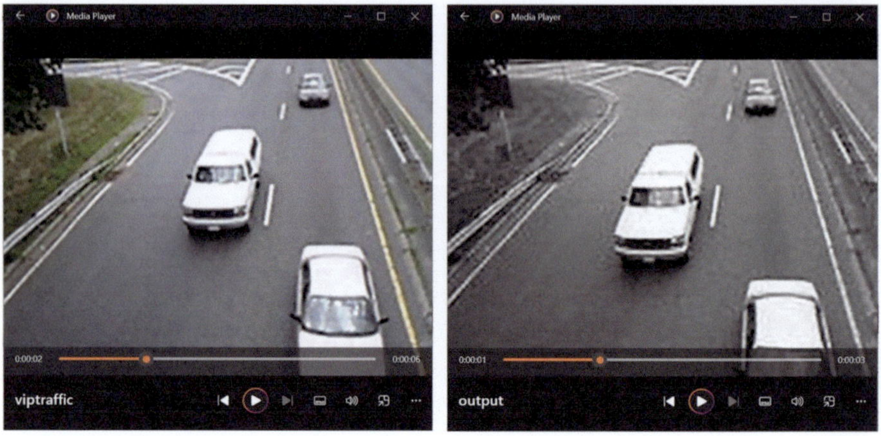


**Fig. 5.39** Reading a video frame-by-frame, displaying the frames, converting the frames to grayscale, and writing them to a new video

```
% closes the videwriter object
close(videoWriter);
```

Now, we can focus on reading, displaying, and potentially writing other types of signals in MATLAB.

I. **Fast Fourier Transform (FFT) and Power Spectral Density (PSD):** One of the common tasks in signal processing is finding the frequencies of a signal along with their relative magnitudes, which can be done using **FFT** or **PSD** (using the **pwelch** command) commands as shown below for a composite signal made of 3 harmonics plus noise. Results are shown in Fig. 5.40.

```
fs = 1000; % Sampling frequency (Hz)
t = 0:1/fs:10-1/fs; % Time vector (10 seconds of data)

% Create a composite signal with three sinusoids and noise
f1 = 50; % Frequency of first sinusoid (Hz)
f2 = 120; % Frequency of second sinusoid (Hz)
```

```
f3 = 300; % Frequency of third sinusoid (Hz)
noisePower = 0.5; % Noise power
signal = sin(2*pi*f1*t) + sin(2*pi*f2*t) + sin(2*pi*f3*t) + sqrt(noisePower)*randn(size(t));


% Compute the FFT of the signal
N = length(signal);
signal_fft = fft(signal);
f_fft = (0:N-1)*(fs/N);
signal_fft_magnitude = abs(signal_fft)/N;

% Estimate and plot the power spectral density using Welch's method
[pxx, f_psd] = pwelch(signal, hamming(1024), 512, 1024, fs);

subplot(2,1,1);
plot(f_fft(1:N/2+1), 2*signal_fft_magnitude(1:N/2+1));
title('FFT of Composite Signal');grid on;
xlabel('Frequency (Hz)');ylabel('Magnitude');
subplot(2,1,2);
plot(f_psd, 10*log10(pxx));
title('Power Spectral Density of Composite Signal');grid on;
xlabel('Frequency (Hz)');ylabel('Power/Frequency (dB/Hz)');
```

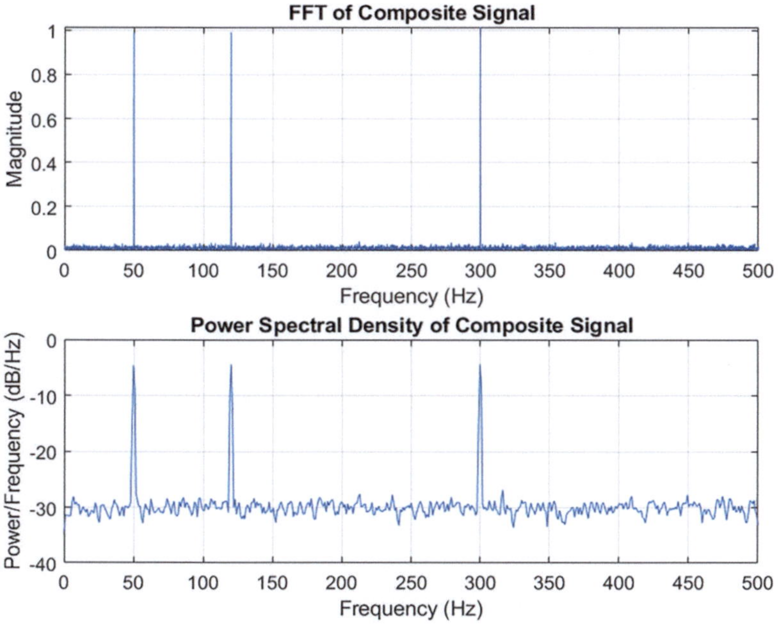


**Fig. 5.40** FFT and PSD plots for a composite signal made of 3 harmonics

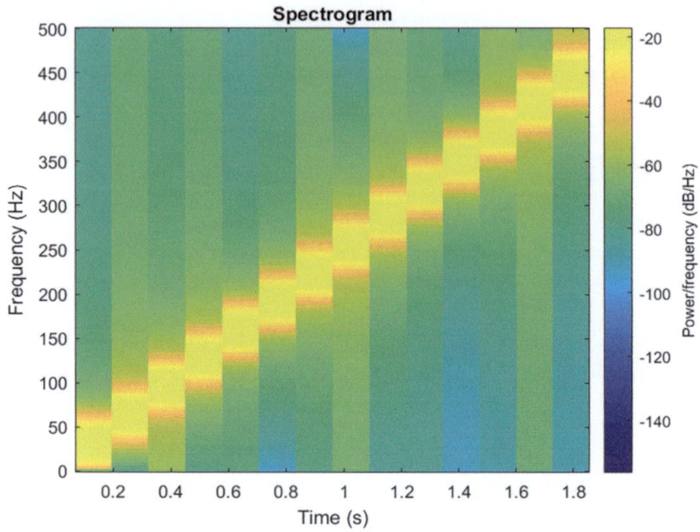


**Fig. 5.41** Spectrogram for a chirp signal

II. **Spectrogram:** When the frequency content of a signal varies over time, such as a **chirp** signal with linear frequency increase as time goes on, a **spectrogram** (time–frequency) plot can be used. Such a plot is created in Fig. 5.41 using the **spectrogram** command for a chirp signal generated using the **chirp** command with the code below:

```
fs = 1000; % Sampling frequency
t = 0:1/fs:2; % Time vector
signal = chirp(t, 0, 2, 500); % chirp signal
spectrogram(signal, 256, [ ], [ ], fs, 'yaxis');
title('Spectrogram');
```

III. **Finding peaks in a signal:** When the peak values in a signal are to be extracted and plotted, command **findpeaks** can be really useful. The following code demonstrates using this command for a harmonic with a high frequency that is deteriorated by noise. Figure 5.42 shows the original signal with its peaks overlaid.

```
fs = 100;
t = 0:1/fs:2-1/fs;
signal = sin(2*pi*50*t) + 0.5*randn(size(t));

% Find and plot peaks
[pks, locs] = findpeaks(signal, 'MinPeakHeight', 0.5);
plot(t, signal);hold on;
plot(t(locs), pks, 'r*', 'MarkerSize', 10);
title('Signal with Peaks');legend('Signal', 'Peaks');
```

```
xlabel('Time (s)');ylabel('Amplitude');
```

IV. **Plotting the scalogram of a signal with continuous wavelet transform:** command **cwt** can be used to obtain the continuous wavelet transform of a signal and plot its scalogram as shown in Fig. 5.43 for a speech signal using the code below.

```
load mtlb;
cwt(mtlb);
```

V. **Plotting the Short-Time Fourier Transform of a signal with a Waterfall Plot:** command **stft** can be used to determine the sinusoidal frequency and phase

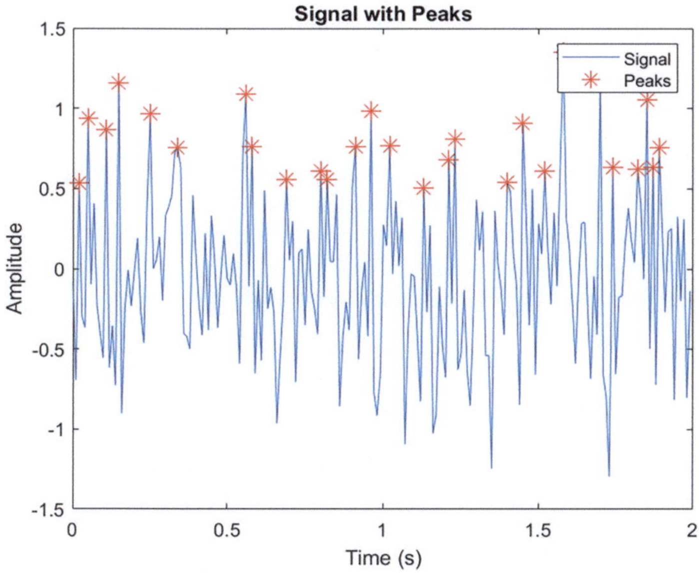


**Fig. 5.42** Finding and plotting peaks in a signal

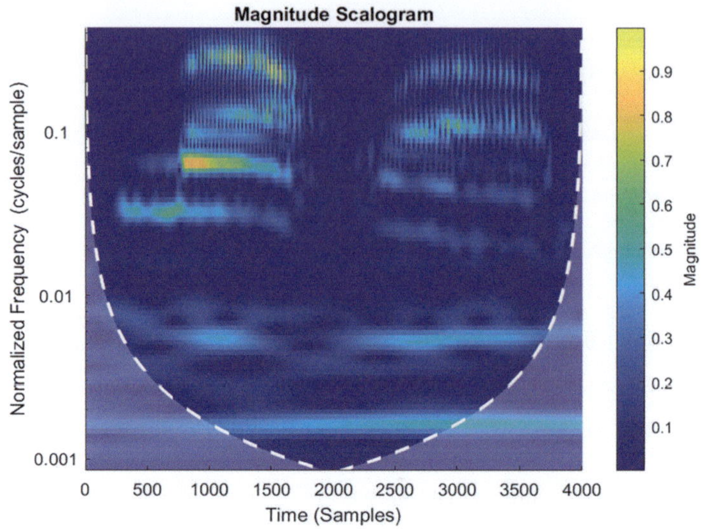


**Fig. 5.43** Scalogram for a speech signal using continuous wavelet transform

content of local sections of a signal as it changes over time (used to create spectrograms). Command **waterfall** creates a waterfall plot, which is a mesh plot with a partial curtain along the y dimension. In the following code, these commands are used to calculate the short-time Fourier transform (using a periodic Hamming window of length 128 and an overlap length of 50 samples) on a composite signal made up of 3 chirp signals:

The first channel consists of a concave quadratic chirp with instantaneous frequency 100 Hz at t = 0 and crosses 300 Hz at t = 1 s. It has an initial phase equal to 45 degrees.

The second channel consists of a convex quadratic chirp with instantaneous frequency 100 Hz at t = 0 and crosses 500 Hz at t = 1 s.

The third channel consists of a logarithmic chirp with instantaneous frequency 300 Hz at t = 0 and crosses 500 Hz at t = 1 s.

The plot of the results as can be seen in Fig. 5.44.

```
fs = 1e3;
t = 0:1/fs:1-1/fs;
x = [chirp(t,100,1,300,'quadratic',45,'concave');chirp(t,100,1,500,'quadratic',[],'convex');...
chirp(t,300,1,500,'logarithmic')]';
[S,F,T] = stft(x,fs,'Window',hamming(128,'periodic'),'OverlapLength',50);
waterfall(F,T,abs(S(:,:,1))');
xlabel('Time (seconds)');ylabel('Frequency (Hz)');
title('Input Channel: 1');
```

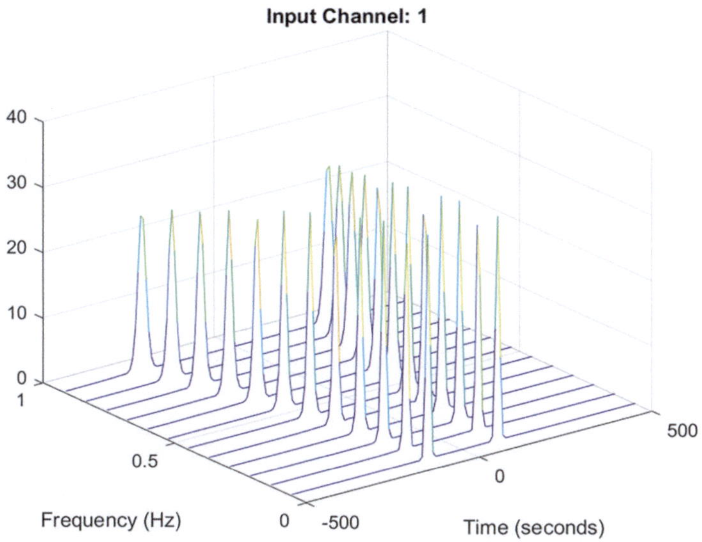


**Fig. 5.44** Waterfall plot for the short-time Fourier transform applied to a composite signal made up of 3 chirps

## Practice Problems

1. Write a script to visualize the bending moment distribution along the length of a simply supported beam subjected to a uniformly distributed load.

### Beam Specifications

- Length of the beam: $L = 10\ m$
- Uniformly distributed load: $\omega = 5\ kN/m$

**Bending Moment Calculation:** The bending moment at any point x along the beam can be calculated using the formula:

$$M(x) = \omega\left(Lx - x^2\right)/2 \tag{5.1}$$

where:

- **M(x)** is the bending moment at position **x** (in kN·m),
- **L** is the length of the beam,
- **w** is the uniformly distributed load.

**Output Plot:** Plot the bending moment *M(x)* along the length of the beam from $x = 0$ to $x = L$.

- Use different markers and colors for the plot.
- Annotate the maximum bending moment on the plot using a **textarrow**.
- Include a **Greek letter** for $\omega$ in the title.

- Customize the plot with grid lines and specify line colors and marker styles.
- Write a script to visualize the performance of a heat exchanger using various engineering plots. The heat exchanger is modeled by a system of nonlinear equations representing the temperature distribution and the heat transfer rate. You will use MATLAB's **fimplicit** function to plot the temperature distribution, create **subplots** to show different aspects of the heat exchanger's performance, and use a **yyaxis** plot to compare the heat transfer rate and temperature gradient on the same graph.

### Problem Requirements

**Heat Exchanger Model:** The heat exchanger can be modeled by the following system of nonlinear equations:

$$T_h(x, y) = T_{h0}e^{-ax}\cos(by),\ T_c(x, y) = T_{c0}e^{-cx}\cos(dy) \tag{5.2}$$

where:

- $T_h(x,y)$ and $T_c(x,y)$ are the temperatures of the hot and cold fluids, respectively, at any point (x,y) in the heat exchanger.
- $T_{h0} = 200°C$ and $T_{c0} = 50°C$ are the inlet temperatures of the hot and cold fluids, respectively.
- a, b, c, and d are constants related to the heat exchanger design.

### Heat Transfer Rate

The heat transfer rate Q between the hot and cold fluids can be modeled by:

$$Q = k[T_h(x, y) - T_c(x, y)] \tag{5.3}$$

where $k = 5$ is a heat transfer coefficient.

### Plotting Requirements

- Use fimplicit to plot the temperature distribution curves for $T_h(x,y) = 100°C$ and $T_c(x,y) = 75°C$ on the same graph.
- Create a subplot to compare the temperature distributions of the hot and cold fluids along the length of the heat exchanger (x-axis).
- Use a yyaxis plot to display both the heat transfer rate Q(x) and the temperature gradient $\nabla T(x) = T_h(x,y) - T_c(x,y)$ on the same graph.
- Write a script to analyze vibration data collected from a rotating machine. The machine is equipped with a sensor that records vibration amplitudes at regular time intervals. Your goal is to analyze the frequency content of the vibrations using the Fast Fourier Transform (FFT) and to visualize the distribution of vibration amplitudes using a histogram.

**Vibration Data**

- The vibration data is provided as a time series representing the vibration amplitude at each time step.
- The data is stored in a vector **vibrationData**, and the time interval between samples is given as **dt**.
- Try the following code to generate artificial **vibrationData** for testing:

```
% Given data and parameters
fs = 1000;  % Sampling frequency in Hz
dt = 1/fs;  % Time interval between samples
t = 0:dt:1-dt;  % Time vector
vibrationData = 0.5 * sin(2 * pi * 50 * t) + 0.2 * sin(2 * pi * 120 * t) + 0.1 * randn(size(t));
```

**Data Analysis**

- **Frequency Analysis:** Perform an FFT on the vibration data to identify the dominant frequencies. Plot the amplitude spectrum showing the magnitude of each frequency component.
- **Amplitude Distribution:** Create a histogram of the vibration amplitudes to visualize their distribution.

**Plotting and Output Requirements:** Plot the amplitude spectrum with proper axis labels and a title on **subplot(2,1,1)**. Create a histogram of the vibration amplitudes with appropriate bin size, axis labels, and a title on **subplot(2,1,2)**. Print the dominant frequency in the command window.

4. Write a script to visualize the electric field generated by two point charges in a 2D plane. Your goal is to compute the electric field vectors at various points in the plane and visualize the field using the quiver command.

**Point Charges:** Consider two point charges located at different positions in a 2D plane. Each charge $q_i$ is located at coordinates $(x_i,)$ and has a magnitude $q_i$.

**Electric Field Calculation:** The electric field $E$ at a point $(x,)$ due to a single point charge.

*qi* located at (*xi*,) is given by:

$$E_i(x, y) = \frac{kq_i}{r_i^2}\hat{r}_i \tag{5.4}$$

where:

$k$ is Coulomb's constant, $r_i$ is the distance between the point $(x,)$ and the charge $q_i$, and $\hat{r}_i$ is the unit vector pointing from $q_i$ to $(x,)$.

The total electric field at $(x,)$ due to the two charges is the vector sum of the fields due to each charge:

$$E(x, y) = E_1(x, y) + E_2(x, y) \tag{5.5}$$

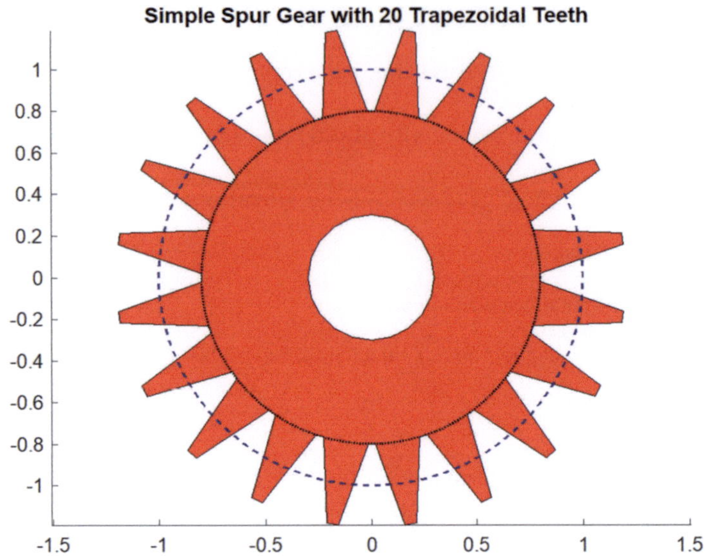


**Fig. 5.45** Plot of a simple spur gear using the patch command

**Visualization:** Use the **quiver** command to visualize the electric field vectors at a grid of points in the 2D plane. Customize the appearance of the vectors (e.g., color, scale) to clearly depict the electric field. Add texts showing the names of point charges and their values at their locations (Fig. 5.45).

5. Write a script that plots the gear in using the patch command and show the addendum, dedendum, and pitch circles given the parameters on the next page:

```
% Gear Parameters
N = 20;            % Number of teeth
rp = 1;            % Pitch radius in meters
ra = 0.2 * rp;     % Addendum (radial distance from pitch to outer)
rd = 0.2 * rp;     % Dedendum (radial distance from pitch to root)
ro = rp + ra;      % Outer radius
rr = rp - rd;      % Root radius
rh = 0.3;          % shaft hole radius
```

6. The stress concentration around a circular hole in an infinite plate under uniform tension can be analyzed. The stress at any point around the hole is a function of the angle θ. The stress distribution $\sigma_\theta$ around the hole can be given by:

$$\sigma_\theta = \sigma_\infty\left(1 - \frac{r_0^2}{r^2}\right) + 2\sigma_\infty\cos(2\theta)(\frac{r_0^2}{r^2}) \tag{5.6}$$

where $\sigma_\infty$ is the applied stress at infinity, $r_0$ is the radius of the hole, $r$ is the radial distance from the center and $\theta$ is the angular position.

Write a script that uses a **polar plot** to visualize the stress distribution around a circular hole in a flat plate for the parameters given below:

```
% Parameters
sigma_inf = 100;  % Applied stress at infinity (in MPa)
r0 = 1;           % Radius of the hole (in meters)
r = r0;           % Distance from the center to the edge of the hole
```

7. You are an engineer tasked with analyzing the temperature distribution on a thin, rectangular metal plate. The plate is heated at 5 different points given below, and you want to write a script to visualize the temperature distribution across the surface using the **surf** command to identify hot spots and temperature gradients. The plate has dimensions of 10 cm by 10 cm. The temperature at specific points on the plate is known. Interpolate the temperature across the surface by the distance they have from the heated points.

```
% (x, y, temperature)
heat_sources = [2, 2, 75;
                8, 2, 100;
                5, 5, 150;
                2, 8, 90;
                8, 8, 120];
```

8. Write a script that plots a spring with the parameters given below using the command **plot3** and annotate it as shown in Fig. 5.46.

```
n_coils = 10;          % Number of coils
radius = 1.5;          % Radius of the spring (helix diameter/2)
pitch = 1.0;           % Pitch of the spring (distance between coils)
wire_thickness = 6;  % Thickness of the spring wire
```

9. Suppose you are an engineer analyzing the vibration response of a mechanical structure. The structure is excited by a force, and its vibration response is measured using an accelerometer. The data collected is in the form of an acceleration-time signal. Write a script to identify the peaks in the signal to analyze the resonant frequencies of the structure. To do so, you need the **FFT** command and the **findpeaks** command. Use the following artificial data for the acceleration signal:

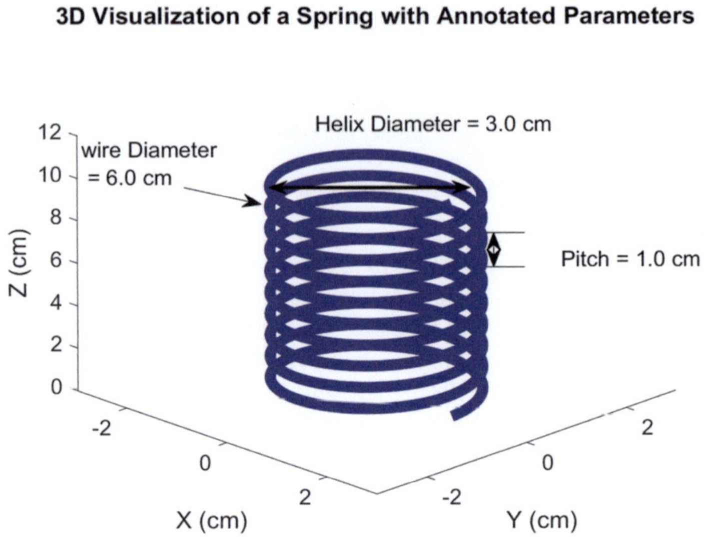


**Fig. 5.46** Plot of a simple spring with annotations using plot3 command

```
% Simulated vibration response data (acceleration in m/s^2)
Fs = 500;           % Sampling frequency in Hz
t = 0:1/Fs:2;        % Time vector from 0 to 2 seconds
N = length(t);       % Number of samples

% Simulated signal: combination of sine waves with a few dominant frequencies
f1 = 10;            % Frequency of the first component in Hz
f2 = 25;            % Frequency of the second component in Hz
f3 = 40;            % Frequency of the third component in Hz

acceleration_signal = 2*sin(2*pi*f1*t) + 1.5*sin(2*pi*f2*t) + sin(2*pi*f3*t) +
0.5*randn(size(t));
```

# Chapter 6
# Creating Apps with App Designer

## 6.1 Design and Deployment of GUI Applications

### *6.1.1 Initializing a New App*

One of MATLAB's many tools is App Designer, which is an interface that assists the user in creating their own applications. The App Designer tool generates the code needed to correctly render the application windows, displays the application for testing, and then exports the program to an ***.exe** file that can be run on systems without MATLAB. While the App Designer tool is not required to program applications using MATLAB, the tool makes the process quicker and easier.

To launch App Designer, simply type **appdesigner** into MATLAB's console. The tool opens in a separate window, shown in Fig. 6.1, and gives users the option to start from scratch, use some existing templates, or to view some example applications. Additionally, previously created app files can be opened using the panel on the left of the screen.

A new blank application, created by selecting the "Blank App" button, is shown in Fig. 6.2, where existing MATLAB GUI objects are in a library on the left and properties of the selected object (in this case the app itself) are in the panel on the right. The middle panel displays the app's current displayed window, but this can be switched to show the MATLAB code responsible for generating the current display by clicking the Display View/Code View toggle above and to the right of the application window.

As can be seen by switching to Code View (Fig. 6.3), MATLAB's applications use object-oriented programming (covered in Chap. 3). The app itself is a subclass of MATLAB's built-in **MATLAB.apps.AppBase** class, where the names of the class object and the ***.mlapp** file are the same. Clicking the Run button will launch the app in a separate window. The app will run in the exact same way if the code was copied out of App Designer into a regular ***.m** file.



M. Andalibi and S. A. Cherkauer, *Advanced MATLAB for Engineers*,
https://doi.org/10.1007/978-3-032-10677-3_6

**Fig. 6.1** App designer's home screen

**Fig. 6.2** New blank application

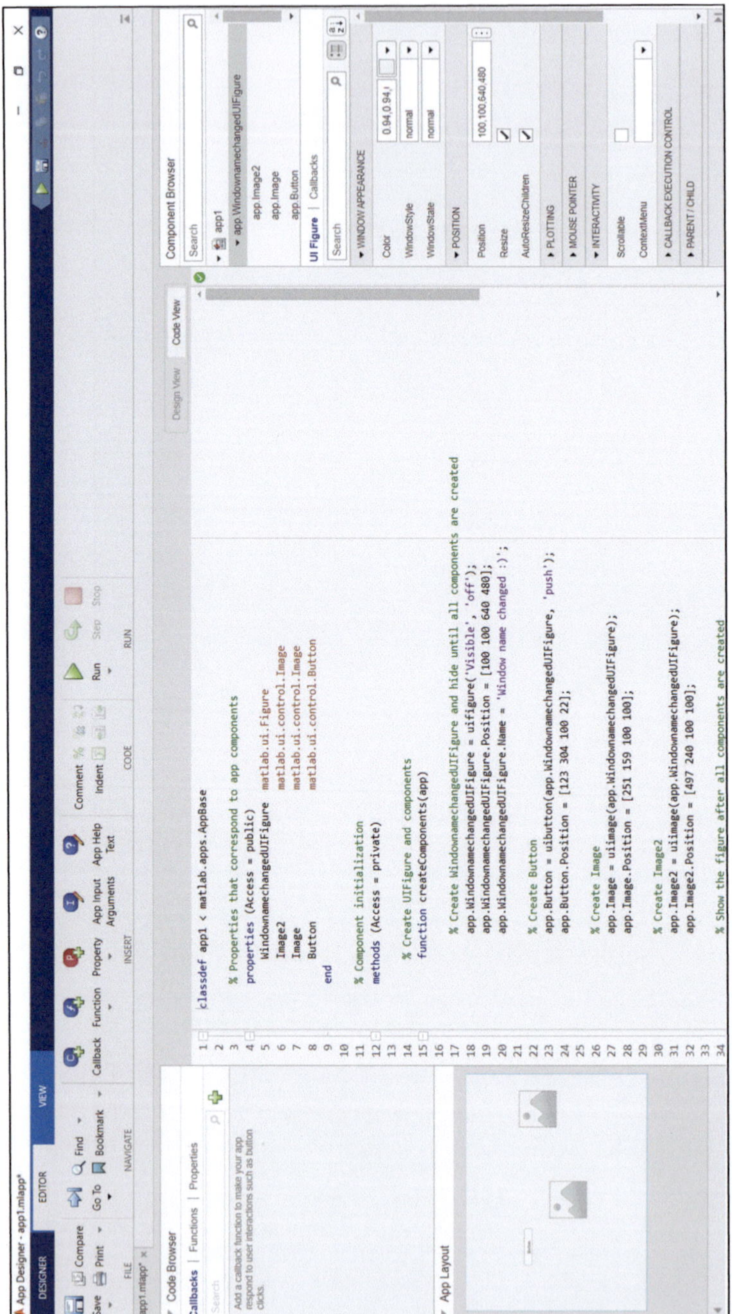


**Fig. 6.3** App designer code view

### 6.1.2 Adding and Editing New Components

By using the App Designer tool, users do not need to remember class names for every possible app component. Instead, all of MATLAB's built-in components can be found in the Component Library on the left of the designer window. The components from the library can be dragged and dropped into a desired location in the Design View window and MATLAB will automatically update the application's code. Notice that every new component added to the app corresponds to a new property in the app's class, where each of these properties is a new class instance. In Fig. 6.3, two images and a button component are added to the app, which in turn adds the **Image**, **Image2**, and **Button** properties with corresponding **MATLAB.ui.control.Image** and **MATLAB.ui.control.Button** datatypes to the app class. Every time one of the existing components is moved around, MATLAB automatically updates the object's **Position** property, which is set in the **createComponents** method.

When in App Designer, most of the code is grayed out, meaning it cannot be directly edited. Values in the code can be changed using the Component Browser panel on the right, however. For example, selecting the **app.UIFigure** component, scrolling down to the Identifiers tab, and changing the name of the component change the value assigned to the **app.UIFigure.Name** property in the app constructor (this property changes the name in the upper left corner of the window when the app is running). Notice how changing settings away from their default in the Components Browser adds additional component property definitions to the code in the **create-Components** method, such as changing the name property of the **UIFigure** object. Removing components is as simple as selecting the component in the Design View or Component Browser and then pressing backspace or delete.

### 6.1.3 Adding New Methods and Properties

Many applications will need additional variables and functions to save inputs, process inputs, and update the interface. MATLAB's ***mlapp** files, just like ***.m** files, can access other function and class files, and these files will be compiled with the rest of the app for distribution. In addition, methods can be created in the app class by switching to Code View, selecting the Functions tab in the Code Browser on the left, and clicking the plus sign. Similarly, class properties can be added by selecting the Properties tab in the Code Browser and clicking the plus sign (Fig. 6.4). Once the first method or property is created, others can be added with the same process or by typing into the white, unlocked code editor section like normal. Method and property blocks with different attributes can be created using the second method, since the default generated attribute is always **(Access = private)**. The **MATLAB.app.AppBase** class is a handle class, so methods do not return an instance of the altered class, but they do take the class handle, which MATLAB represents with the **app** variable, as the first input.

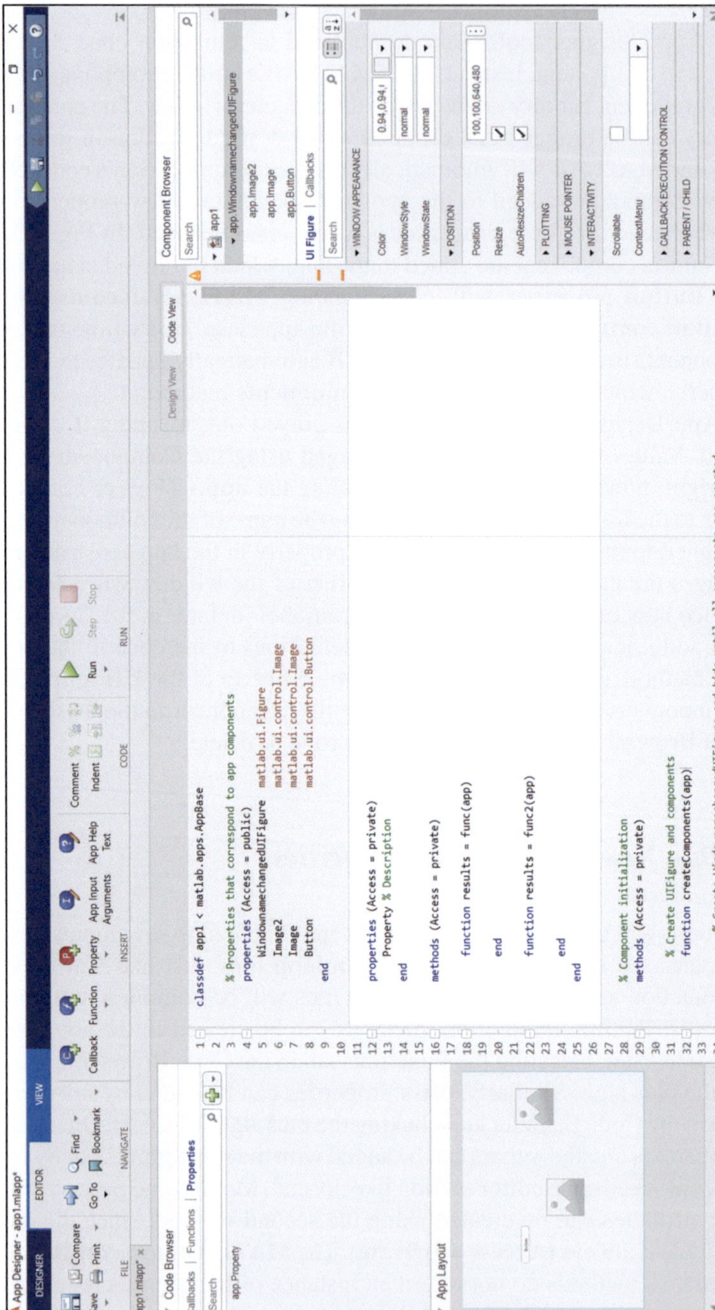


**Fig. 6.4** New class property and methods

### 6.1.4 Component Callbacks

MATLAB's user interface components can also call functions, known as callbacks, when certain events occur. One such case is initiating a callback function once a button is clicked. Callbacks in App Designer are created the same way as methods and properties using the Callbacks tab in the Code Browser. When creating a new callback, MATLAB will ask for the specific component to assign the function to, which event to trigger the callback on, and what to name the callback function (Fig. 6.5). The first two sections are dropdown menus that contain all of the possible choices, and all of the callback event options are named clearly. Newly created callbacks are just private access methods, just like any other class method, but they do not return anything and cannot be renamed or moved in the code editor. The App Designer tool keeps callback functions and other methods separated so that only specially created callback functions can be used as callbacks. Callback functions can be renamed or deleted by right-clicking on their names in the Code Browser. One special callback, tied to the app class itself, is the startup callback, which is a function called when the app is first run.

### 6.1.5 Compiling an Application

Once an application has been completed and thoroughly tested, it can be compiled into a Windows ***.exe** installer file, web app, or MATLAB app for sharing and distribution. This process is started by selecting the Designer tab at the top of the App Designer window, clicking on the Share dropdown, and selecting one of the three compiling options as shown in Fig. 6.6. The MATLAB application installer will add the program to the App tab in the MATLAB editor, the web app compiler will upload the program to an existing website, but the standalone installer will export the program to an ***.exe** that will allow the program to be installed and run on any Windows machine without needing to install or even own MATLAB.

The next popup will ask for an app name, author, description, and other information, unless the program is being deployed as a web app. Figure 6.7 shows the popup when the standalone desktop app option is clicked. In this case, the installer file can be named by typing into the text box in the packaging options section after choosing whether or not to include MATLAB Runtime in the installer package. MATLAB Runtime is the standalone software that executes the compiled MATLAB code without a license, and including Runtime in the installer package can speed up installation at the cost of making the installer file larger. Once done inputting information, click Package and MATLAB will open the location of the installer files once compilation is completed.

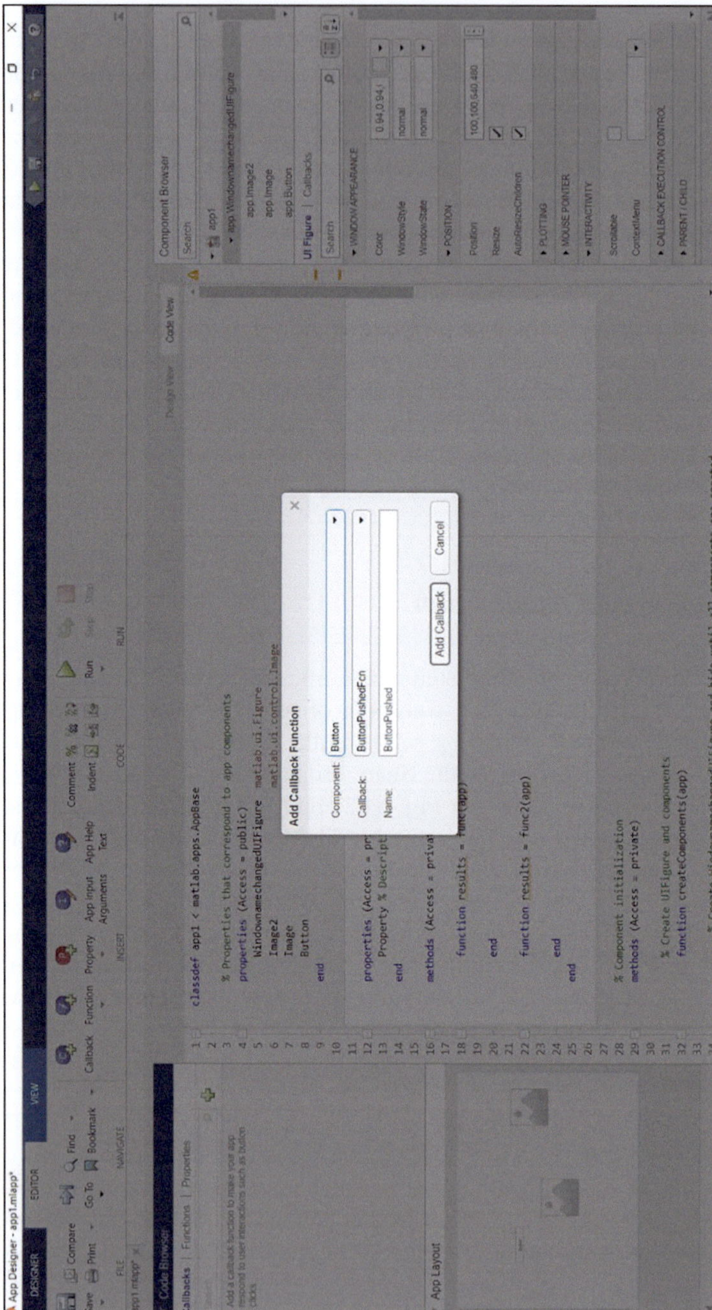


**Fig. 6.5** New callback function popup

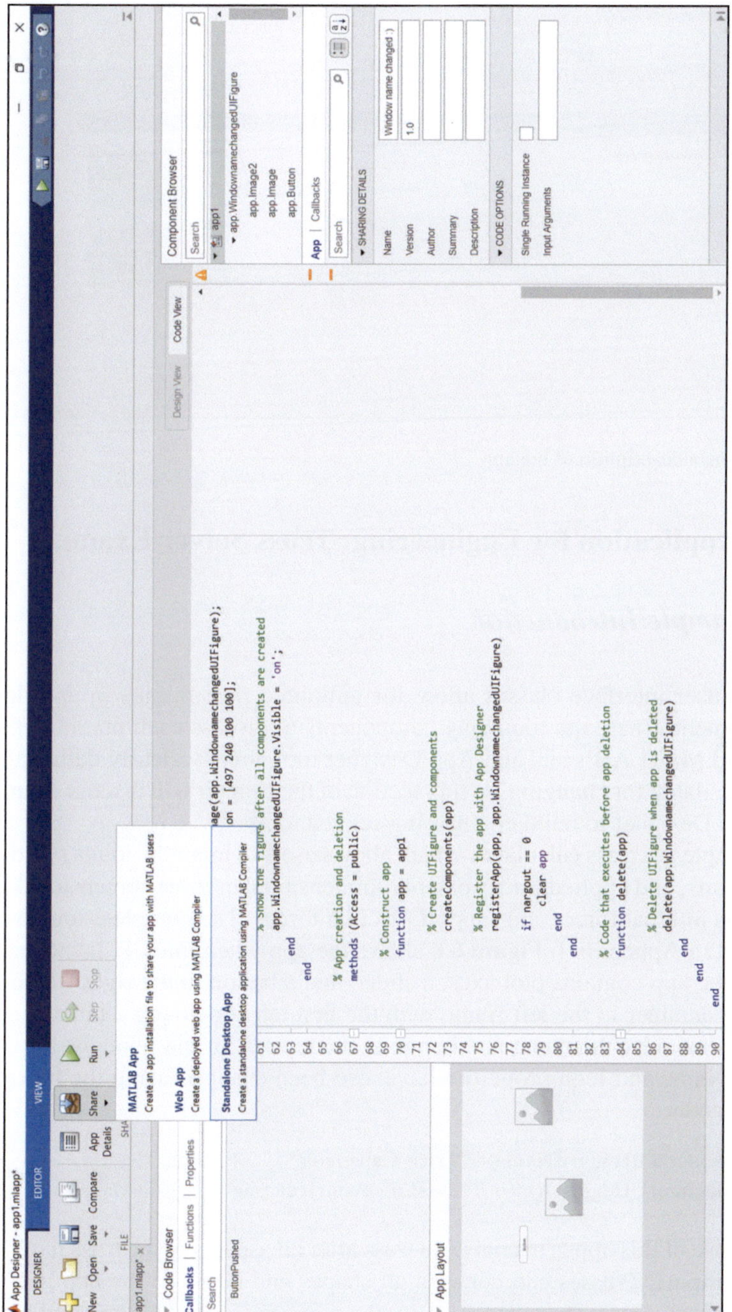


**Fig. 6.6** Start compiling an app for distribution

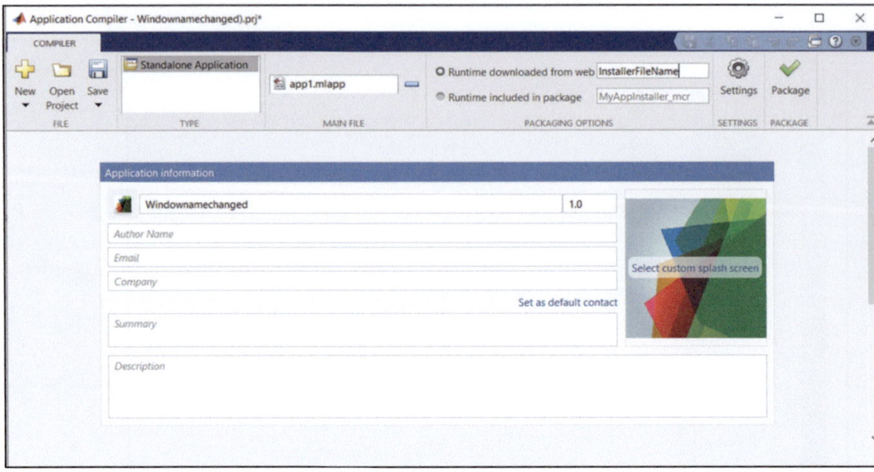


**Fig. 6.7** Adding a description of the app

## 6.2 An Application for Engineering: Truss Solver Example

### *6.2.1 Example Introduction*

MATLAB's user interface classes allow for unlimited possibilities in app design and development. There are too many components to discuss each one in depth in this book and MATLAB's site and App Designer tool already clearly define all the properties available for changing, so this section of the chapter will discuss examples of using App Designer to build engineering-related tools.

This example is a truss calculator, which allows users to input the joints (vertices), beams, supports, and applied forces of a two-dimensional or three-dimensional truss and obtain its internal forces and support reaction forces. The complete source code can be found in Appendix I. Figure 6.8 shows the app when initially launched. On this screen, the app contains plot axes, a slider, and a button in the right frame, and a tab group container in the left frame with the first tab containing a table and two buttons. As shown by the name and image at the top left of the window, the app's **UIFigure**'s Name and Icon properties have also been changed using the following two lines of code:

```
app.TrussCalculatorUIFigure.Name = "Truss Calculator"
app.TrussCalculatorUIFigure.Icon = "TrussCalculatorIcon.png"
```

The purpose of this app is to analyze a truss after all aspects of the truss have been provided as inputs. Trusses can come in all shapes and sizes, so the app needs an input method that allows the user to specify the number of truss vertices and beams, the location of the truss vertices, and how all the vertices are connected together

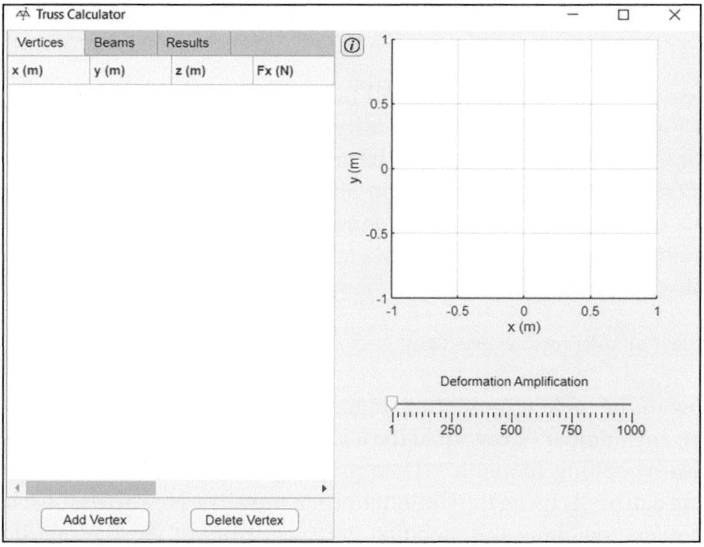


**Fig. 6.8** Newly launched truss calculator

by beams. Additionally, the user needs to specify any external forces and supports applied to any of the vertices and cross-sectional characteristics of the truss beams. In terms of outputs, the app needs to be able to display the calculated internal forces and support reaction forces and be able to render the deformed shape of the truss. The input and output requirements of the app help to decide how the app should work and what components each input and output should use, so outlining them in a table is a great starting point in app development (Table 6.1).

**Table 6.1** App requirements

| Truss calculator requirements | |
|---|---|
| Input requirements | Output requirements |
| • Number of vertices<br>• Number of beams<br>• Locations of vertices<br>• Vertex connections<br>• External forces<br>• Locations of supports (constraints)<br>• Beam characteristics (Cross-sectional area and elastic modulus) | • Internal forces<br>• Reaction forces<br>• Deformed shape |

### 6.2.2 UITables and Containers

This truss calculator uses MATLAB's **UITable** components to accept an unspecified number of inputs in a structured and easily understandable format. Separate tables are used for the truss vertices and truss beams, and these tables are stored in separate tabs of a container to organize the app and avoid crowding the screen. Inserting components into containers (or the container's tab in this case) is done by dragging the table component onto the container's tab in App Designer or initiating the table component with the container's tab as the specified parent object:

```
app.UITable = uitable(app.VerticesTab);
```

Each row of the table represents a single vertex or beam, so the table needs to be able to have any number of rows that the user can add or remove. Rows can be added to **UITables** by setting the table's Data property, and the allowable data types for each column can be set using the ColumnFormat property. New rows must have blank cells or contents matching the specified datatype. Both of these properties are cell arrays, where each element represents a table column. Each column can accept the datatypes of **'char'**, **'numeric'**, or **'logical'**, where **'char'** columns can be defined by a nested cell array to create dropdown menus and **'logical'** columns appear as check boxes. In the truss calculator example, the vertex row inputs accept character vectors for the Cartesian coordinates and force vectors so that mathematical functions such as **sqrt** and **log** could be inputted and converted to numbers (using the **'numeric'** datatype prevents all letters and special characters other than decimals and negative signs from being inputted). Boolean (**'logical'**) datatypes are used to specify which axes are fixed (Fig. 6.9):

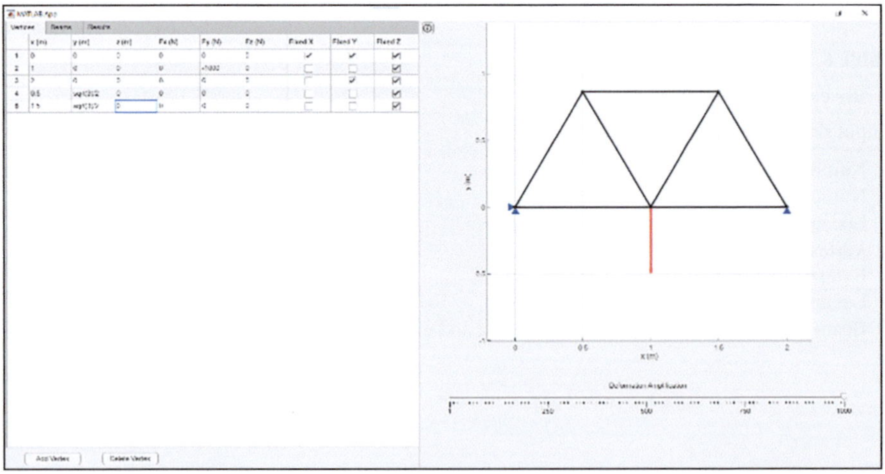

**Fig. 6.9** Completed vertices tab

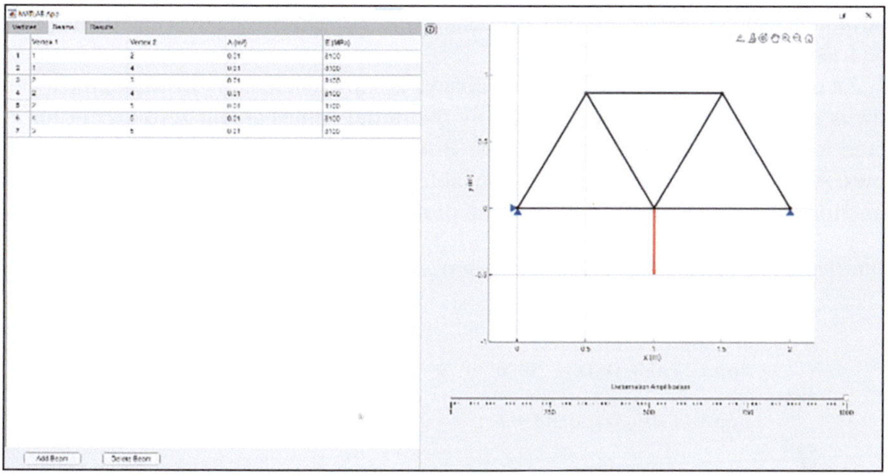

**Fig. 6.10** Completed Beams tab

```
app.UITable.ColumnFormat = {'char' 'char' 'char' 'char' 'char' 'char' ...
                            'logical' 'logical' 'logical'};
app.UITable.Data(end + 1, :) = {'0' '0' '0' '0' '0' '0' 0 0 1};
```

The beam table rows similarly accepted character vectors for the beam characteristics, but then used nested cell arrays of character vectors to select connecting vertices from a dropdown list (Fig. 6.10). The column format for the beam table is updated every time the vertex table is updated so that the dropdown list always contains every existing vertex. The dropdowns are defined by creating a cell array of character vectors and inserting them into the ColumnFormat cell array, as shown below:

```
function updateDropdowns(app)

    % Make the 2 vertices of the beam table dropdowns (dropdowns can only contain
    % strings)
    options = cellstr(string(1:1:height(app.UITable.Data)));
    app.UITable_2.ColumnFormat = {options options 'char' 'char'};
end
```

### 6.2.3 *UIButtons*

Most applications need buttons in one form or another to trigger events, and this example is no different. This app uses buttons to add and remove table rows, open help menus, and turn the inputs into calculated outputs using the buttons' callback functions. The add and remove buttons on both the Vertices and Beams tabs update the tables' Data properties, update the beam table's dropdown menus, and then update the truss plot.

An example of one of the button callbacks is shown below. In this function, a new row is added to the Vertices table, where the initial values are all zero except for all three Booleans in the first row and the Boolean along the Z-axis in the following rows. After adding a new row, the Beam table's dropdown format is updated using the function shown previously, and then the plot is updated in the **updatePlot** function.

```
function AddVertexButtonPushed(app, event)

        % Add a new table row
        if isempty(app.UITable.Data)
                app.UITable.Data = {'0' '0' '0' '0' '0' '0' 1 1 1};
        else
                app.UITable.Data(end + 1, :) = {'0' '0' '0' '0' '0' '0' 0 0 1};
        end

        app.updateDropdowns()
        app.updatePlot()
end
```

## 6.2.4 *UIAxes*

As will be covered later in Chap. 7, MATLAB's plotting capabilities are a great way for displaying and animating information. The **UIAxes** component is App Designer's equivalent to plotting figures, and updating **UIAxes** plots is no different than any other plotting in MATLAB. After creating a new **UIAxes**, all plotting functions can be used to update that specific plot by providing the plot's handle as the first argument. For example, the following code changes the **UIAxes** plot to a three-dimensional view and then plots each vertex in the Vertices table sequentially as solid black dots:

```
view(app.UIAxes, 3)
vertexTable = app.convert(app.UITable.Data);
hold(app.UIAxes, 'on');
for index = 1:1:height(vertexTable)
        row = vertexTable(index, :);
        plot3(app.UIAxes, row(1), row(2), row(3), 'Marker', 'o', 'MarkerSize', 4, ...
                'MarkerFaceColor', 'k', 'MarkerEdgeColor', 'k');
end
hold(app.UIAxes, 'off');
```

### *6.2.5 UISliders and UITextareas*

Once all of the vertices and beams are entered, and calculations are completed by pressing the Calculate button on the Results tab, the results need to be output in a readable and copyable format. This action could be performed by many components, but most notably by **UILabels**, **UIEditfields**, and **UITextareas**. This example uses **UITextareas** because they are large enough to show multiple rows and can easily accommodate a changing number of outputs. While both **UIEditfields** and **UITextareas** are usually used to accept inputs, they can be used to display outputs by setting their Editable properties to **'off'** (or **false**) and then assigning the output to their Value properties. Keep in mind that while a **UITextarea** can display multiple lines, its Value property can only be one string, so the multiple line output must be converted all into a single string using **'\n'** as shown below:

In the following code, **R** is a 3xn array collection of column vectors each containing the reaction force on one vertex. Within the **sprintf** function, the column number is placed in brackets followed by the contents of the column separated by tabs (**'\t'**) and then followed by an end of line command (**'\n'**). The results of this code are shown in Fig. 6.11 in the Reaction Forces text area.

```
app.ReactionForcesNTextArea.Editable = 'off';
app.ReactionForcesNTextArea.Value = sprintf("[%d]\t" + strjoin(repmat("%.4f", ...
                                    1, height(R)), ', \t') + "\n", [1:1:width(R); R]);
```

Pressing the Calculate button also updates the plot to show the undeformed and deformed trusses. The slider below the axes can be used to change the amplification of the deformation so that it can more easily be seen. The plot is updated every time the slider is moved using the **ValueChangingFcn** callback. This callback takes the

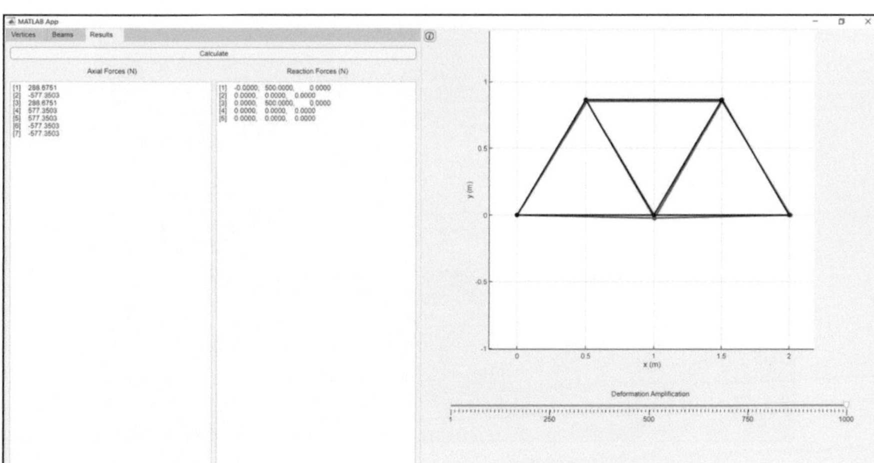


**Fig. 6.11** Outputs for the inputted truss

class handle and event information as inputs, which MATLAB calls **app** and **event**, respectively. The current position of the slider can be used by reading the value of **event.Value**.

### 6.2.6 *Finishing Touches*

Once all major functions of the application are completed, it is time to add finishing touches that help the end user to navigate the app and use it correctly. These helpful little additions can be done in many ways using many different components, but some especially useful components are tooltips and **UIAlerts**. Every component in MATLAB has a Tooltip property that contains a character cell array. When a mouse hovers over a component, that component's respective tooltip will appear, and can be used to inform the user what the component does or how it can be used.

```
app.AddBeamButton.Tooltip = {'Adds a new row to the table.'};
```

Well-made applications are able to prevent a wide range of user errors from crashing the program and can provide messages informing the user what went wrong. **UIAlerts** are a useful tool for dealing with user error, since they create popups that can explain the problem. For example, the Truss Calculator aborts solving and opens an alert when a table cell is left empty (Fig. 6.12).

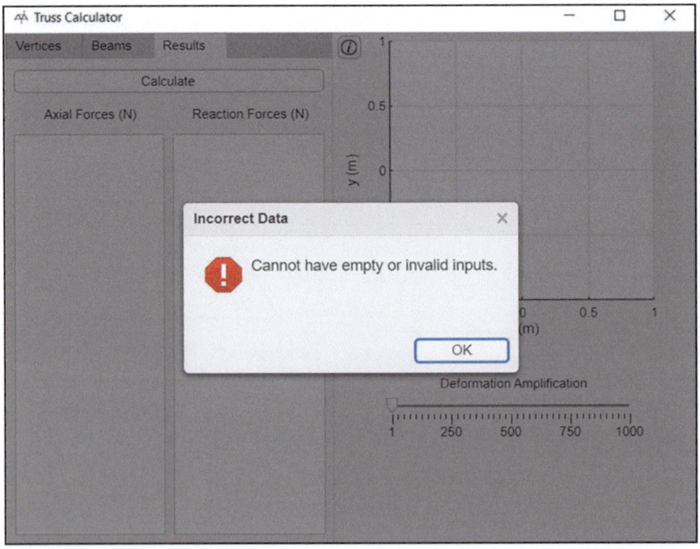


**Fig. 6.12** Empty cell error popup

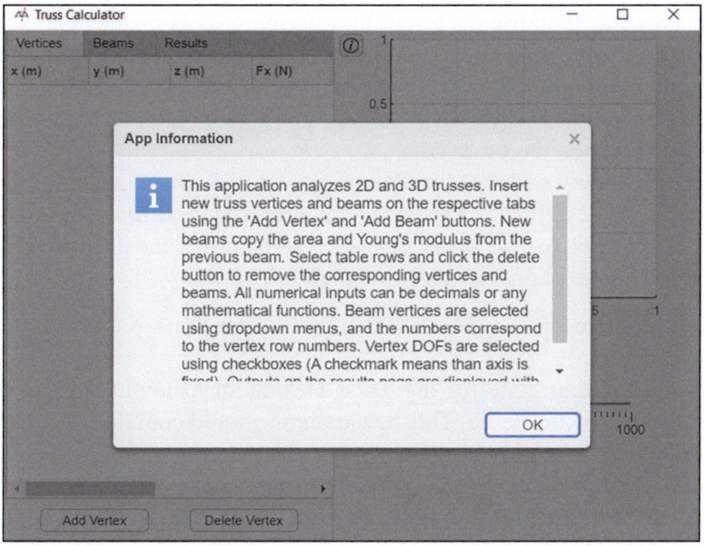


**Fig. 6.13** Help information popup

```
uialert(app.TrussCalculatorUIFigure, 'Cannot have empty or invalid inputs.', ...
        'Incorrect Data')
```

Additionally, **UIAlerts** can be used for information popups and help pages to explain more about certain aspects of the application. The calculator example uses a button with a different Icon property setting to open a help window (Fig. 6.13).

```
app.InfoButton.Icon = 'InfoIcon.png';
uialert(app.TrussCalculatorUIFigure, "This application analyzes 2D and 3D trusses. Insert
new truss vertices and beams on the respective tabs using the 'Add Vertex' and 'Add Beam'
buttons. New beams copy the area and Young's modulus from the previous beam. Select
table rows and click the delete button to remove the corresponding vertices and beams. All
numerical inputs can be decimals or any mathematical functions. Beam vertices are
selected using dropdown menus, and the numbers correspond to the vertex row numbers.
Vertex DOFs are selected using checkboxes (A checkmark means than axis is fixed).
Outputs on the results page are displayed with the corresponding vertex or beam
number.", 'App Information', 'Icon', 'info');
```

## Practice Problems

(1) Using the existing code for the Truss Calculator example, add number labels to each vertex and beam on the plot axes corresponding to their row numbers. Add two checkboxes below the deformation slider that toggle the vertex and beam numbering on and off and update the plot whenever the checkbox values are changed.

(2) Using the existing code for the Truss Calculator example, change the two output text areas to two output tables. The output tables' contents should be un-editable,

and each table should have a tooltip describing its contents. The axial forces table should have only one column, and the reaction forces table should have three columns, one for each force component. Each row in the tables should represent one beam or vertex with the same row number as the corresponding beam or vertex on the other tabs.

(3) Using the existing code for the Truss Calculator example, add vertex and beam numbering as described in Problem 1, but use two toggle tools in a toolbar rather than two checkboxes. To assist app users, add tooltips and custom images to the toggle tools to explain the buttons' functions.

**Hint**: The value of a toggle tool is stored in the **State** property rather than a **Value** property.

(4) Using the existing code for the Truss Calculator example, add an additional column to the beams table. This new column should contain a dropdown menu that allows the selection of different line thicknesses, which is then applied to the plot (the deformation plotting function can remain unchanged). Set the first row default value of this dropdown to a line width of 2 but have all future rows' defaults copy the value from the row above. Remember to label the new column!

(5) In the current version of the Truss Calculator, the program throws an error when a number not available in the beam vertex dropdown is inputted (For example, typing 12 into the cell when there are only three vertices). Using the existing code for the Truss Calculator example, add something to catch or prevent this error and open a **UIAlert** to inform the user of the problem.

# Chapter 7
# Animation Techniques in MATLAB

## 7.1 Techniques for Creating Animations in MATLAB

### *7.1.1 Using Plotting Commands, Drawnow and Pause*

Command **pause(n)** will stop the execution of the remaining code for **n** seconds allowing the user to see the animation. For instance, the code below gradually constructs a sine curve for which three snapshots at different time points are shown in Fig. 7.1:

```
x = linspace(0, 2*pi, 100);
y = sin(x);
for k = 1:length(x)
        plot(x(1:k), y(1:k));
        axis([0 2*pi -1 1]);
        pause(0.1);
end
```

Command **drawnow** updates figures and processes any pending callbacks, so it can be used in animations when their data is updated. For instance, in the code below, a sine curve is plotted, then a secondary plot handle for a red circle moving on this curve is created, and finally the circle's location on the curve is updated using the commands **set(h,'XData',x(ii)), set(h,'XData',x(ii)),** and **drawnow**. Three snapshots for this animation at different time points are shown in Fig. 7.2:



M. Andalibi and S. A. Cherkauer, *Advanced MATLAB for Engineers*,
https://doi.org/10.1007/978-3-032-10677-3_7

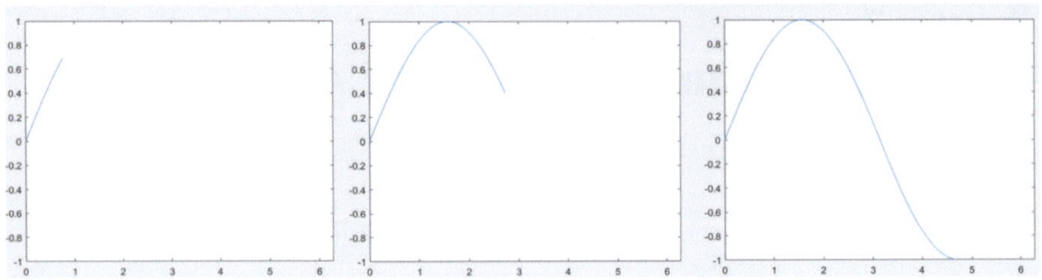

Fig. 7.1 Animation of a sine curve gradually constructed using pause

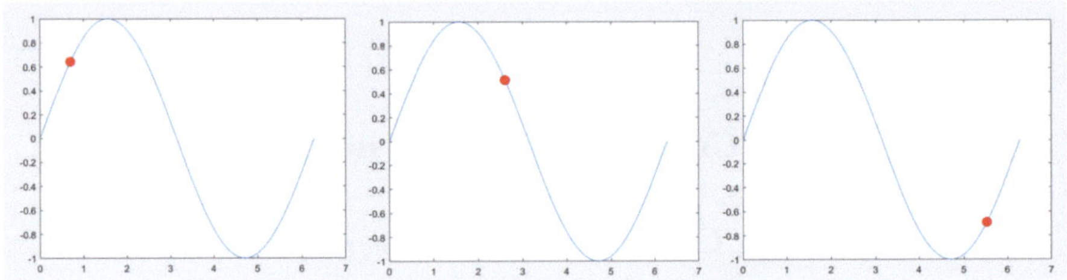

Fig. 7.2 Animation of a particle moving on a sine curve using drawnow

```
x = linspace(0, 2*pi, 100);
y = sin(x);
plot(x,y);
hold on;
h=plot(x(1),y(1),'ro','MarkerSize',10,'MarkerFaceColor','r');
for ii = 2:length(x)
        set(h,'XData',x(ii));
        set(h,'YData',y(ii));
        drawnow;
        pause(0.04);
end
```

To make the motion smoother, use more points in the sin curve and reduce the pause time.

### *7.1.2 Using Getframe, Movie, and Videowriter to Convert Captured Frames and Create a Movie*

Command **getframe** captures the current axes as it appears on the screen as a movie frame, i.e., it can be used as **getframe(fig)** or **getframe(ax)** too.

The resulting structure from this command has a field called **cdata** that contains the returned frame. Using the command **frame2im**, the frame can be converted into an image that can then be written to the hard drive using the command **imwrite** as shown below:

```
frame = getframe(fig);
im = frame2im(frame);
imwrite(im,"myframe.png");
```

However, since the goal in this chapter is to make animations, we can store these frames in a (movie) variable, such as **mov**, that can then be played in MATLAB **n** times a frame rate of **fps** using the syntax **movie(mov,n,fps)** or be written to an external movie file, such as an **.avi** file using the **VideoWriter** command.

The code below creates a 3D sine wave, captures the frames, stores them in a movie variable, plays the movie in MATLAB, and finally writes the movie to an external **.avi** file:

```
t=0:0.1:10;
[x,y]=meshgrid(-2*pi:pi/10:2*pi);
z=cos(t(1))*sin(x).*sin(y);
figure;
h=surf(x,y,z);
set(gca,'Zlim',[-1 1]);
m=1;
mov(m)=getframe;
for ii = 2:length(t)
        z=cos(t(ii))*sin(x).*sin(y);
        set(h,'ZData',z);
        drawnow;
        pause(0.1);
        m=m+1;
        mov(m)=getframe;
end
movie(mov,1,15);
v=VideoWriter('2D_sine_wave.avi');
open(v);
writeVideo(v,mov);
close(v);
```

Three frames of this movie are shown in Fig. 7.3. Note that when the external movie file is written using the **writeVideo** command, you need to close the **VideoWriter** object **v** to release the resources.

### *7.1.3 Using comet and comet3*

Command **comet(x,y)** displays a comet plot of **y** versus **x** while command **comet3(x,y,z)** displays a comet plot of **z** versus **x** and **y**. In a 2D comet plot, a

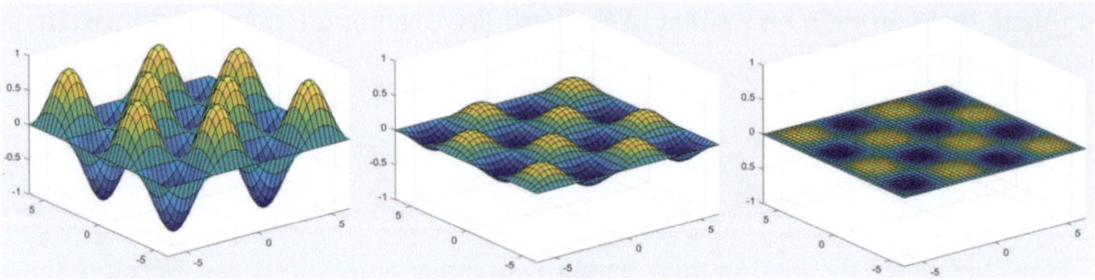

**Fig. 7.3** Animation of a 3D sine wave and converting it to a movie

**marker/particle** is leading the curve being constructed as can be seen in Fig. 7.4, which shows three snapshots taken from the results of running the code below for a sine wave:

```
x = linspace(0, 2*pi, 100);
y = sin(x);
comet(x,y);
```

A 3D comet is an animation of a **marker (head)** and a **line (tail)** tracing a growing line over the data points. The tail is a solid line that traces the entire function as can be seen in Fig. 7.5 for three snapshots taken from the results of running the code below for the **peaks** surface:

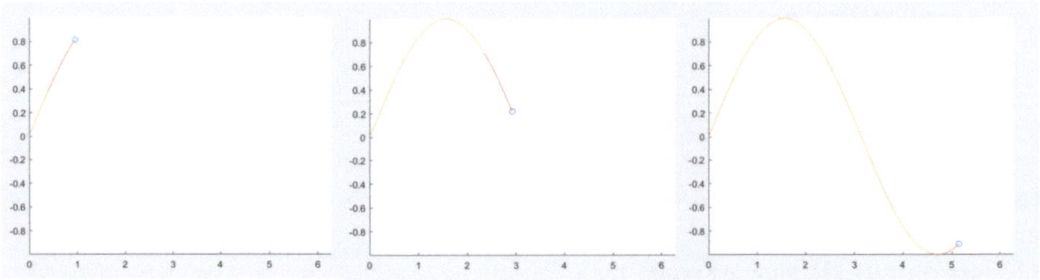

**Fig. 7.4** 2D comet plot for a sine wave

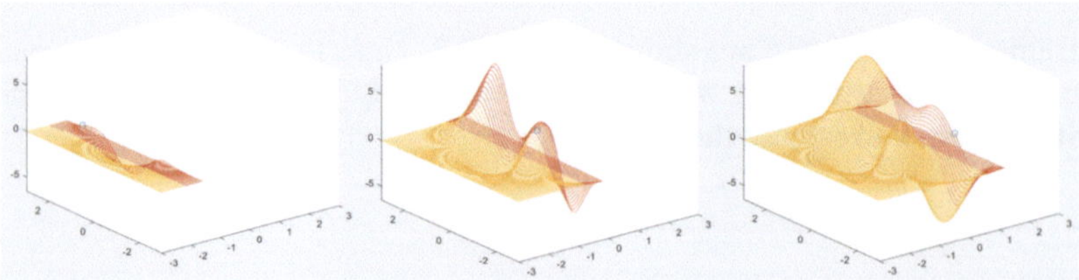

**Fig. 7.5** 3D comet plot for the peaks surface

```
[xmat,ymat,zmat] = peaks(100);
xvec = xmat(:);
yvec = ymat(:);
zvec = zmat(:);
comet3(xvec,yvec,zvec)
```

### *7.1.4 Using Animatedline*

Command **animatedline(x,y)** or **animatedline(x,y,z)** creates an animated line with initial data points defined by **x** and **y** (and **z**) as shown in the code below for a sine curve:

```
x = linspace(0, 2*pi, 100);
y = sin(x);
h = animatedline('Color','r','LineWidth',3);
axis([0 2*pi -1 1]);
for k = 1:length(x)
        addpoints(h, x(k), y(k));
        drawnow;
end
```

Note that **animatedline** basically creates a figure handle but it has a different class name. Also, similar to the method in Sect. 7.1.1 and unlike the comet command, **animatedline** requires a **for** loop to change the plot handle data, which is done using the command/method **addpoints**. Snapshots of the resulting animation can be seen in Fig. 7.6.

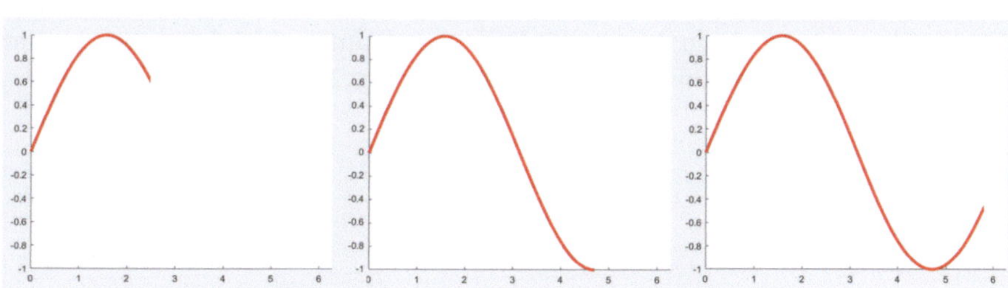

**Fig. 7.6** Using animatedline to create animation for a 2D sine curve

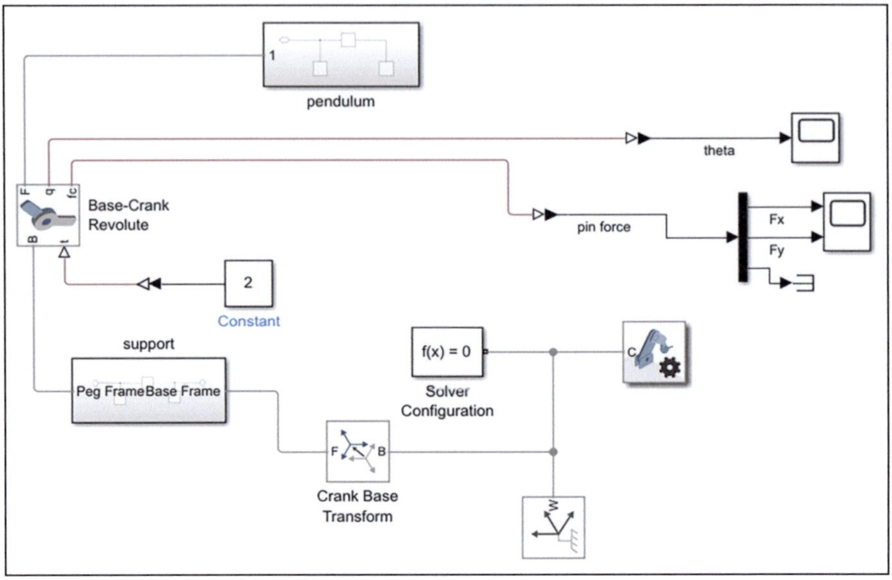


**Fig. 7.7** Simulink diagram for a pendulum made using Simscape/Multibody environment

### *7.1.5 Using Simulink Simscape Multibody*

While Simulink will be discussed in Chap. 8, here, we would like to show that the **Multibody** environment inside **Simulink/Simscape** is capable of creating animation/movies from the assembled mechanical elements made/exported into this environment.

To see an example, first type the constant values below in the command window (these values are needed for the dimensions of the rigid bodies in the Simulink file), then download and run the Simulink file simulating a pendulum from the link provided at the bottom of this page.[1] and shown in Fig. 7.7.

[1] https://drive.google.com/file/d/1rAjBA6ZNJtxKn3icrchHQh2EnJHTE_ty/view?usp=sharing

```
% Body Geometry Parameters
l = 20; % Hole-to-hole distance
w = 2; % Link width
d = 1.2; % Hole diameter
t = 1; % Link thickness

% Main Solid Cross-section:
A = linspace(-pi/2,pi/2)';
B = linspace(pi/2,-pi/2)';
csRight = [l/2+w/2*cos(A) w/2*sin(A)];
csLeft = [-l/2 w/2; -l/2 + d/2*cos(B) d/2*sin(B); -l/2 -w/2];
csMain = [csRight; csLeft];

% Hole Solid Cross-section:
C = linspace(pi/2,3*pi/2)';
D = linspace(3*pi/2,pi/2)';
csHole = [w/2*cos(C) w/2*sin(C);
d/2*cos(D) d/2*sin(D)];
```

The generated pendulum in Simulink can be observed in Fig. 7.9, where the icon highlighted in a red box in this figure can convert the resulting simulation into an external animation video. Figure 7.8 shows the options for creating the video file using this icon.

In this Simulink model, a constant torque is applied to the pendulum, causing it to spin constantly in the same direction. Three snapshots are shown in Fig. 7.10.

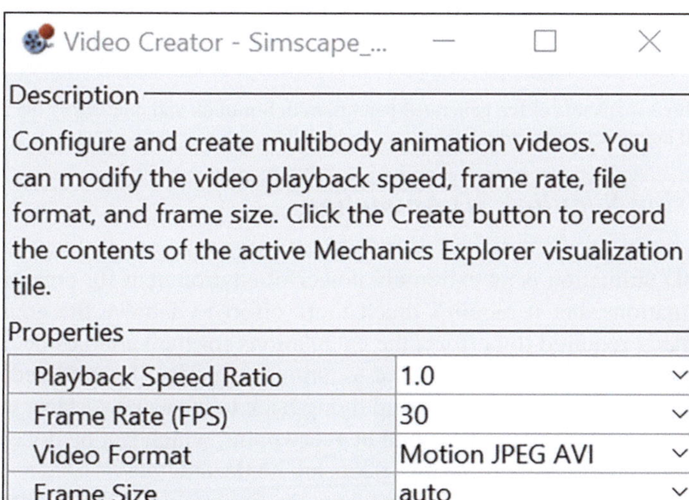


**Fig. 7.8** Options for creating an animation video file in Simscape/Multibody environment

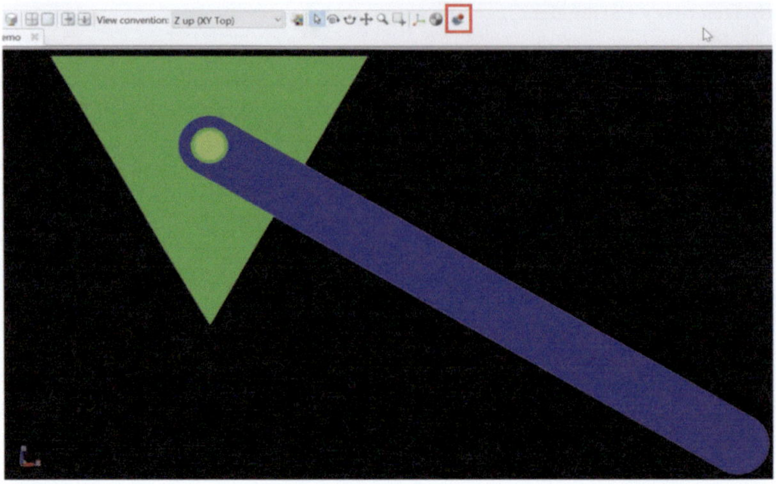


**Fig. 7.9** Generated pendulum in Simulink and converting the simulations to an external animation video file

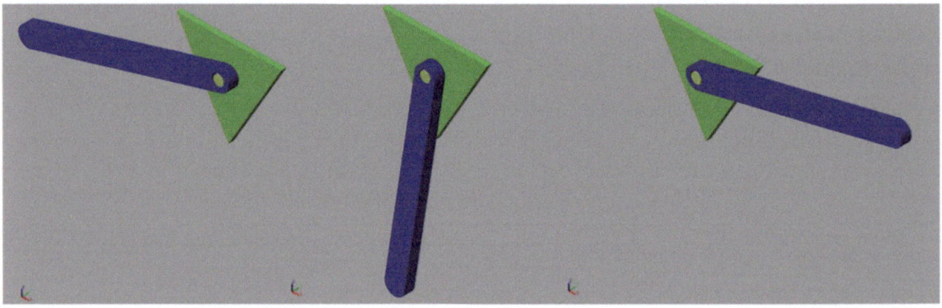

**Fig. 7.10** Three snapshots of the generated pendulum in Simulink and converting the simulations to an external animation video file

## *7.1.6 Using Simulink 3D Animation*

Simulink 3D animation is an extremely powerful environment for creating sophisticated animations, but it requires much more effort to achieve the goal. Here, a Simulink file is required to perform the calculations for the variables necessary for the animation. For instance, if we want to animate a projectile, rendered as a ball, being launched from a fixed surface and flying back to the same surface, we need to calculate the (x,y,z) location of the ball at every point, which can be done using the diagram shown in Fig. 7.11 if the air effects are neglected.

The Simulink file can be downloaded from the link provided at the bottom of this page.[2] Next, we need to create a Virtual Reality (VR) environment for the graphics.

[2] https://drive.google.com/file/d/1duHOR46hU_ir5GpLGw0rytbiWxZ8ErMH/view?usp=sharing

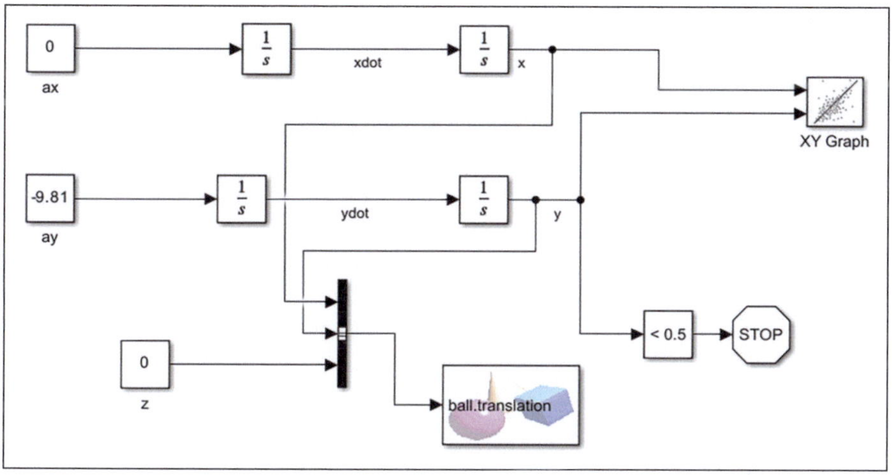


**Fig. 7.11** Simulink diagram for a projectile simulation neglecting the air effects

To do that, we first need to install the VR editor in MATLAB using the following command:

```
vrinstall -install editor
```

Next, from under **Simulink 3D animation branch/Utilities**, drag and drop the **VR Sink** block to the Simulink file and attach it to the **Bus** block that gathers the (x,y,z) location of the ball as can be seen in Fig. 7.11 (the block is renamed as **ball.translation**).

Next, as shown in Fig. 7.12, you need to click on the **VR Sink** block to select an existing VR environment or click on New to create a file.

When creating a new VR environment, the objects created are called **Nodes**, and they will be added under the **ROOT** which is seen on the left window as shown in Fig. 7.13. To create these objects, we can insert basic shapes from the component library or import previously designed parts as an STL, XML, SDF, or other types of compatible files.

In Fig. 7.13, three nodes are created for the ball (projectile), the fixed surface (ground), and the sky and ground (background). You can download the VR (x3d) file for this environment from the link at the bottom of this page.[3] Note that for the objects/nodes, you can rename them and alter their properties, including the geometry (size), location, orientation, scale, location of the center of gravity, appearance (color, texture, material), etc. Even for texture and material, there are libraries under **Nodes/Insert from**.

[3] https://drive.google.com/file/d/14Z8raDUx2gJBd7htdJVAWRblat-jbXgq/view?usp=sharing

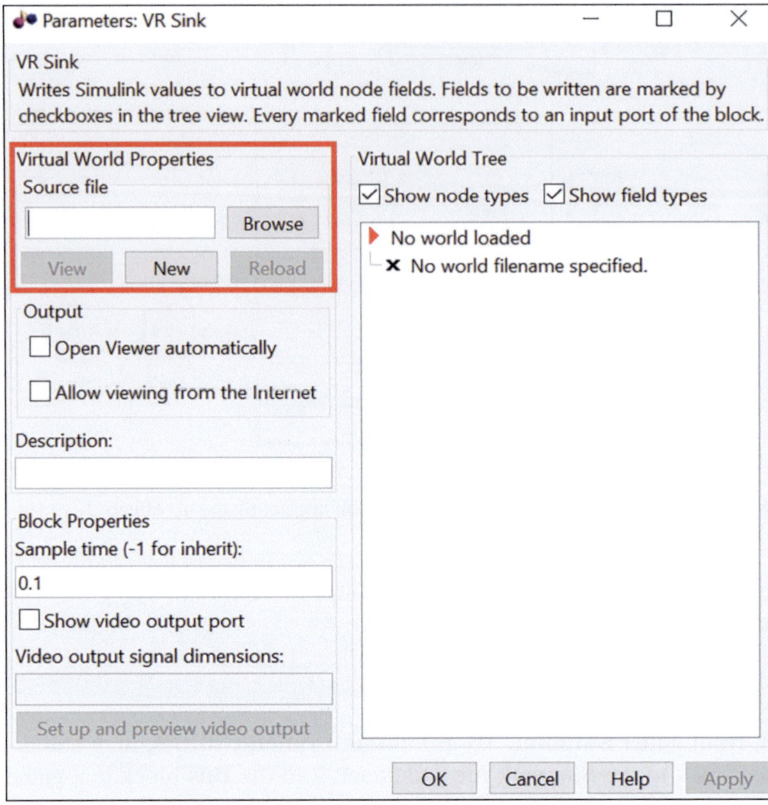


Fig. 7.12 Selecting or creating the VR environment inside the VR Sink

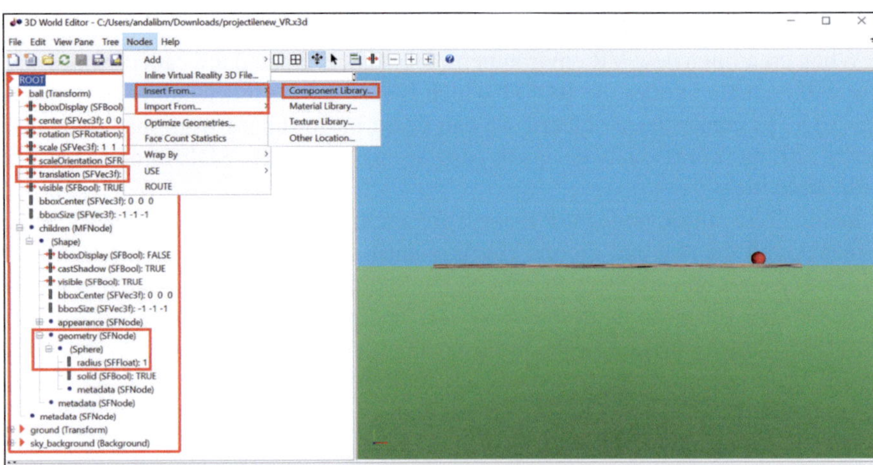


Fig. 7.13 VR Editor environment and nodes for the projectile animation

Once the VR file is created, saved, and selected in the **VR Sink** block, we need to connect the location of the ball from simulations to the **VR Sink**, as it does not have an input port.

To do so, we double-click and open the **VR Sink** block, go under **Simulation/ Block parameters**, and then expand the **ball** and select **translation** under the **ROOT** in the right window to create an input port for a translation vector as can be seen in Fig. 7.14.

Note that if you want to modify the VR file, you can open the **VR Sink** block and then open the VR environment in the VR Editor by navigating to **File/Open in Editor**.

Figure 7.15 shows three snapshots of the resulting animation overlaid in the same picture, where the ball is shown in three different positions.

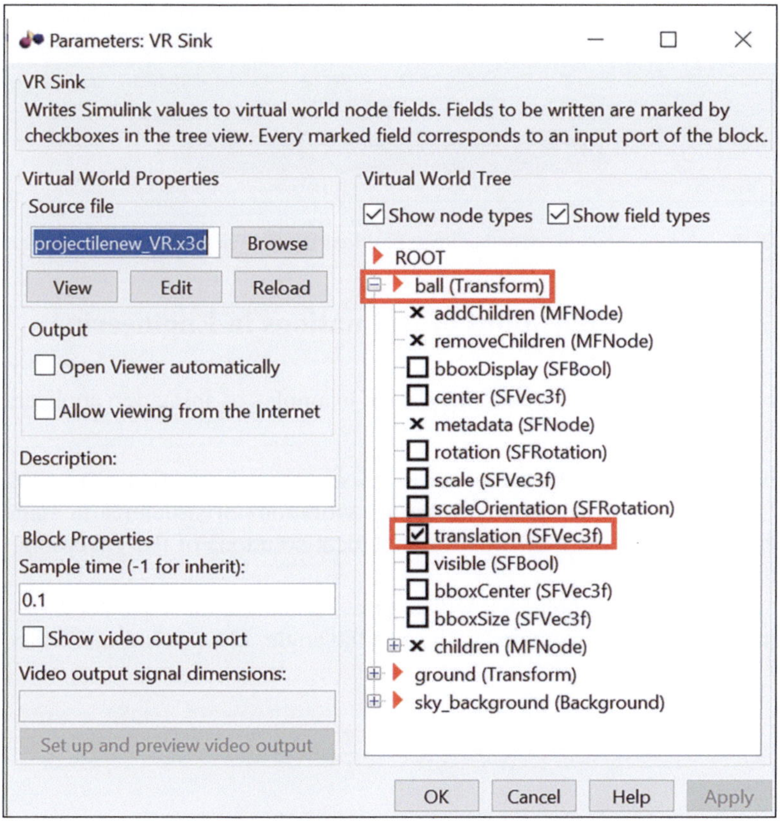


**Fig. 7.14** Creating an input port for the VR Sink block by selecting the ball translation

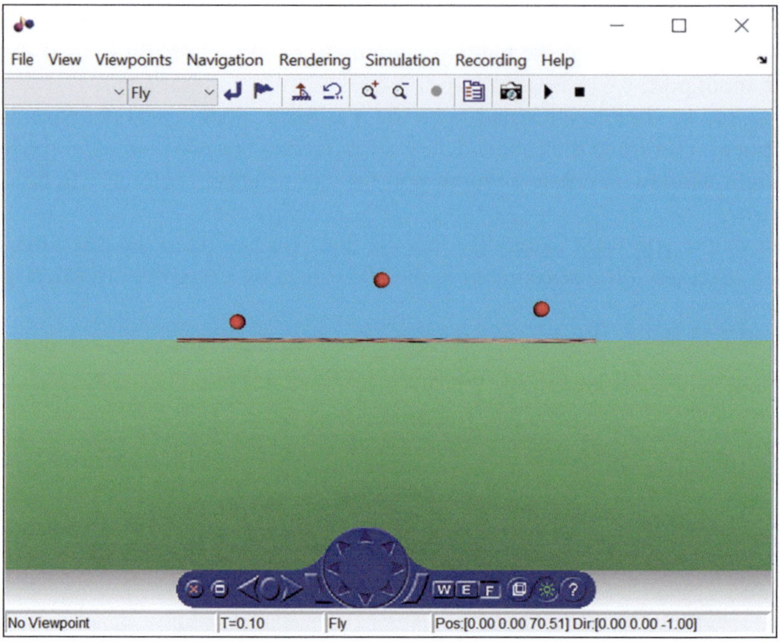


**Fig. 7.15** Resulting animation for ball at three positions overlaid in the same picture

## 7.2 Practical Applications of Animations in Engineering

In this section, we are going to look at some examples of animation applications in engineering disciplines.

(I) **Engineering Mechanics**: Here, we can visualize the motion of moving parts in dynamics and the deformations and vibrations of structures in statics and vibrations using the numerical or analytical solutions of the governing ODEs and PDEs.

As the first example, we will look at the animation code for a nonlinear pendulum released from rest with an initial angle. The governing ODE is given by Eqn. 7.1:

$$\frac{d^2\theta}{dt^2} = -\frac{g}{L}\sin(\theta) \tag{7.1}$$

```
L = 1; % Length of the pendulum (m)
g = 9.81; % Acceleration due to gravity (m/s^2)
theta0 = pi/4; % Initial angle (rad)
omega0 = 0; % Initial angular velocity (rad/s)

t_end = 10; % Simulation end time (s)
dt = 0.01; % Time step (s)
t = 0:dt:t_end; % Time vector

% Pre-allocate arrays for efficiency
theta = zeros(size(t));
omega = zeros(size(t));

% Initial conditions
theta(1) = theta0;
omega(1) = omega0;
```

```
% Numerical integration using Euler's method
for ii = 1:length(t)-1
        omega(ii+1) = omega(ii) - (g/L)*sin(theta(ii))*dt;
        theta(ii+1) = theta(ii) + omega(ii)*dt;
end

% Animation Loop
figure;
hold on;
x = L*sin(theta(1));
y = -L*cos(theta(1));
h1=plot([0 x], [0 y], 'r', 'LineWidth', 3);
h2=plot(x, y, 'bo', 'MarkerSize', 10, 'MarkerFaceColor', 'b');
axis equal;
axis([-1.5 1.5 -1.5 0.5]);
for ii = 2:length(t)
        x = L*sin(theta(ii));
        y = -L*cos(theta(ii));
        set(h1,'XData',[0 x]);
        set(h1,'YData',[0 y]);
        set(h2,'XData',x);
        set(h2,'YData',y);
        drawnow;
end
```

Figure 7.16 shows three snapshots of the resulting animation, where the numerical solution of the governing ODE was obtained using Euler's method and a small sample time to be sufficiently accurate.

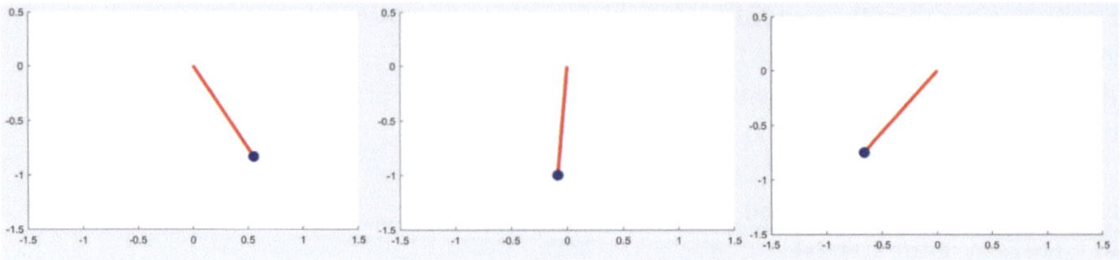

**Fig. 7.16** Animation for a nonlinear pendulum released from rest and $\theta_0 = \frac{\pi}{4}$

The second example is about the free vibrations of a cantilevered beam given by Eqn 7.2:

$$\frac{\partial^4 y}{\partial x^4} = -\beta^4 \frac{\partial^2 y}{\partial t^2} \tag{7.2}$$

The solution to the governing PDE is acquired via the **Finite Difference method**. Also, stability of the solution is checked when the time increment and number of nodes are provided by the user.

```
% User Data
L=0.5; % total length of the beam (m)
dx=0.025; %step size for the length (m)
x=0:dx:L; % locations of nodes
N=length(x); % number of nodes
tf=2; % final computation time
dt=0.000004; % time increment
t=0:dt:tf; % time points
M=length(t); % number of time steps used
A=0.05*0.2; % m^2: rectangular cross section W=20cm, H=5cm
ro=2700; % kg/m: aluminum beam
E=69e9; % Pa : aluminum beam
I=0.2*(0.05^3)/12; % m^4: rectangular cross section W=20cm, H=5cm
beta=((ro*A)/(E*I))^0.25;
disp('alpha must be less than or equal 1 for stability')
alpha=dt/((dx^2)*(beta^2))

% Boundary Conditions and Initial Condition
Y=zeros(N,M); % initialization of displacements for all nodes at all time points
% Boundary conditions
Y(1,:)=0;
Y(2,:)=Y(1,:);
% initial conditions
P=3*E*I*0.1/(L^3); % force for a 10-cm upward displacement at the end
Y(:,1)=P*(x.^2).*(3*L-x)/(6*E*I); % initial displacements at all nodes, q(x)
Y(:,2)=Y(:,1);
ymax=max(abs(Y(:,1)));
```

```
% Solution
for ii=3:M % loop over time points
        for jj=3:N-2 % loop over nodes
                Y(jj,ii)=-(alpha^2)*(Y(jj-2,ii-1)-4*Y(jj-1,ii-1)-4*Y(jj+1,ii-1)+...
                Y(jj+2,ii-1)+6*Y(jj,ii-1))-Y(jj,ii-2)+2*Y(jj,ii-1);
        end
        % Boundary conditions
        Y(N-1,ii)=2*Y(N-2,ii)-Y(N-3,ii);
        Y(N,ii)=3*Y(N-2,ii)-2*Y(N-3,ii);
end
% visualization
h=plot(x,Y(:,1),'-','LineWidth',10);
xlabel('nodal locations');
ylabel('displacements');
title('free vibrations of a beam')
axis([0 1.1*L -1.1*ymax 1.1*ymax]);
for kk=2:200:M
        set(h,'YData',Y(:,kk));
        drawnow;
end
```

Figure 7.17 shows three snapshots of the resulting animation. Note that the resulting animation is fast and can be slowed down by reducing the number of skipped time points (here 200) or adding a pause command in the last **for** loop.

(II) **Thermo-Fluid Mechanics**: Here, we can animate the fluid flow in pipes, around objects, or in open channels, see the path of particles within a fluid flow, and visualize velocity fields, pressure distribution, and turbulence, where the results are obtained via common techniques in Computational Fluid Dynamics (CFD).

For the first example, we will study the heat transfer via convention problem (defined by the custom **heatpde** function) within a 1D medium with fixed temperatures at both end points given by the governing PDE of $\alpha \frac{\partial^2 T}{\partial x^2} = \frac{\partial T}{\partial t}$. The problem is solved using the built-in command **pdepe**, which expects a standard equation format given by Eqn. 7.3:

$$c\left(x, t, u, \frac{\partial u}{\partial x}\right)\frac{\partial u}{\partial t} = x^{-m}\frac{\partial}{\partial x}\left(x^m f\left(x, t, u, \frac{\partial u}{\partial x}\right)\right) + s\left(x, t, u, \frac{\partial u}{\partial x}\right) \quad (7.3)$$

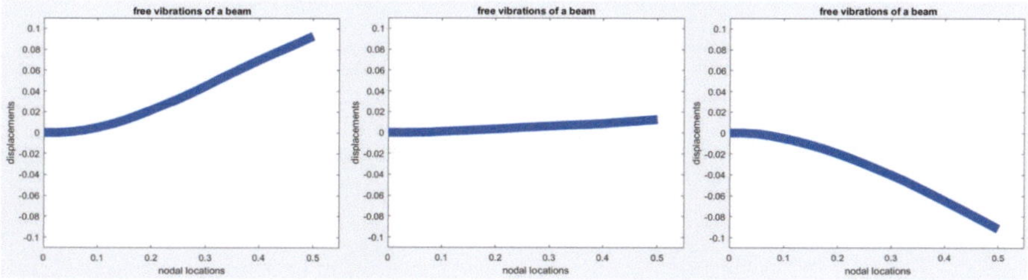


**Fig. 7.17** Animation for free vibrations of a cantilevered beam

An initial harmonic temperature distribution is being considered, which satisfies the fixed temperatures at the ends. These conditions are defined using the custom function **initial_condition.**

If the end points are denoted by $\boldsymbol{x_l}$ and $\boldsymbol{x_r}$, boundary conditions must be defined by Eqn. 7.4:

$$p(x, t, u) + q(x, t)f\left(x, t, u, \frac{\partial u}{\partial x}\right) = 0 \tag{7.4}$$

These boundary conditions are defined by the custom **boundary_conditions** function at the end of the script. The outputs $\boldsymbol{p_l}$ and $\boldsymbol{q_l}$ correspond to ***p(x,t,u)*** and ***q(x,t)*** for the left boundary and the outputs $\boldsymbol{p_r}$ and $\boldsymbol{q_r}$ correspond to ***p(x,t,u)*** and ***q(x,t)*** for the right boundary.

```
% Parameters
L = 1; % Length of the rod (m)
T0 = 100; % Temperature at x = 0 (°C)
TL = 0; % Temperature at x = L (°C)
n = 50; % Number of spatial points
t_end = 200; % End time for simulation (s)
dt = 0.01; % Time step (s)
alpha = 1e-2; % Thermal diffusivity (m^2/s)

% Spatial and time discretization
x = linspace(0, L, n);
t = 0:dt:t_end;

% PDE solver
m = 0; % Symmetry parameter for slab (0 means Cartesian coordinates)
sol = pdepe(m, @heatpde, @initial_condition, @boundary_conditions, x, t);
% Extract solution
T_sol = sol(:,:,1);

% Animation setup
figure;
h = plot(x, T_sol(1,:), 'b-', 'LineWidth', 2);
xlabel('Position (m)');ylabel('Temperature (°C)');
ylim([TL, T0]);
```

```
% Time-stepping loop for animation
for ii = 1:10:length(t)
        set(h, 'YData', T_sol(ii,:));
        title(['Temperature Distribution at t = ', num2str(t(ii), '%.2f'), ' s']);
        drawnow;
end

% PDE function
function [c, f, s] = heatpde(x, t, T, dTdx)
        alpha = 1e-4; % Thermal diffusivity (m^2/s)
        c = 1;
        f = alpha * dTdx;
        s = 0;
end

% Initial condition function
function T_0 = initial_condition(x)
        L = 1;
        T_0 = 50 * sin(pi * x / L); % Initial temperature distribution
end

% Boundary conditions function
function [pl, ql, pr, qr] = boundary_conditions(xl, Tl, xr, Tr, t)
        T0 = 100; % Temperature at x = 0 (°C)
        TL = 0; % Temperature at x = L (°C)
        pl = Tl - T0;
        ql = 0;
        pr = Tr - TL;
        qr = 0;
end
```

Figure 7.18 shows the initial, an intermediate, and the final temperature distributions within the time frame of 200 s, in parts (a), (b), and (c), respectively. As can be seen, the temperature distribution approaches the steady-state solution, which is linear since:

$$t \to \infty : \frac{\partial T}{\partial t} \to 0 : \frac{\partial^2 T}{\partial x^2} \to 0 \Rightarrow T = Ax + B \tag{7.5}$$

For the second example, we will look at the 1D steady, laminar, fully-developed flow of an incompressible viscous fluid in a pipe, for which the governing Navier–Stokes equation is given by Eqn. 7.6:

$$\frac{1}{r}\frac{\partial}{\partial r}\left(r\frac{\partial u}{\partial r}\right) = \frac{1}{\mu}\frac{\partial P}{\partial x} \tag{7.6}$$

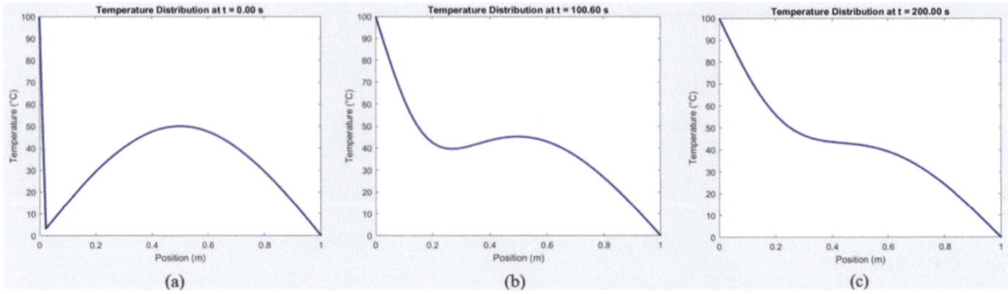


**Fig. 7.18** Initial (**a**); an intermediate (**b**); and the final (**c**) temperature distributions with a time frame of 200 s in a 1D medium heat transfer/conduction problem with temperatures at the ends and a harmonic initial temperature distribution

In Eqn. 7.6, **r** is the radial distance from the centerline of the pipe, **x** is the distance along the pipe, **u(r)** is the particle velocity, and $\mu$ is the dynamic viscosity of the fluid.

The boundary conditions are given by:

$$u(R) = 0 : \textit{no slip condition}, \frac{\partial u}{\partial r}(r = 0) = 0 : \textit{symmetry condition} \tag{7.7}$$

where **R** is the radius of the pipe.

Integrating the governing equation with respect to ***r*** and applying the symmetry condition gives:

$$\frac{\partial}{\partial r}\left(r\frac{\partial u}{\partial r}\right) = \frac{r}{\mu}\frac{\partial P}{\partial x} \Rightarrow r\frac{\partial u}{\partial r} = \frac{r^2}{2\mu}\frac{\partial P}{\partial x} + C_1, \frac{\partial u}{\partial r}(r = 0) = 0 \Rightarrow C_1 = 0 \tag{7.8}$$

Simplifying and integrating the governing equation with respect to ***r*** again gives:

$$\frac{\partial u}{\partial r} = {}_{\{}^{r}2\mu\frac{\partial P}{\partial x} \Rightarrow u = \frac{r^2}{4\mu}\frac{\partial P}{\partial x} + C_2, u(R) = 0 \Rightarrow C_2 = -\frac{R^2}{4\mu}\frac{\partial P}{\partial x}$$

$$\Rightarrow u(r) = \frac{-\left(R^2 - r^2\right)}{4\mu}\frac{\partial P}{\partial x} \tag{7.9}$$

In the steady fully developed flow, the pressure drop/gradient is going to be a fixed value and given by $\frac{\Delta P}{L}$, where **L** is the length of the pipe, so the velocity profile is going to be:

$$u(r) = u_{max}\left(1 - \frac{r^2}{R^2}\right) \textit{where } u_{max} = \frac{-R^2}{4\mu}\frac{\Delta P}{L} \tag{7.10}$$

To animate the trajectories of fluid particles inside the pipe which form a parabolic velocity profile, we will use the following code:

```
% Parameters
L = 1; % Length of the pipe (m)
R = 0.1; % Radius of the pipe (m)
Umax = 1; % Maximum velocity at the center of the pipe (m/s)
n_particles = 20; % Number of particles to animate
n_steps = 200; % Number of animation steps
dt = 0.05; % Time step for animation (s)
trail_length = 50; % Number of steps to show in the trail

% Generate initial positions of particles along the pipe radius
y = linspace(-R, R, n_particles);
% Compute velocity profile (parabolic profile for laminar flow)
velocity_profile = Umax * (1 - (y / R).^2);
% Initialize particle positions along the pipe length
x = zeros(n_particles, 1);
% Initialize trail history
x_trail = nan(n_particles, trail_length);
y_trail = repmat(y', 1, trail_length);

% Animation setup
figure;
hold on;
pipe = plot([-0.1 L+0.1], [R R], 'k', 'LineWidth', 2);
plot([-0.1 L+0.1], [-R -R], 'k', 'LineWidth', 2);
particles = scatter(x, y, 'filled');
trails = plot(x_trail', y_trail', 'b');
xlabel('Pipe Length (m)');
ylabel('Pipe Radius (m)');
xlim([-0.1, L+0.1]);ylim([-R-0.05, R+0.05]);
title('1D Laminar Flow in a Pipe');
% Animation loop
for step = 1:n_steps
        % Update particle positions
        x = x + velocity_profile' * dt;
        % Check for particles reaching the end of the pipe and stop the animation
        if any(x >= L)
                break;
        end
        % Update trail history
        x_trail = [x, x_trail(:, 1:end-1)];
        % Update plot
        set(particles, 'XData', x, 'YData', y);
        for ii = 1:n_particles
```

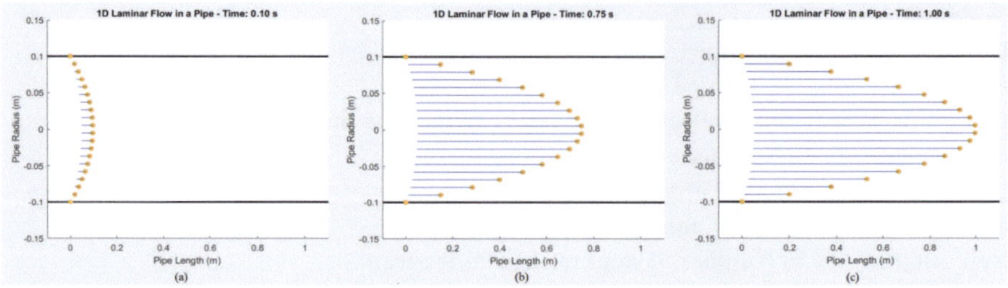


**Fig. 7.19** Initial (**a**); an intermediate (**b**); and the final (**c**) velocity profiles of a steady fully developed laminar flow for a viscous incompressible fluid inside a 1D pipe

```
            set(trails(ii), 'XData', x_trail(ii,:), 'YData', y_trail(ii,:));
        end
        % Update title with time
        title(['1D Laminar Flow in a Pipe - Time: ', num2str(step * dt, '%.2f'), ' s']);
        % Draw and pause for animation effect
        drawnow;
        pause(0.1);
end
hold off;
```

Snapshots of the parabolic velocity profile at the beginning, in the middle, and at the end of the particles' motion inside the pipe can be seen in Fig. 7.19.

(III) **Electrical Engineering**: We can use MATLAB animation to animate the behavior of electrical circuits, showing current and voltage variations, the distribution of electromagnetic fields around conductors and antennas, and the effects of signal processing techniques on waveforms and spectra.

As our first example, we will animate the light brightness in an RC circuit with a DC voltage source, as can be seen in part a of Fig. 7.20, where the capacitor is charged for half of the simulation time by closing the $S_1$ switch (part b of Fig. 7.20) and then discharged by opening the $S_1$ switch and closing the $S_2$ switch to turn on the green light. The light brightness will fade because of dropping voltage/current (parts c and d of Fig. 7.20).

The code below generates the animation:

```
% Parameters
R = 1e3; % Resistance in ohms
C = 500e-6; % Capacitance in farads
V0 = 5; % Initial voltage in volts
t_charge_end = 5; % End time for charging in seconds
t_discharge_end = 5; % End time for discharging in seconds
dt = 0.01; % Time step for simulation in seconds
% Time vectors for charging and discharging
t_charge = 0:dt:t_charge_end;
t_discharge = 0:dt:t_discharge_end;
% Calculate voltage across the capacitor during charging
charging_voltage = V0 * (1 - exp(-t_charge / (R * C)));
% Calculate voltage across the capacitor during discharging
discharging_voltage = V0 * exp(-t_discharge / (R * C));
% Combine charging and discharging voltages and time
combined_voltage = [charging_voltage, discharging_voltage];
```

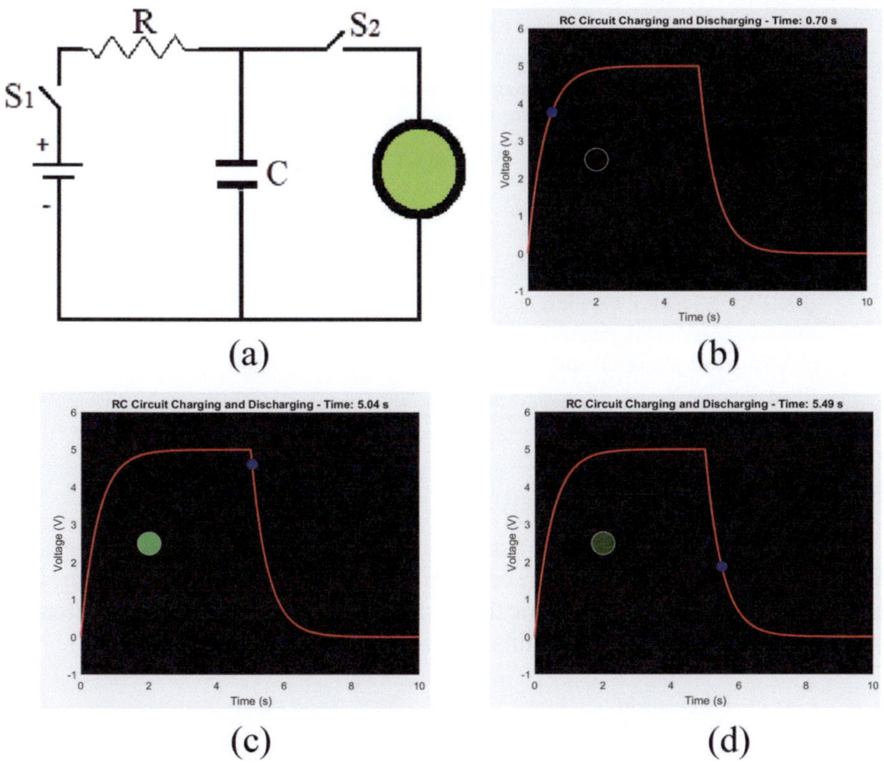


**Fig. 7.20** RC circuit with a light and a DC voltage source (**a**); capacitor being charged (**b**); capacitor being discharged, the light is bright in the beginning of voltage drop (**c**); capacitor being discharged, the light is bright in toward the end of voltage drop (d)

```
combined_time = [t_charge, t_charge_end + t_discharge];

% Animation setup
figure;
hold on;
h = plot(combined_time, combined_voltage, 'r', 'LineWidth', 2);
charge_point = plot(0, 0, 'bo', 'Markersize',10,'MarkerFaceColor', 'b');
xlabel('Time (s)');ylabel('Voltage (V)');
ylim([-1, V0+1]);
title('RC Circuit Charging and Discharging');
% Setup for the light representation
light_pos = [mean(combined_time)-3, V0 / 2]; % Position of the light in the plot
light = scatter(light_pos(1), light_pos(2), 500, 'filled', 'MarkerEdgeColor', 'w'); %
Initial light (off)
set(gca, 'Color', 'k'); % Set background color to black
% Animation loop
for ii = 1:5:length(combined_time)
        % Update the point on the voltage curve
        set(charge_point, 'XData', combined_time(ii), 'YData', combined_voltage(ii));
        % Update the light brightness based on discharging voltage only
        if combined_time(ii) > t_charge_end
                light_intensity = discharging_voltage(ii - length(t_charge)) / V0;
                set(light, 'CData', [0 1 0] * light_intensity); % Green color, scaled by
                intensity
        else
                set(light, 'CData', [0 0 0]); % Light off during charging
        end
        % Update title with time
        title(['RC Circuit Charging and Discharging - Time: ',
        num2str(combined_time(ii), '%.2f'), ' s']);
        % Draw and pause for animation effect
        drawnow;
        pause(0.03);
end
hold off;
```

For the second example, we use animation to visualize the variations in the electric field around a dipole as the distance between the conductors is increased. Figure 7.21 shows three snapshots of this electric field, where the separation distance is increasing from top to bottom. The code for this animation is provided below:

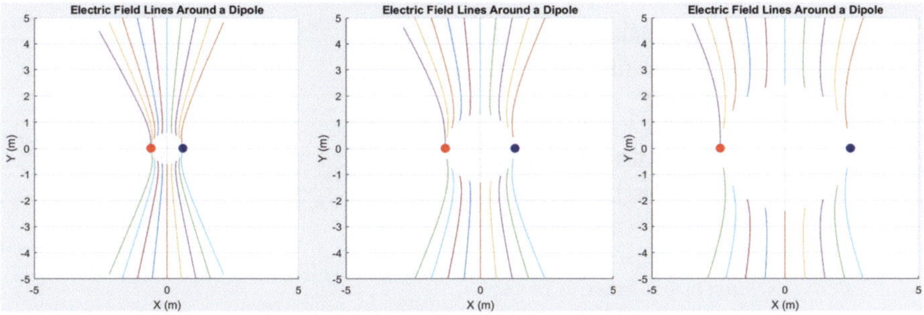


**Fig. 7.21** Electric field around a dipole with increasing distance between the conductors

```
% Parameters
q = 1; % Charge (Coulombs)
d_start = 1; % Initial separation distance (meters)
d_end = 5; % Final separation distance (meters)
d_steps = 100; % Number of steps in the animation
N = 20; % Number of field lines to plot
r_min = 0.1; % Minimum radius for the field lines (meters)
r_max = 5; % Maximum radius for the field lines (meters)
n_points = 100; % Number of points for each field line

% Create figure
figure;
hold on;
axis equal;
xlim([-r_max, r_max]);
ylim([-r_max, r_max]);
xlabel('X (m)');ylabel('Y (m)');
title('Electric Field Lines Around a Dipole');
grid on;
% Colors for the field lines
colors = lines(N);
% Generate angles for field lines
angles = linspace(0, 2*pi, N+1);
angles(end) = [];

% Animation loop
for step = 1:5:d_steps
        % Clear the plot
        clf;
        hold on;
        axis equal;
        xlim([-r_max, r_max]);
        ylim([-r_max, r_max]);
        xlabel('X (m)');
        ylabel('Y (m)');
        title('Electric Field Lines Around a Dipole');
        grid on;
        % Calculate current separation distance
        d = d_start + (d_end - d_start) * (step - 1) / (d_steps - 1);
        % Generate field lines
        for ii = 1:N
```

```
            % Starting points for the field lines
            theta = angles(ii);
            x_start = d/2 * cos(theta);
            y_start = d/2 * sin(theta);
            % Initialize field line coordinates
            x_line = zeros(1, n_points);
            y_line = zeros(1, n_points);
            x_line(1) = x_start;
            y_line(1) = y_start;
            % Generate field line points
            for jj = 2:n_points
                % Compute the electric field at the current point
                r1 = sqrt((x_line(jj-1) - d/2)^2 + y_line(jj-1)^2);
                r2 = sqrt((x_line(jj-1) + d/2)^2 + y_line(jj-1)^2);
                E1 = q / r1^2;
                E2 = q / r2^2;
                Ex = E1 * (x_line(jj-1) - d/2) / r1 + E2 * (x_line(jj-1) + d/2) / r2;
                Ey = E1 * y_line(jj-1) / r1 + E2 * y_line(jj-1) / r2;
                % Normalize the electric field vector
                E = sqrt(Ex^2 + Ey^2);
                Ex = Ex / E;
                Ey = Ey / E;
                % Compute the next point along the field line
                x_line(jj) = x_line(jj-1) + Ex * r_max / n_points;
                y_line(jj) = y_line(jj-1) + Ey * r_max / n_points;
            end
            % Plot the field line
            plot(x_line, y_line, 'Color', colors(ii,:));
        end
        % Draw the conductors (charges)
        plot(-d/2, 0, 'ro', 'MarkerFaceColor', 'r', 'MarkerSize', 8); % Positive charge
        plot(d/2, 0, 'bo', 'MarkerFaceColor', 'b', 'MarkerSize', 8); % Negative charge
        % Update the plot
        drawnow;
        pause(0.05);
end
hold off;
```

(IV) **Control Systems**: We can animate the response of control systems to various inputs or disturbances and visualize the effects of controller parameters on system performance.

For the first example, we will look at using a PID controller to control the angle of an inverted pendulum on a cart, where the goal is to push the system from an initial condition of $x(0) = 0, \theta(0) = \frac{5\pi}{6}$ and rest to an upright position for the pendulum ($\theta_{ss} = \pi$). We investigate the effect of the integral gain, whose job is to reduce the

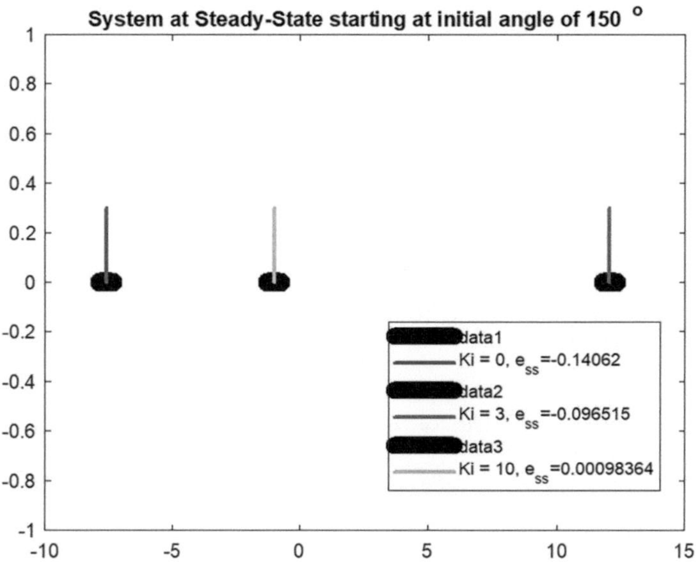


**Fig. 7.22** Effect of integral gain of a PID controller on steady-state error in an inverted pendulum on the cart

steady-state error, by plotting the system in the steady-state position overlaid by the steady-state angular error in degrees. As can be seen in Fig. 7.22, where three gains are used, a larger integral gain reduces this error. Note that caution must be taken to avoid instability of the system at large integral gains. The code for this simulation and plot is provided below:

```
% Parameters
m = 0.2; % mass of the pendulum (kg)
M = 0.5; % mass of the cart (kg)
L = 0.3; % length of the pendulum (m)
g = 9.81; % acceleration due to gravity (m/s^2)
d = 0.1; % damping coefficient (N/m/s)

% PID controller gains
Kp = 20; % proportional gain
Kd = 5; % derivative gain
Ki_values = [0, 3, 10]; % integral gains to be tested
```

```
% Simulation parameters
tspan = 0:0.01:10; % time vector
theta0 = 150 * pi / 180; % initial pendulum angle (radians)
x0 = [0; 0; theta0; 0]; % initial state [cart position; cart velocity; pendulum angle;
pendulum angular velocity]
```

```
% Plot setup
figure;
colors = lines(length(Ki_values));
for ii = 1:length(Ki_values)
    Ki = Ki_values(ii);
    % Define the PID controller
    controller = @(t, e, de, ie) Kp * e + Kd * de + Ki * ie;
    % Solve the system
    [t, y] = ode45(@(t, x) inverted_pendulum_dynamics(t, x, controller), tspan, x0);
    % Extract the cart position and pendulum angle at the last time step (steady-
    state)
    cart_position = y(end, 1);
    pendulum_angle = y(end, 3);
    % Plot the steady-state position of the cart and pendulum
    plot([-0.25 0.25] + cart_position, [0 0], 'k', 'LineWidth', 10); % Cart
    hold on;
    plot([cart_position, cart_position + L * sin(pendulum_angle)], ...
    [0, -L * cos(pendulum_angle)], 'Color', colors(ii, :), 'LineWidth', 2, ...
    'DisplayName', ['Ki = ' num2str(Ki) ', e_{ss}=' num2str((pi-
    pendulum_angle)*180/pi)]);
end
ylim([-1 1]);
legend('Location','best');
hold off;
title('System at Steady-State starting at initial angle of 150^{o}');

% Dynamics of the inverted pendulum on the cart system
function dxdt = inverted_pendulum_dynamics(~, x, controller)
    m = 0.2; % mass of the pendulum (kg)
    M = 0.5; % mass of the cart (kg)
    L = 0.3; % length of the pendulum (m)
    g = 9.81; % acceleration due to gravity (m/s^2)
    d = 0.1; % damping coefficient (N/m/s)
    % States
    pos = x(1);
    vel = x(2);
    theta = x(3);
    omega = x(4);
    % Control error and integrals
    persistent int_error;
    if isempty(int_error)
        int_error = 0;
    end
```

```
        % Compute control input
        error = pi-theta;
        derror = -omega;
        int_error = int_error + error * 0.01; % integrate error over time
        u = controller(0, error, derror, int_error);
        % Equations of motion
        S = sin(theta);
        C = cos(theta);
        D = m * L^2 * (M + m * (1 - C^2));
        dxdt = zeros(4, 1);
        dxdt(1) = vel;
        dxdt(2) = (1 / D) * (-m^2 * L^2 * g * C * S + m * L^2 * (m * L * omega^2 * S - d
        * vel)) + m * L^2 * (1 / D) * u;
        dxdt(3) = omega;
        dxdt(4) = (1 / D) * ((M + m) * m * g * L * S - m * L * C * (m * L * omega^2 * S -
        d * vel)) - m * L * C * (1 / D) * u;
end
```

```
% Parameters for the mass-spring-damper system
m = 1; % mass (kg)
k = 10; % spring constant (N/m)
c = 1; % damping coefficient (N·s/m)
% PID controller gains
Kp = 100; % proportional gain
Kd = 20; % derivative gain
Ki = 10; % integral gain

% Disturbance force parameters
disturbance_time = 4; % time when disturbance occurs (s)
disturbance_force = 8; % magnitude of the disturbance force (N)

% Simulation parameters
tspan = 0:0.01:10; % time vector
x0 = [0; 0]; % initial state [position; velocity]
% Animation setup
figure;
hold on;
xlim([-0.5 10]);ylim([-0.5 4]);
xlabel('Time (s)');ylabel('Position (m)');
title('Disturbance Rejection in a Controlled System');
% Define the PID controller
controller = @(t, e, de, ie) Kp * e + Kd * de + Ki * ie;
% Solve the system
[t, y] = ode45(@(t, x) mass_spring_damper_dynamics(t, x, controller, disturbance_time,
disturbance_force), tspan, x0);
% Extract position
position = y(:, 1);
```

For the second example, we will look at the effect of disturbance on a mass-spring-damper system that is controlled by a PID controller for a set-point position of $x = 3$ m. The disturbance is an external force with a magnitude of 8 Newtons that is applied at t = 4 s and lasts for 2 s. The position change of the mass is animated using the code below:

```
% Animation loop
for ii = 1:10:length(t)
        % Plot the mass
        plot(t(ii), position(ii), 'bo-', 'MarkerFaceColor', 'b');
        % Update title with time
        title(['Disturbance Rejection in a Controlled System - Time: ', num2str(t(ii),
        '%.2f'), ' s']);
        % Draw and pause for animation effect
        drawnow;
        pause(0.01);
end
hold off;

% Dynamics of the mass-spring-damper system with disturbance
function dxdt = mass_spring_damper_dynamics(t, x, controller, disturbance_time,
disturbance_force)
        m = 1; % mass (kg)
        k = 10; % spring constant (N/m)
        c = 1; % damping coefficient (N·s/m)
        % States
        pos = x(1);
        vel = x(2);
        % Control error and integrals
        persistent int_error;
        if isempty(int_error)
                int_error = 0;
        end
        % Desired position is 0 (setpoint)
        setpoint = 3;
        error = setpoint - pos;
        derror = -vel;
        int_error = int_error + error * 0.01; % integrate error over time
        u = controller(t, error, derror, int_error);
        % Disturbance force
        if (t >= disturbance_time)&(t<= disturbance_time+2)
                disturbance = disturbance_force;
        else
                disturbance = 0;
        end
        % Equations of motion
        dxdt = zeros(2, 1);
        dxdt(1) = vel;
        dxdt(2) = (u - k * pos - c * vel + disturbance) / m;
end
```

As can be seen in Fig. 7.23, position of the mass goes through a sudden change for the duration of the disturbance and then goes back to the set-point after the disturbance is removed. This change is not significant, which shows a good disturbance rejection for the PID controller.

```
load exampleMaps
map = occupancyMap(simpleMap,10);
ss = stateSpaceSE2;
sv = validatorOccupancyMap(ss);
sv.Map = map;
sv.ValidationDistance = 0.01;
ss.StateBounds = [map.XWorldLimits; map.YWorldLimits; [-pi pi]];
planner = plannerRRT(ss,sv,MaxConnectionDistance=0.3);
start = [0.5 0.5 0];
goal = [2.5 0.2 0];
[pthObj,solnInfo] = plan(planner,start,goal);

% Extract path
path = pthObj.States(:, 1:2);

% Robot parameters
robotRadius = 0.2;
sampleTime = 0.1;
tVec = 0:sampleTime:10;

% Differential-drive robot
robot = differentialDriveKinematics('TrackWidth', 0.4, 'VehicleInputs',
'VehicleSpeedHeadingRate');
% Initial conditions
initPose = [start(1:2), 0]; % [x, y, theta]

% Simulation setup
goalRadius = 0.1;
controlRate = robotics.Rate(1/sampleTime);

% Animation setup
figure;
hold on;
show(map);
plot3(path(:,1), path(:,2), zeros(size(path, 1), 1), 'r--', 'LineWidth', 2);
```

```
plot3(start(1), start(2), 0, 'go', 'MarkerFaceColor', 'g', 'MarkerSize', 10); % Start
position
plot3(goal(1), goal(2), 0, 'rx', 'MarkerFaceColor', 'r', 'MarkerSize', 10); % Goal
position
robotHandle = plot3(initPose(1), initPose(2), 0, 'bo', 'MarkerFaceColor', 'b');
% Path following controller
controller = controllerPurePursuit;
controller.Waypoints = path;
controller.DesiredLinearVelocity = 0.5;
controller.MaxAngularVelocity = 2;
controller.LookaheadDistance = 0.5;
% Simulation loop
while norm(initPose(1:2) - goal(1:2)) > goalRadius
    % Compute control inputs
    [v, omega] = controller(initPose);
    % Simulate the robot using the control inputs
    vel = derivative(robot, initPose, [v omega]);
    initPose = initPose + vel' * sampleTime;
    % Update robot position
    set(robotHandle, 'XData', initPose(1), 'YData', initPose(2), 'ZData', 0);
    % Pause for animation effect
    waitfor(controlRate);
    % Redraw the robot
    drawnow;
end
hold off;
```

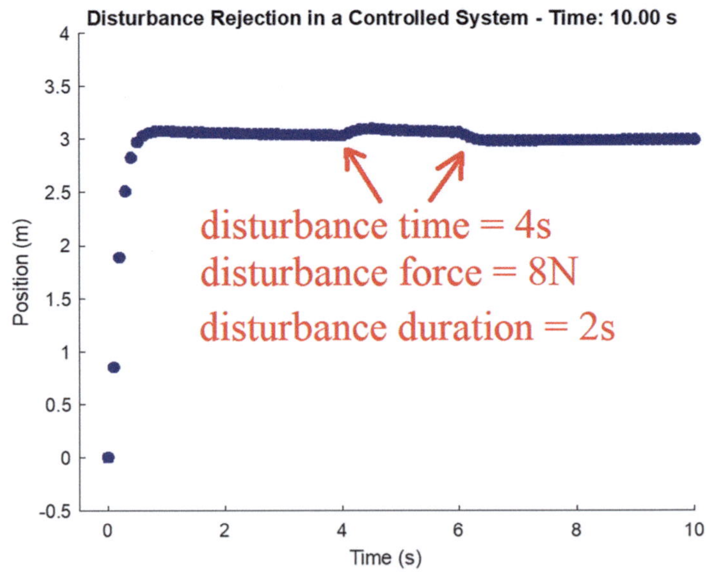


**Fig. 7.23** Effect of disturbance on a mass-spring-damper system that is controlled by a PID controller for a set-point position

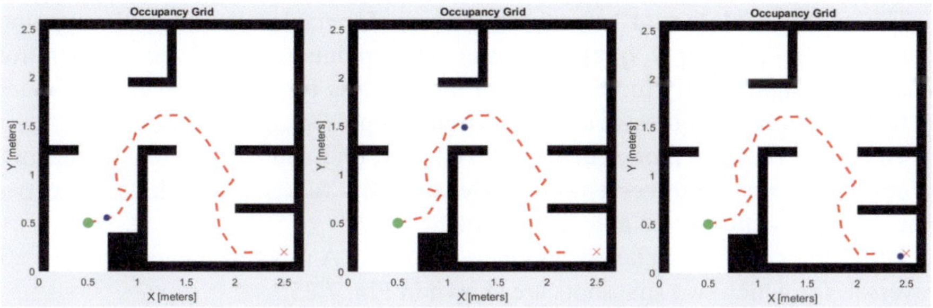


**Fig. 7.24** Path planning for a differential-drive robot using RRT planner in a binary occupancy grid map

(V) **Robotics**: We can animate the planned path of a robot in its environment, visualize sensor data and robot perception in real-time, and show the movement of kinematic chains and the effect of different configurations.

For the first example, we look at the path planning problem for a differential-drive robot going through obstacles that are defined by a binary occupancy grid map, as can be seen in Fig. 7.24. In the following code that animates the robot path traversal, MALTAB built-in functions for the RRT planner (e.g., **plannerRRT** and **plan**) as well as occupancy grid maps from the navigation toolbox (from the "exampleMaps" file) have been used. Also, built-in functions for the differential-drive robot kinematics model and control (e.g., **differentialDriveKinematics** and **controller**) have been utilized to make the code given below more compact:

Figure 7.24 shows three snapshots for the robot doing path traversal in the beginning (green circle is the start point), in the middle, and toward the end of the path (red cross is the goal point).

For our second example, we will look at animating a SCARA robot as it goes through predefined waypoints to write something in the workspace. Here, **.stl** files for SCARA parts are read into MATLAB using the utility function**stlread** written by MathWorks. These **.stl** files, the utility function **stlread**, and the **.mat** file for the waypoints can be downloaded from the link at the bottom of the page.[4]

[4] https://drive.google.com/file/d/1jvewQVdMdqDrMm_kmIUHktS4KMiUe59o/view?usp=sharing

The **.mat** file **"data.mat"** contains three sets of variables: the original waypoints represented by **[q_x, q_y, q_z]**, interpolated waypoints from trajectory generation which have significantly more points for smoothness of path traversal represented by **task_point**, and their associated time points represented by **time**.

The utility function **plot_scara.m** reads the **.stl** files and plots them with the joint values returned by the **inverse_kinematics.m** utility function, which itself has two other utility functions **rotz.m** and **trans.m**.

The entire code shown below animates the SCARA robot as it performs the path traversal, for which two snapshots are shown in Fig. 7.25.

```
% loading way points and time data
load('data.mat');

% animation loop
figure;
for ii = 1:length(time)
        set(gca,'XLim',[-0.2,0.2],'YLim',[-0.2,0.2],'Ydir','reverse','Xdir','reverse')
        cla;
        hold on
        plot3(q_x(:,1),q_y(:,1),q_z(:,1),'k');
        plot3(task_points(:,1),task_points(:,2),task_points(:,3),'rx');
        calcd_angs(ii,1:4) = inverse_kinematics([q_x(ii,1);q_y(ii,1);q_z(ii,1)],0);
        plot_scara(calcd_angs(ii,1),calcd_angs(ii,2),calcd_angs(ii,3),calcd_angs(ii,4));
        pause(0.01);
end

% inverse kinematics function
function [results] = inverse_kinematics(r, psi)
        a1 = 0.111;
        a2 = 0.109;
        d1 = 0.149;
        results = ["NaN"; "NaN"; "NaN"; "NaN"].';
        c2 = (r(1,1)^2 + r(2,1)^2 - a1^2 - a2^2)/(2*a1*a2);
```

```
        if(c2 <= 1 && r(3) <= 0.105)
                theta2 = atan2(sqrt(1-c2), c2);
                theta1 = atan2(r(2),r(1)) - atan2(a2*sin(theta2), a1 + a2*cos(theta2));
                theta3 = theta1 + theta2 - psi;
                d3 = 0.105 - r(3);
                results = [theta1; theta2; theta3; d3].';
        end
end
% function for reading stl files of SCARA and plotting the robot
function plot_scara(theta1, theta2, theta3, d3)
        a1 = 0.111;
        a2 = 0.109;
        d1 = 0.149;
        [Base, Base_f, n, c, stltitle] = stlread('YK220x_Base.stl');
        [Link1, Link1_f, n, c, stltitle] = stlread('YK220x_Link1.stl');
        [Link2, Link2_f, n, c, stltitle] = stlread('YK220x_Link2.stl');
        [Link3, Link3_f, n, c, stltitle] = stlread('YK220x_Link3.stl');
        view(220,50);
        Base_v = Base';
        Link1_v = trans(0,0,d1)*rotz(theta1)*[Link1,ones(length(Link1),1)]';
        Link2_v =
        trans(0,0,d1)*rotz(theta1)*trans(a1,0,0.022)*rotz(theta2)*[Link2,ones(length(Lin
        k2),1)]';
        Link3_v = trans(0,0,d1)*rotz(theta1)*trans(a1,0,0.022)*rotz(theta2)*trans(a2,0,-
        0.039-d3)*rotz(theta3)*[Link3,ones(length(Link3),1)]';
        patch('Faces',Base_f,'Vertices',Base_v','EdgeColor','k','FaceColor','r');
        patch('Faces',Link1_f,'Vertices',Link1_v(1:3,:)','EdgeColor','k','FaceColor','y');
        patch('Faces',Link2_f,'Vertices',Link2_v(1:3,:)','EdgeColor','k','FaceColor','b');
        patch('Faces',Link3_f,'Vertices',Link3_v(1:3,:)','EdgeColor','k','FaceColor','g');
end

% utility functions for the plot_scara function
function [translation] = trans(dx, dy, dz)
        translation = [1 0 0 dx; ...
        0 1 0 dy; ...
        0 0 1 dz; ...
        0 0 0 1 ];
end
function [rotation] = rotz(q)
        rotation = [cos(q) -sin(q) 0 0; ...
        sin(q) cos(q) 0 0; ...
        0 0 1 0; ...
        0 0 0 1];
end
```

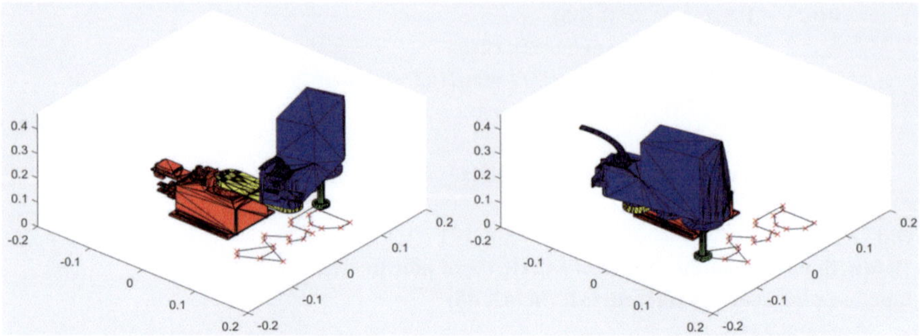

Fig. 7.25 Animating a SCARA robot writing a text in the workspace using mesh files

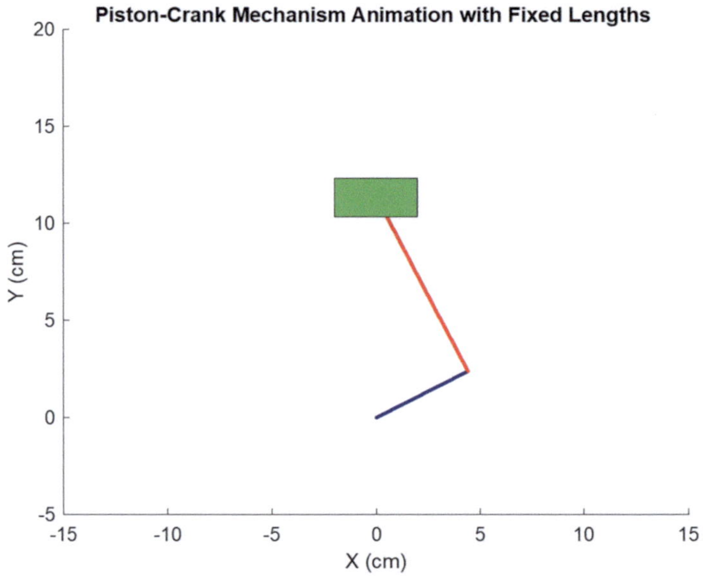


Fig. 7.26 One frame from a slider-crank mechanism animation

## Practice Problems

(1) Write a script that animates a slider-crank mechanism with a crankshaft length of 10 cm and a connecting rod of length 5 cm for one full rotation of the crankshaft. The piston moves up and down as shown in Fig. 7.26. Use 100 time points for the animation and utilize the command **drawnow** and a **for** loop.

(2) Write a script that animates deflection of a cantilever beam under a time-varying load applied at its free end as the load increases linearly from 0 to a maximum value over time. Deflection of a cantilevered beam with a load **F** at the end is given by Eqn. 7.11:

$$y(x,t) = \frac{F(t)x^2}{6EI}(3L - x) \tag{7.11}$$

In this equation, **E** is the young modulus, **I** is the section moment of area, and **L** is the beam length.

Apply the following numerical values:

```
L = 10;            % Length of the beam (in meters)
E = 210e9;         % Young's modulus (in Pascals, for steel)
I = 8.33e-6;       % Moment of inertia (in m^4)
F_max = -1000;     % Maximum load (in Newtons)
n_frames = 100;    % Number of frames in the animation
```

(3) Write a script to animate the heat transfer by conduction through a one-dimensional rod of length **L** with an initial uniform temperature. The ends of the rod are exposed to constant different temperatures. Over time, heat conduction causes the temperature distribution along the rod to change according to the update equation (Eqn. 7.12):

$$T_i^{n+1} = T_i^n + \frac{\alpha \Delta t}{(\Delta x)^2}(T_{i+1}^n - 2T_i^n + T_{i-1}^n) \tag{7.12}$$

In this equation, $\mathbf{T_i^n}$ is the temperature of node**i** at iteration/time **n**, $\mathbf{\Delta t}$ is the time step, $\mathbf{\Delta x}$ is the distance between the nodes, and $\alpha$ is thermal diffusivity.

Use the numerical values for the parameters given below:

```
L = 0.5;            % Length of the rod (meters)
nx = 50;            % Number of spatial points
alpha = 1e-4;       % Thermal diffusivity (m^2/s)
dx = L / (nx - 1);  % Spatial step size
dt = 0.1;           % Time step size (seconds)
nt = 1000;          % Number of time steps
T_initial = 100;    % Initial temperature of the rod (K)
T_left = 400;       % Temperature at the left end (K)
T_right = 100;      % Temperature at the right end (K)
```

(4) Write a script that animates the charging and discharging behavior of a capacitor in an RC (resistor–capacitor) circuit. The circuit consists of a resistor **R** and a capacitor **C** connected in series with a voltage source. You want to animate the voltage across the capacitor over time during the charging and discharging phases.

Equations of capacitor voltage during charging and discharging are given by Eqn. 7.13:

$$charging: V_c(t) = V_0(1 - e^{-\frac{t}{T}}),\ discharging: V_c(t) = V_0 e^{-t/T} \tag{7.13}$$

In this equation, **T = RC** is the time constant of the system.

Use the following numerical values:

```
R = 1000;           % Resistance in ohms
C = 100e-6;          % Capacitance in farads
V0 = 5;             % Initial voltage (in volts)
tau = R * C;         % Time constant
t_max = 5 * tau;      % Maximum time for the simulation
dt = 0.01;           % Time step size (in seconds)
n_frames = round(t_max / dt);  % Number of frames
```

(5) Write a script that analyzes the vibrational modes of a circular membrane, such as a drumhead. The membrane is fixed at its boundary and can vibrate in different modes depending on its initial conditions. You want to animate the time evolution of the membrane's vibration pattern when it is excited in one of its fundamental modes. The membrane's displacement can be described by the Bessel function of the first kind, which is typical for circular membranes. The fundamental mode of vibration can be represented by the $J_0$ Bessel function, which shows radial symmetry. The animation will use the **surf** command to show the time evolution of the membrane's displacement.

For a circular membrane of radius **R** and fixed at the boundaries, the membrane's vertical displacement at radial distance **r** and for mode shape **n** is given by Eqn. 7.14:

$$z(r, \theta, t) = J_0(\frac{n\pi r}{R})\cos(\omega t) \tag{7.14}$$

In MATLAB, you need to use the function **besselj(m, n)**.

Use the following numerical values:

```
L = 0.5;              % Length of the rod (meters)
nx = 50;              % Number of spatial points
alpha = 1e-4;            % Thermal diffusivity (m^2/s)
dx = L / (nx - 1);        % Spatial step size
dt = 0.1;             % Time step size (seconds)
nt = 1000;             % Number of time steps
T_initial = 100;          % Initial temperature of the rod (K)
T_left = 400;           % Temperature at the left end (K)
T_right = 100;           % Temperature at the right end (K)
```

Use a black color for the fixed boundary to distinguish that from the rest of the membrane as shown in Fig. 7.27.

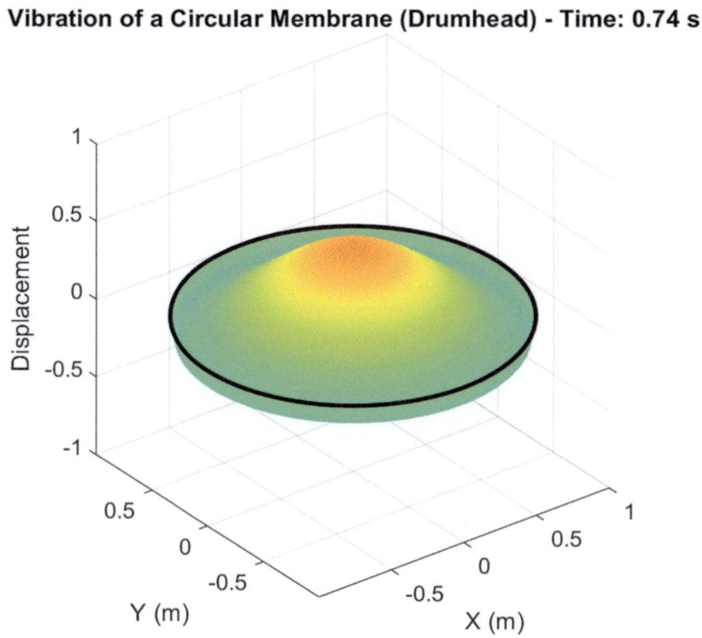


**Fig. 7.27** One frame from animation of a circular membrane's vibration

# Chapter 8
# Simulink for Engineers

## 8.1 Introduction to Simulink for System Simulation

### *8.1.1 What is Simulink?*

Simulink is a graphical programming environment for modeling, simulating, and analyzing dynamic systems. Developed by MathWorks, it allows users to create models using block diagrams, where each block represents a specific function, process, or system component. This block-oriented approach makes it highly intuitive, as users can drag, drop, and connect blocks in a visual workspace to build complex systems without needing to write extensive code.

Simulink supports both continuous- and discrete-time simulations, making it versatile for a wide range of applications. Furthermore, Simulink is integrated with MATLAB, allowing users to write scripts, manipulate data, and execute models seamlessly between both environments.

To open Simulink, click on the Simulink button  inside the MATLAB "Home" tab. Figure 8.1 shows a Simulink file simulating a nonlinear pendulum with an external torque for which the governing differential equation is numerically solved over a duration of 10 s.

Simulink libraries can be opened and browsed by clicking on the "Library Browser" button as can be seen on the top left of this figure.

Simulation results can be observed by double-clicking on the "scope" block in the model (which stands for oscilloscope) or by logging the output (theta) signal and clicking on the "Data Inspector" button as will be explained in further detail later in this chapter.



M. Andalibi and S. A. Cherkauer, *Advanced MATLAB for Engineers*,
https://doi.org/10.1007/978-3-032-10677-3_8

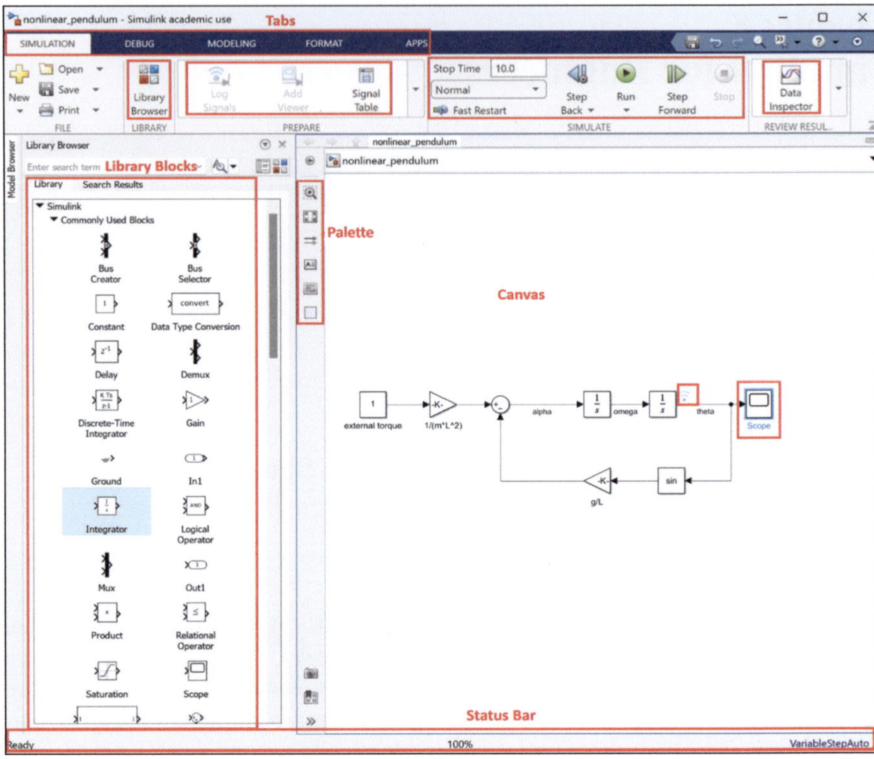


**Fig. 8.1** Components of a typical Simulink model

## *8.1.2 Applications of Simulink in Engineering and Industry*

Simulink is a powerful tool widely used across numerous engineering disciplines and industries for its ability to simulate and model complex systems. Some of the key applications include the following:

i. **Control Systems Design**: Simulink is widely used in control systems engineering for designing and testing controllers like PID (proportional-integral-derivative) controllers, state-space models, and other control architectures. Automotive companies, for example, use Simulink to model systems such as engine control, cruise control, and autonomous driving features before implementing them in real vehicles.

ii. **Signal Processing and Communications**: Simulink is often employed to model, simulate, and analyze signal processing algorithms. It is used in designing filters, modulators, and signal encoding/decoding systems. In the telecommunications industry, Simulink helps in simulating wireless communication systems, including aspects like modulation, channel modeling, and error correction.

iii. **Mechanical System Simulation**: Simulink can integrate with Simscape, a library used for modeling physical systems. Engineers use Simulink to model multi-domain systems like robotics, mechatronics, and electromechanical systems, where the mechanical, electrical, and control aspects of the design can be simulated together.
iv. **Electrical and Power Systems**: Simulink is also prevalent in electrical engineering for simulating circuits, power electronics, and electrical distribution networks. For instance, companies in the renewable energy sector use Simulink to model and optimize power grids and inverters for solar and wind energy systems.
v. **Aerospace and Defense**: In aerospace, Simulink helps engineers design and simulate control systems for aircraft, drones, and spacecraft. It is used to model flight dynamics, sensor integration, autopilot systems, and fault-tolerant control systems, ensuring that designs meet strict safety standards.
vi. **Automotive Industry**: Simulink is crucial for automotive engineers to design, simulate, and validate systems like ABS (anti-lock braking system), electronic stability control, and Advanced Driver Assistance Systems (ADAS). It is often paired with hardware-in-the-loop (HIL) simulations to test control algorithms on physical hardware before implementation in vehicles.

### *8.1.3 Difference Between Simulink and MATLAB*

Simulink and MATLAB are complementary tools, both developed by MathWorks, but they serve different purposes and have distinct features:

i. **Nature of the Environment**: Simulink is a graphical environment for system-level modeling, where users interact with blocks to create models and simulations. It is more focused on designing and analyzing dynamic systems through visual diagrams, which can represent processes like signal flow, system behavior, and physical phenomena. MATLAB is a textual programming language and environment focused on matrix-based computations, numerical analysis, algorithm development, and data visualization. MATLAB allows users to write scripts and functions to perform complex calculations, process data, and develop algorithms.
ii. **Modeling versus Programming**: Simulink provides a drag-and-drop, visual approach to modeling systems. Users can assemble systems with predefined blocks (e.g., integrators, gains, and summations) and simulate the behavior of dynamic systems over time. This is particularly useful for users who prefer a block diagram-based approach for system modeling. MATLAB involves coding to develop mathematical models and perform simulations. MATLAB excels at solving complex mathematical problems, processing large datasets, and performing symbolic computations through scripting.
iii. **Time-Based Simulations**: Simulink specializes in time-based simulations, allowing users to simulate systems over continuous or discrete time. This

makes it ideal for real-world dynamic systems where time plays a crucial role, such as control systems, electrical circuits, and mechanical systems. While MATLAB can perform simulations, it is not specifically designed for time-based or dynamic system simulation. MATLAB is more focused on static computations and numerical solutions.

iv. **Application Domains**: Simulink primarily used for dynamic system modeling, including applications in control systems, signal processing, communications, and embedded systems. It is often integrated with hardware and used for real-time system testing and verification. MATLAB is used in a broader range of applications, including data analysis, algorithm development, machine learning, optimization, and computational mathematics. MATLAB's flexibility makes it useful for researchers and engineers working in varied fields from finance to robotics.

## 8.2 Simulink Basics

When the Simulink button is pressed, the Simulink Start Page shown in Fig. 8.2 appears, and a new model can be created (a blank model, subsystem, library, fixed-step size simulation, project, etc.).

On this page, we can open our own template models, open templates from different toolboxes, such as Simscape, DSP System Toolbox, or Aerospace Blockset, and learn from related examples.

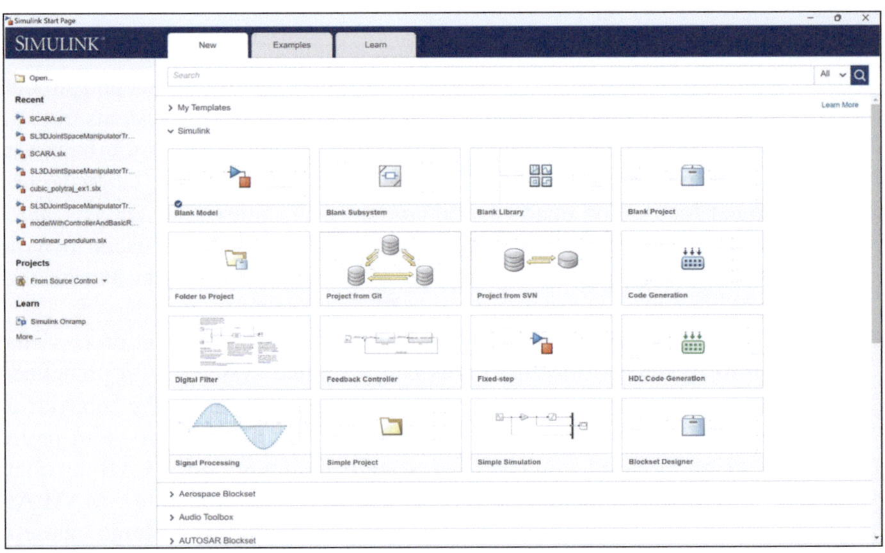


**Fig. 8.2** Simulink start page

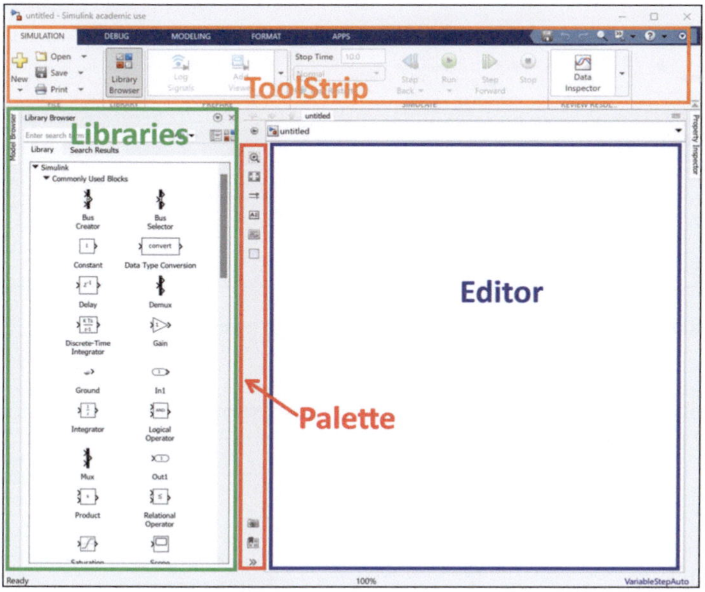


**Fig. 8.3** Components of the model editor window

To start, we create a "Blank Model" and then click on the "Library Browser" icon  to get access to Simulink libraries of blocks. The "Model Editor" window shown in Fig. 8.3 allows the user to access the Libraries, Palette, ToolStrip, and Editor.

Note that the first model you open in a MATLAB session takes longer to open than subsequent models. By default, MATLAB does not start Simulink until you open the first model to reduce MATLAB startup time and avoid unnecessary system memory use. To speed up opening the first model, you can configure MATLAB startup to also start Simulink.

The Library Browser has blocks from different toolboxes as mentioned earlier. For simple models, most of the required blocks can be found under the "Simulink" branch, specifically under "Commonly Used Blocks", "Dashboard", "Logic and Bit Operations", "Math Operations", "Matrix Operations", "Ports and Subsystems", "Sinks", "Sources", and "User-Defined Functions".

To demonstrate Simulink basics, let's create a simple series RC circuit with an AC voltage source from which the output voltage is acquired across the capacitor. The governing equation for this system is given in Eq. 8.1:

$$R\frac{dI}{dt} + \frac{I}{C} = \frac{dV_{in}}{dt}, V_{out} = \frac{\int Idt}{C} \tag{8.1}$$

When solving an Ordinary Differential Equation (ODE) in Simulink, we typically rearrange the governing ODE with the highest order derivative on the left side and all other terms on the right side. Then we try to use "Integrator" blocks found under

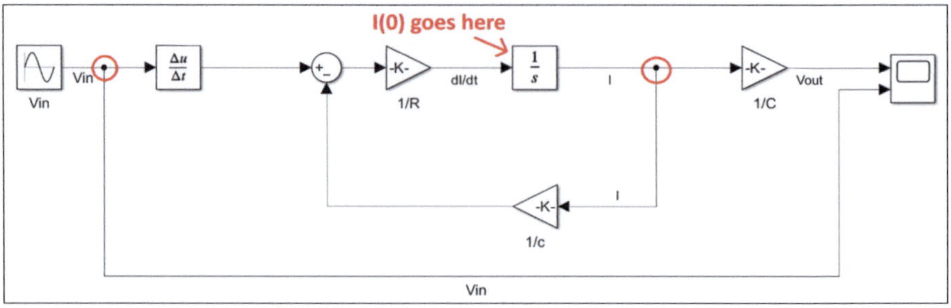


**Fig. 8.4** Model of a simple RC circuit with an AC voltage source

"Commonly Used Blocks" or "Continuous" and external inputs using blocks under "Sources" to form terms on the right side:

$$\frac{dI}{dt} = \frac{\left(\frac{dV_{in}}{dt} - \frac{I}{C}\right)}{R} \tag{8.2}$$

From Eq. 8.2, we can see that in order to form the first derivative of the current (**I**), we need to have the **I** term, which itself is obtained by integrating this derivative one time (forming a loop). We also need to multiply the current by a constant **−1/C** (which can be done in Simulink using a "Gain" block), take a time derivative of the AC input voltage function (a "Sine Wave" block under "Sources") using the "Derivative" block under "Continuous", add these two terms together using a "Sum" block, and finally multiply the right-hand side by a constant term **1/R** using a "Gain" block as shown in Fig. 8.4. To visualize the input and output voltages versus time, we can use a "Scope" block which is found under "Sinks".

Note that in order to add any of the blocks mentioned above to the blank model, we can either drag and drop that block into the model, right-click on it and use "Add Block to Model XX", or double-click anywhere in the blank model and search for the block's name (if we already know it) and Simulink will add it to the model once it is found.

Once the required blocks are added, they need to be connected/wired together. This can be done by holding the left mouse key and dragging it from either the output of one block to the input of the other block or vice versa. You can also hold down the Ctrl key and by selecting the related blocks, Simulink will connect them together.

In some places in the model diagram, we need to have wiring branches, marked in Fig. 8.4 by red circles. Creating a branch from a signal can be done by either wiring backward from a block to a branch (e.g., from the "**1/c** Gain block" to the **I** signal or from the "Scope" block on the right to the "Vin" block) or by holding down the Ctrl button and wiring forward from the branch to the destination block.

Most Simulink blocks have internal parameters that must be provided as **numerical values**. For instance, a gain block expects the value of the gain, the type of multiplication (element-wise product **K.*u**, left Matrix product **K*u**, or right Matrix product **u*K**), the type and limits of the block output, the "Integer Rounding Mode", etc. The numerical values must be either directly typed into the block window (Fig. 8.5a) or be provided from the MATLAB workspace, which is shared with Simulink. In the second case, only the name of the parameter in the MATLAB workspace will be entered in the block (Fig. 8.5b). If that parameter does not exist in the MATLAB workspace, the parameter color will be red inside the block. In the RC circuit example, we used the values of **C = 0.001** Farad and **R = 100** ohms.

When solving an Nth-order ODE that involves time (initial value problem or IVP), we need to provide **N** initial conditions, which will be entered inside the **N** "Integrator" blocks (or given from **N** external inputs to the integrator blocks as shown in Fig. 8.5c). In this example, we used a value of **I(0) = 0.1** amp. Integration operations can be reset if needed, which will be discussed later.

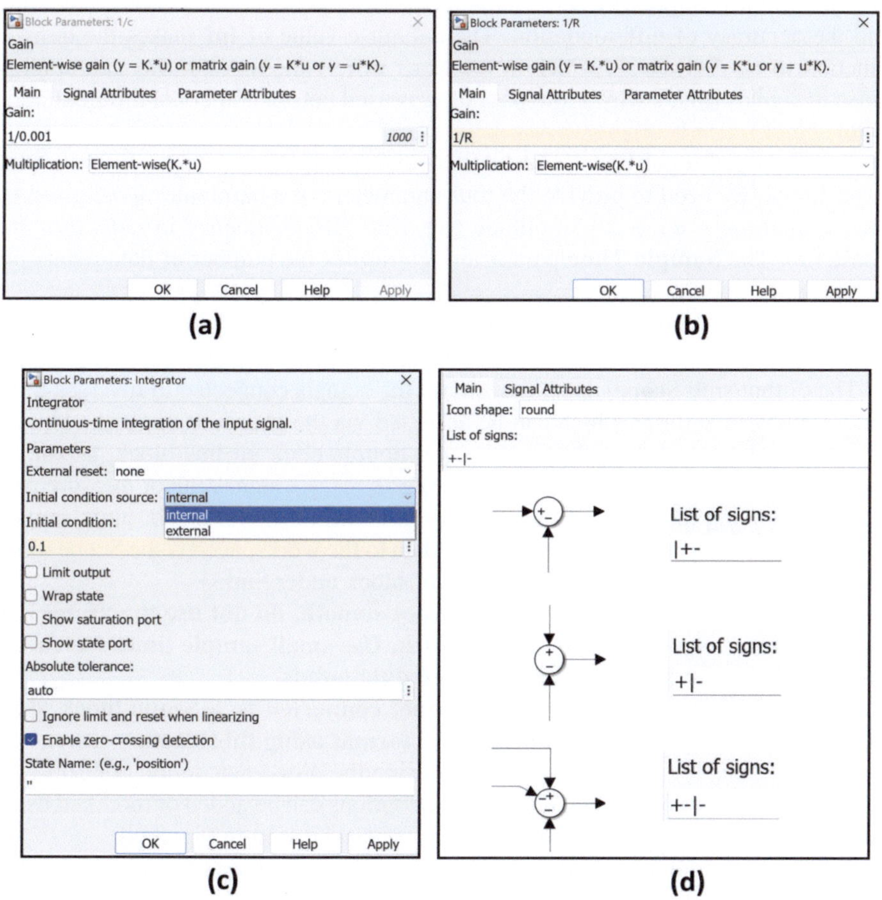


**Fig. 8.5** **a** Parameters of a Gain block (numerical value directly entered); **b** Parameters of a Gain block (numerical value provided in the workspace); **c** Parameters of an Integrator block; **d** Parameters and three configurations of a Sum block

A "Sum" block can add or subtract any number of signals together. The | character and the order of + and − signs determine the location of the input ports with respect to the output port. Figure 8.5d shows three different combinations of these signs.

Note that any block can be renamed by double-clicking on its name and providing the new name, and any signal/wire can be named by a similar action. Furthermore, once a signal is named, if you double-click anywhere else along the signal the name will appear at that location as well.

Another important consideration is that integration and differentiation in Simulink are both done numerically and not analytically. While the step size of Simulink determines the accuracy of numerical integration (setting the type and value for the step size will be discussed later), for the "Derivative" block, the transfer function is given by $G(s) = \frac{s}{cs+1}$, which is a high-pass filter (a derivative operation after a low-pass filter is applied) where the parameter **c** affects the bandwidth of the filter and the accuracy of differentiation. The default **c** value of **inf** makes the transfer function to be $G(s) = \frac{1}{s}$, which is the exact differentiation, but this can amplify noise in signals (that's why a low-pass filter is used before numerical differentiation in this block).

Each external input block will have its own parameters. In the case of a Sine Wave block, we need to provide the four parameters of a harmonic signal given by $x(t) = A\sin(\omega t + \varphi) + B$: Amplitude (**A**), Bias (**B**), Frequency in rad/s ($\omega$), and Phase ($\varphi$). The **Sample Time** parameter determines the number of data points per unit of time.

The total simulation time (**Stop Time**) is provided in the middle of ToolStrip. The simulation can be started using the "Run" button on the top .

The output/sink Scope block, that shows the signals connected to it versus time, has a variety of settings which can be accessed via the gear icon on the top left of its window (check Fig. 8.6a) when you double-click on this block, as can be seen in Fig. 8.6b. It also has tools for zooming , signal/cursor measurement , and running the Simulink. These settings control the axes labels, units, ranges, legends, grid, sample time, and logging of data to the workspace via the Scope block (it can also be done using a "To Workspace" block under Sinks).

**If the curves on the Scope block are not smooth, do not use the "Inherited Sample Time" or (−1) for the Scope block. Use small sample times instead to increase the number of generated/plotted data points.**

**Note that if there is more than 1 signal connected to a Scope block, data logging cannot be done using an "Array" format using this block.**

Color options for axes can be modified using the dropdown option next to the gear icon as can be seen in Fig. 8.6c. Subplot options can be added or modified using the dropdown option next to the gear icon as can be seen in Fig. 8.6d.

To zoom in and out of the Simulink model, you can use the mouse scroll button, or the "Zoom" tool and "Fit to View" tool on the Palette on the left (the Space key from the keyboard can also be used to perform the "Fit to View" operation).

To invert a block, we can use the keyboard shortcut of Ctrl + I. To rotate a block 90° clockwise each time, we can use the keyboard shortcut of Ctrl + R.

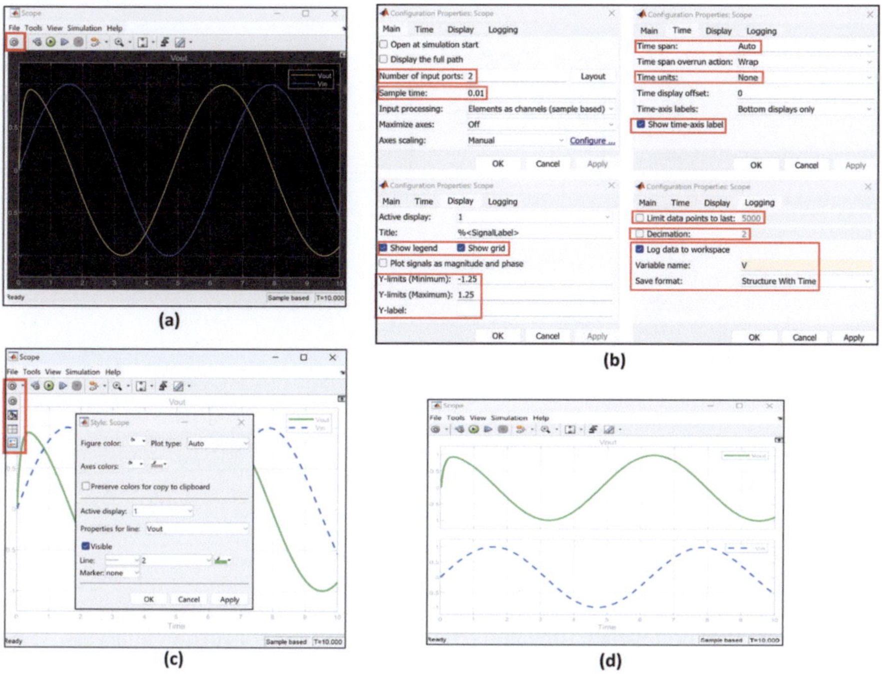


**Fig. 8.6** **a** Signal visualization using the Scope block and its settings; **b** Parameters of a Scope block; **c** Color options for the Scope block; **d** Subplot option for a Scope block

Blocks can be "Commented Out" and made ineffective (their color will turn into gray) like lines of code in a MATLAB script using the keyboard shortcut of Ctrl + Shift + X.

To replicate a block, other than copy/paste, we can hold down the Ctrl key and drag/drop that block.

Annotations , pictures , and area highlights ☐ tools from the Palette can be added to a model for further illustrations and explanations as can be seen in Fig. 8.7.

While a Simulink model can be self-sufficient in terms of the parameters, where all of its input/output signals are kept within the model, it can also obtain the parameters and inputs from the MATLAB workspace or external files, such as **.mat** files and Excel spreadsheets. Additionally, Simulink models can send the output signals to the workspace or the same types of external files. These IO blocks are found under "Commonly Used Blocks/Sinks" and "Commonly Used Blocks/Sources".

Running a Simulink model can also be done from the MATLAB command window or a script/function using the command **sim** instead of clicking on the model "Run" button as we will see next.

One major application of the topic mentioned above is parameter tuning, where the effects of certain parameters on the system performance are investigated using a script/function that runs the Simulink model in a loop, providing different sets of

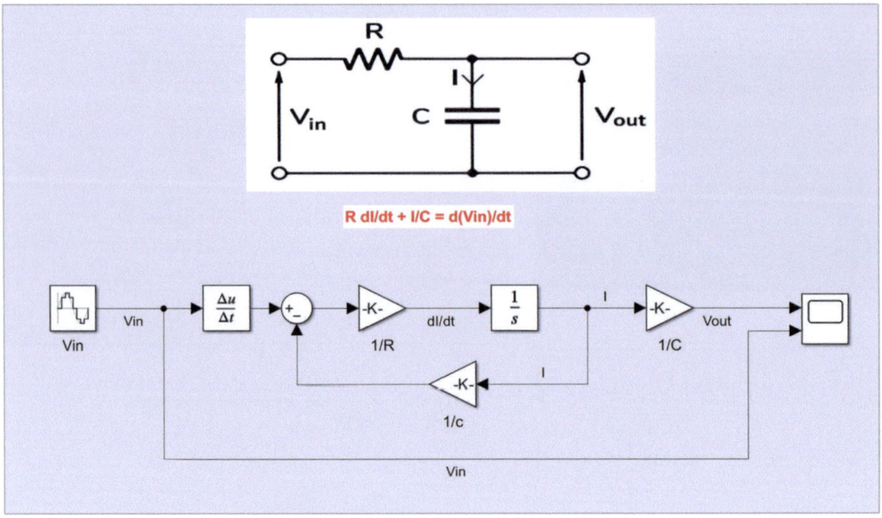


**Fig. 8.7** Adding pictures, annotations, and area highlights to a model

parameters to the model each time and receives its outputs for further analysis. Then, the designer can pick an optimal parameter value.

To demonstrate this topic, we will look at the effect of the damping coefficient of a linear viscous damper in a mass-spring-damper system on the mass position response (overdamped, critically damped, and underdamped responses).

In this system, we will have no external forces. The governing ODE is given by Eqn. 8.3:

$$m\frac{d^2x}{dt^2} + c\frac{dx}{dt} + Kx = 0 \tag{8.3}$$

and we will use the following numerical values for parameters and initial conditions:

$$m = 1kg,\, K = 9\frac{N}{m},\, x(0) = 2m,\, v(0) = 0\frac{m}{s}$$

We will pick the damping coefficient from the 10 equally spaced values in the range of [0 9] using a **for** loop in the MATLAB script shown below:

```
% parameters
m=1;
K=9;
C=0:1:9;

% initial conditions
x0=2;
v0=0;

% final time
tf=10;

% legends
lables=cell(1,length(C));

figure; hold on;

% main loop
for ii=1:length(C)
        c=C(ii);
        out=sim('msd_noforce.slx'); % running the Simulink model
        t=out.x.time;
        x=out.x.signals.values;
```

```
plot(t,x,'LineWidth',2);
lables{ii}=sprintf('damping coeff=%4.2f',c/(2*sqrt(K*m)));
end

xlabel('t(s)');ylabel('x(m)');
title('effect of damping coefficient on mass position response');
legend(lables,'Location','best');
```

Note that the command **sim('filename')**, which is used in the code above to run the Simulink model, can return an output, which is generally a structure with the fields shown below:

```
>> out = Simulink.SimulationOutput:

                    tout: [1007x1 double]

                       x: [1x1 struct]


    SimulationMetadata: [1x1 Simulink.SimulationMetadata]

            ErrorMessage: [0x0 char]


>> out.x


                    struct with fields:


                    time: [1001×1 double]

                    signals: [1×1 struct]

                    blockName: 'msd_noforce/To Workspace'


>> out.x.signals
```

```
                    struct with fields:


                    values: [1001×1 double]

                    dimensions: 1

                    label: 'x'
```

If there are several signals that are logged to the workspace, they will be found under the "Out" structure as separate fields. If you prefer to log these variables as separate variables on the workspace and not combine them under the name of one structure, click on the "Model Setting" icon  under the "Modeling" tab. Next, click on the "Data Import/Export" tab on the left, and then uncheck the box next to "Single simulation output".

The Simulink model used for this example, the resulting plot, and the settings for the "To Workspace" block are shown in parts (a), (b), and (c) of Fig. 8.8, respectively.

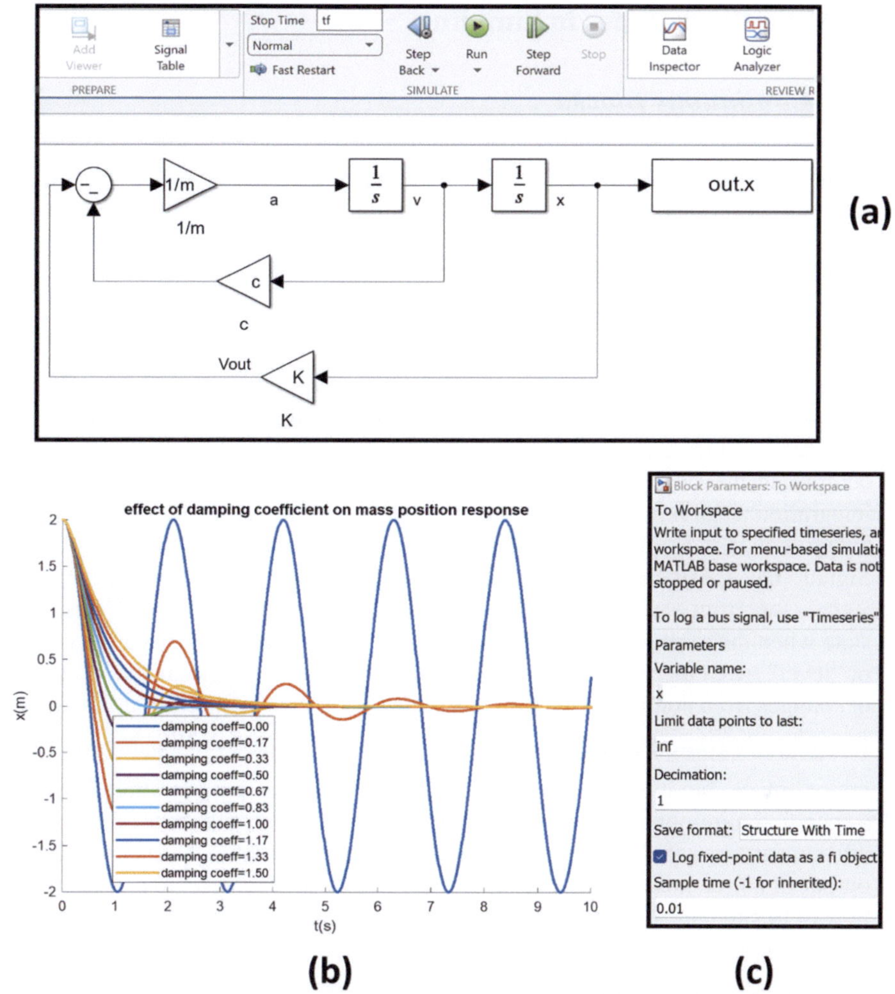


**Fig. 8.8** **a** Simulink model for the mass-spring-damper system with no external force; **b** Resulting plot of the script showing the effect of damping coefficient on the mass position response; **c** Settings for the "To Workspace" block

The values for all parameters and initial conditions are set in the script and sent to the MATLAB workspace, which will then be accessed by Simulink blocks.

Note that the value of **Decimation** that can be seen in many blocks determines the number of data points discarded between two consecutive samples, i.e., a decimation factor of **N** only records every **N**th data point.

## 8.3 Main Block Types in Simulink

### 8.3.1 Continuous Blocks

In Simulink, the "Continuous" library contains blocks used for simulating continuous-time dynamic systems. These blocks are essential for modeling systems whose behavior is described by continuous signals. Here are some of the most important blocks in the "Continuous" library, along with their applications:

- **Integrator Block** 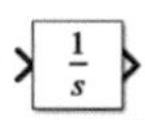 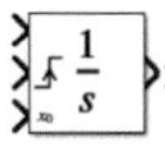: This block computes the integral of an input signal with respect to time. It is used to model systems where integration is required, such as in mechanical systems (e.g., calculating position from velocity), electrical circuits (e.g., calculating charge from current), or control systems (e.g., computing the state of a dynamic system). It takes the "Initial Condition" for integration as an input which can be typed directly into the block or be provided using an external signal. It also can reset the integration if the "Resetting Input Signal" has specific properties, such as "rising" (reset the state when the reset signal rises from a negative or zero value to a positive value), "falling" (reset the state when the reset signal falls from a positive value to a zero or negative value), or "level" (reset the state when the reset signal is nonzero at the current time step or changes from nonzero at the previous time step to zero at the current time step).
- **Integrator Limited Block** 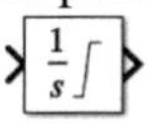: This block is similar to the standard "Integrator" block, but it includes limits on the output. The output is restricted within a specified minimum and maximum range. When the output reaches the limit, it will stay at that value until the conditions change. It is used in systems where the output of an integration process should be limited, for example, to prevent overflow or to simulate a system with physical limitations, such as a battery charge or velocity, that cannot exceed a certain threshold.
- **Integrator Second-Order Block** 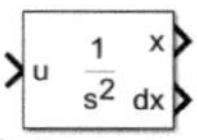: This block is used to model a second-order integration process, which integrates a signal twice. The block represents the dynamics of a system where two levels of integration are required, typically modeled by second-order differential equations. It is commonly used in systems where the output is the double integral of the input. This is especially relevant for mechanical and electrical systems that require modeling of acceleration, position, and velocity, or systems with a second-order response. It has two separate tabs for "**x**" and "**dx/dt**" to take their initial conditions and control them independently.
- **Derivative Block** 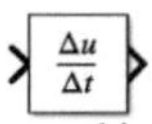: This block computes the derivative of the input signal with respect to time and is used to model systems where differentiation is required, such as in computing velocity from position or analyzing rate-of-change behavior in control systems. Note that differentiation is done numerically using a transfer function $\frac{s}{cs+1}$, which is a high-pass filter. The default value for the coefficient **c** is

infinity, which makes this filter act similar to an actual derivative block. Infinite **c** can amplify noise in its input signal, however, so smaller values of **c** are used for noisy signals to control the bandwidth of the filter and filter out the noise before taking the derivative.

- **Transfer Function Block** 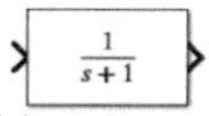

: This block represents a system's transfer function in the form **b(s)/a(s)**, where **a(s)** and **b(s)** are polynomial function of the Laplace variable **s**. It is used to represent linear time-invariant (LTI) systems, such as electrical circuits, mechanical systems, and control systems. It is commonly used to model system dynamics in the frequency domain.
- **State-Space Block** 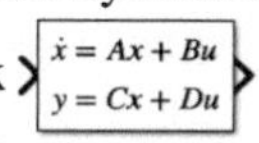

: This block describes a system in state-space form with matrices **A**, **B**, **C**, and **D** representing the system's dynamics, inputs, and outputs. It also takes the initial conditions for the system states. It is used to model systems that are defined by a set of first-order differential equations. It is common in control system analysis and design, especially when dealing with multi-input, multi-output (MIMO) systems.
- **PID Controller Block** 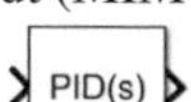

: This block implements a proportional-integral-derivative (PID) controller to regulate the behavior of a system. It is commonly used in control systems to maintain the desired output by adjusting the system's inputs. It is essential for feedback control loops in many applications like temperature control, motor speed regulation, and autopilot systems. Note that the actual transfer function of this block is proportional plus integral plus high-pass filter (instead of derivative) to avoid amplifying noise as explained in the "Derivative" block and sometimes it is called a PIDF (PID + Filter) block.
- **Zero-Order Hold Block** 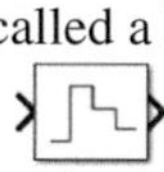: This block converts continuous signals to discrete signals by holding the last value of the input for one sample period and is used in control systems and digital signal processing where a discrete-time implementation is needed, such as in hardware-in-the-loop (HIL) simulation or digital controller systems.
- **First-Order Transfer Function Block** 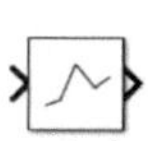: This block generates a continuous piecewise linear approximation of the input signal, and is used to model first-order systems, such as simple RC circuits, low-pass filters, or processes with a single time constant.
- **Transport Delay Block** : This block introduces a time delay to the input signal, meaning the output signal is shifted in time by a specified delay duration. The block can be used for modeling systems where delays exist due to physical characteristics such as signal propagation or processing time. It is common in systems where signal delays are present, such as communication systems, electrical circuits with transmission delays, or processes where a reaction time exists (e.g., delays in a feedback loop). It is also used to simulate the time required for

the signal to propagate through a system. It can also take a "Pade Approximation order" to linearize the nonlinear effect of the transport delay.

### 8.3.2 Discrete Blocks

The "Discrete" library in Simulink contains blocks that model discrete-time systems, where signals are sampled at specific intervals. These blocks are essential for simulating digital systems, sampled-data control systems, and signal processing applications. Below are the most important unique blocks in the "Discrete" library and their applications:

- **Discrete-Time Integrator Block** 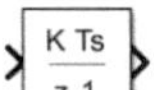

: This block computes the discrete-time integral of the input signal based on the sample time. The output accumulates the input values over discrete steps. It is used in digital control systems and signal processing for cumulative summation or integration, such as position tracking from velocity or implementing an integrator in a digital PID controller.
- **Discrete-Time Derivative Block** 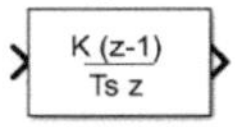

: This block computes the discrete-time derivative of the input signal using the difference between successive input values. It is used to approximate the derivative of a signal in discrete-time systems, such as calculating velocity from position or implementing a derivative action in a PID controller.
- **Discrete Transfer Function Block** 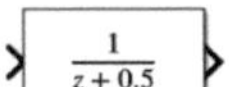

: This block represents a discrete-time system defined by its transfer function in terms of $z^{-1}$. It models linear discrete systems in the z-domain, such as digital filters, control systems, or signal processing algorithms.
- **Discrete State-Space Block** 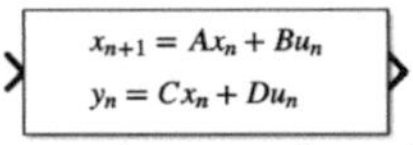

: This block represents a discrete-time system in state-space form using matrices **A**, **B**, **C**, and **D**. It models discrete-time state-space systems, commonly used in digital control and simulation of sampled-data systems.
- **Discrete FIR Filter Block** 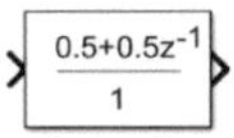

: This block implements a discrete Finite Impulse Response (FIR) filter based on specified coefficients. It is widely used in digital signal processing for tasks such as filtering, smoothing, or noise reduction.
- **Discrete Filter Block** 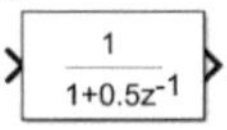

: This block implements a digital filter in terms of numerator and denominator coefficients in the $z^{-1}$ domain. It is used for signal processing, filtering out noise, or implementing controllers in digital systems.
- **Memory Block** 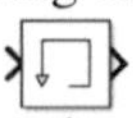: This block holds and outputs the value of the input signal from the previous time step and is used in feedback loops to create delay effects or to implement recursive algorithms. One application of this block is to find the

maximum/minimum of a signal over time in Simulink without recording the data (let's say the maximum height of a projectile during the flight). To achieve this task, we can simply pass the signal of interest (y value for the projectile) to a memory block and compare the output of it (previous y value) to the signal itself using a"Max" block 

. The output of the "Max" block will be the maximum of the y values over the entire simulation time.

- **Delay Block** 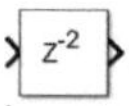

: This block introduces a specified number of discrete-time step delays to the input signal. You can configure the block to delay the input signal by any integer number of sample times. The block allows initialization of the delay with a user-defined initial value. The delay can be fixed or variable. It is used to model systems with inherent delays or memory effects, such as pipeline delays in signal processing, time-lagged feedback in control systems, or propagation delays in communication systems. It is often employed in digital filter design, such as implementing recursive filters or feedback loops.
- **Difference Block** 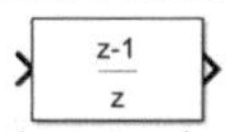

: This block computes the difference between the current input and the previous input signal. This block effectively implements a first-order difference equation, commonly used to approximate derivatives in discrete systems. It is used in signal processing to detect changes between successive samples, such as edge detection in images or transitions in control signals. It is also used in control systems to approximate the derivative of a signal, such as computing velocity from position data in sampled systems.

### *8.3.3 Source Blocks*

Source blocks are used to generate signals or provide input data to a model. These blocks are essential for driving simulations and allowing models to respond to various input conditions.

In this section, we will only look at the source blocks under the "Simulink" tab, but note that there are several other source blocks under other toolboxes, such as DSP systems, Computer Vision toolbox, and Audio toolbox, which can import data from pictures, videos, audio files, hardware interfaces, instruments, etc.

- **Constant Block** 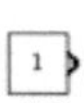

: This block produces a constant signal. It is used to apply a fixed value input, like a constant force or bias, throughout the simulation.
- **Signal Generator Block** 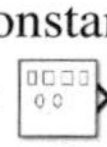: This block generates periodic waveforms such as sine, square, and sawtooth signals. It is useful for testing system responses to oscillatory or repetitive inputs, like testing filters, control systems, and signal processing algorithms. Settings for this block are shown in Fig. 8.9a.
- **Sine Wave Block** 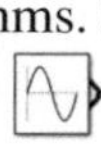: As discussed earlier, this block generates a sinusoidal waveform, and it is commonly used in control systems and signal processing to represent alternating signals or periodic disturbances.

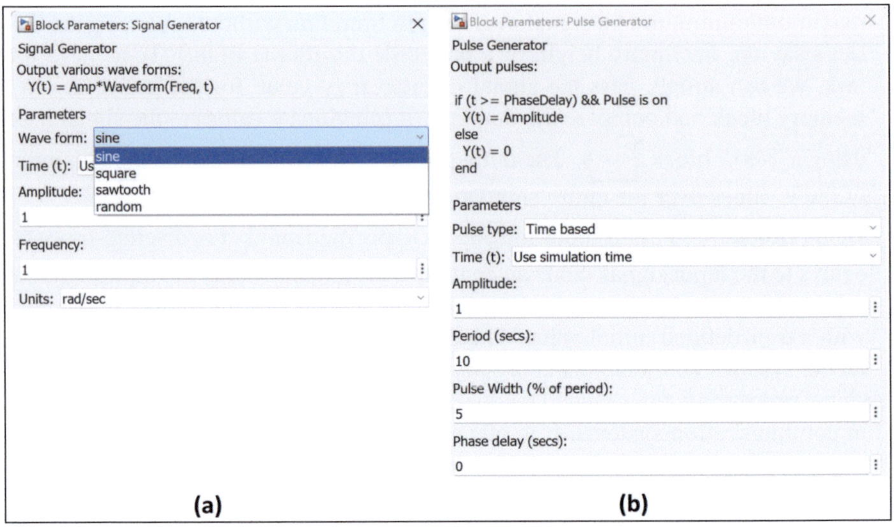


**Fig. 8.9** **a** Settings for the "Signal Generator" block; **b** Settings for the "Pulse Generator" block

- **Step Block**: This block produces a step signal that changes from an "Initial Value" to a "Final Value" at a specified time called the "Step Time". It is often used in control systems to test step responses and analyze transient behavior.
- **Pulse Generator Block**: This block generates square wave pulses with specified width, period, and amplitude. It is useful in digital systems, testing pulse responses (such as pulse width modulation or PWM control for the speed of DC motors), or for triggering events within a system. Settings for this block are shown in Fig. 8.9b.
- **Repeating Sequence Block**: This block generates a repeating sequence of defined "Output Values" that happen at defined "Time Values". It is used to apply recurring patterns of inputs, useful in power systems and signal processing.
- **Random Number Block**: This block produces random numbers from a Gaussian distribution with specific mean and variance values and can be used to simulate noise in systems or model random disturbances.
- **Uniform Random Number Block**: This block is similar to the Random Number block but allows specification of a uniform distribution with defined minimum and maximum values. It is suitable for scenarios requiring bounded random disturbances.
- **Ramp Block**: This block produces a linearly increasing or decreasing signal over time defined by a "Slope" and starting at an "Initial Output" at a "Start Time". It is commonly used in control systems to test responses to linear changes, like simulating a gradual increase in load or temperature.

- **Chirp Signal Block** : This block produces a frequency-swept sinusoidal signal, starting from a defined "Initial Frequency" at a time of **t = 0** s, and ending at the "Final Frequency" at a "Target Time". It is used to test system frequency response, and is particularly useful in analyzing resonant frequencies.
- **Clock Block** **and Digital Clock Block** : These blocks output the current simulation time (second one at a specific rate). They are useful for time-dependent calculations, logging, or triggering events based on simulation time.
- **From Workspace Block** : This block imports data from MATLAB's workspace into Simulink. It is useful for feeding real-world data, pre-computed results, or complex time-series data into the simulation.
- **From File Block** **and From Spreadsheet Block** : These blocks read input data from a **.mat** file and an Excel Spreadsheet, respectively. They allow integration of large datasets or historical data for simulations, ideal for validating models with real data.

### *8.3.4 Sink Blocks*

Sink blocks are used to capture, display, or log simulation outputs. These blocks help in visualizing, storing, or sending output data from the model for further analysis.

In this section, we will only look at the sink blocks under the "Simulink" tab, but note that there are several other sink blocks under other toolboxes, such as DSP systems, Computer Vision toolbox, and Audio toolbox, which can log/send data to pictures, videos, audio files, hardware interfaces, instruments, etc.

- **Scope Block** : This block displays signals in real time as the simulation runs, similar to an oscilloscope. It is useful for viewing the behavior of signals during simulation, such as visualizing waveforms, observing transient responses, and debugging. It was explained earlier in this chapter.
- **Floating Scope Block** : This block is similar to the Scope block, but it can display signals without being directly connected. It is useful in large models where signals need to be visualized at multiple points without additional block connections, making it easier to inspect signals anywhere in the model.
- **To Workspace Block** : This block sends simulation data to MATLAB's workspace. It is ideal for post-processing and further analysis in MATLAB, enabling you to plot, analyze, or manipulate data after the simulation is completed.
- **To File Block** : This block saves data to a file, usually in a format that can be loaded into MATLAB. It is useful for saving large datasets from a simulation to a file for future use, logging, or sharing with others.
- **Display Block** : This block shows the current value of a signal during the simulation in a numerical format. It is handy for monitoring specific signal values or state variables in real time, especially for debugging purposes.

- **XY Graph Block** 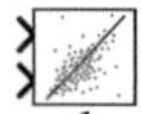: This block plots one signal against another in an X–Y plot. It is commonly used in control and robotics applications to visualize trajectories, phase planes, or other relationships between two variables.
- **Stop Simulation Block** : This block stops the simulation when the input signal to it is nonzero. It is used to terminate the simulation when a certain criterion is met, such as when a trajectory hits the ground.
- **Terminator Block** : This block terminates unconnected output ports. It is used to clean up model warnings or errors about unconnected ports, ensuring all model inputs and outputs are managed properly.

### 8.3.5 *Math, Matrix, Logic, and User-Defined Blocks*

In this section, we will introduce the most used math and logic blocks and avoid explaining all existing blocks in Simulink for these categories.

i. **Math Operations Blocks**

   - **Add/Subtract Blocks**  : As explained earlier, this block adds or subtracts input signals. It is commonly used in control systems and signal processing, such as calculating error signals (difference between setpoint and measurement).
   - **Gain Block** 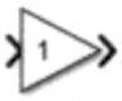: As explained earlier, this block multiplies an input signal by a constant value and is used for scaling signals, implementing proportional gains in control systems, or converting units.
   - **Product Block** : This block multiplies input signals. It is useful in physics-based models, where products of variables (e.g., force = mass * acceleration) are needed.
   - **Divide Block** : This block divides its top input signal by the bottom input signal. It is used for calculating ratios or normalizing values, such as the angle of a projectile, which is the **arctan()** function applied to the ratio of projectile velocity in the y-direction and its velocity in the x-direction.
   - **Math Function Block** : This block performs various mathematical functions (square, exp, log, power, mod, etc.). It is useful in signal processing, control systems, and engineering computations where trigonometric, exponential, or logarithmic functions are required.
   - **Sqrt Block** : This block computes the square root of a signal. It is often used in algorithms involving distance or magnitude calculations, such as calculating the Euclidean distance or norm of a vector like a projectile velocity magnitude.

- **Abs Block** 

: This block computes the absolute value of the input. It is used in algorithms where the magnitude of values is more relevant than their sign, like distance metrics.
- **Sign Block** 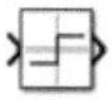: This block computes the sign of the input, i.e., the output is **+1** for positive inputs, **−1** for negative inputs, and **0** if the input is 0. It is used in algorithms where the sign of a signal is important, such as adaptive controls or machine learning.
- **Trigonometric Function Block** 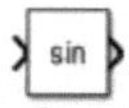

: This block performs trigonometric functions like **sin**, **cos**, **atan**, **atan2**, etc. It is essential for modeling waveforms, oscillations, and whenever trigonometric functions are needed, such as forward kinematics equations of mobile robots.

ii. **Matrix Operations Blocks**

- **Matrix Multiply Block** 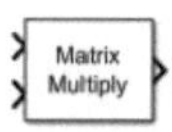

: This block multiplies two matrices. It is commonly used in state-space models, transformations, and kinematic calculations in robotics and mechanics.
- **Transpose Block** 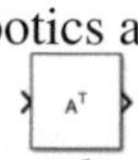: This block transposes a matrix and is used for operations in linear algebra, such as matrix manipulations in control and robotics.
- **Inverse Matrix Blocks** 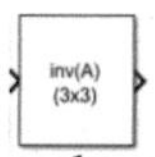

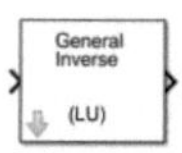

: The first block computes the inverse of a 3 × 3 matrix only and is found under the Aerospace Blockset Toolbox. The second block computes the inverse of an $n \times n$ matrix and is found under the DSP System Toolbox. They are critical in solving systems of linear equations, often used in control systems for calculating system gains.
- **Matrix Determinant Block** 

: This block computes the determinant of a 3 × 3 matrix only and is found under the Aerospace Blockset Toolbox. It is useful in systems that require linear algebra, such as determining the singularity configurations of a robotic arm. For a general $n \times n$ matrix, there is a pre-built block in Simulink, but it can also be done using user-defined functions as we will see later.
- **Matrix Norm Block** 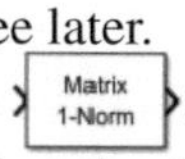

: This block only calculates the 1-Norm of a matrix and is found under the DSP System Toolbox. It can be used in applications, such as path planning, where this norm aka Manhattan norm is used. For other norms of a matrix, there is no pre-built block in Simulink, but they can be calculated using user-defined functions as we will see later.
- **Matrix Concatenate Block** 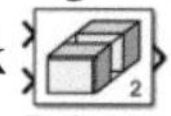

: This block concatenates input matrices along a specified dimension. It is used in data preparation and manipulation, particularly in models requiring stacking or appending matrices.

- **Dot Product and Cross-Product Blocks** 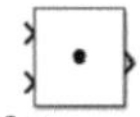 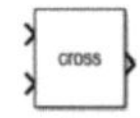

: These blocks compute the dot product and cross product of two vectors. They are often used in physics and engineering for projections, work calculations, torque calculations, and other vector-related operations.

iii. **Logic and Bit Operations Blocks**

- **Logical Operator Block** 

: This block performs logical operations (AND, OR, NOT, XOR). It is commonly used in control systems and decision-making processes, such as setting conditions for switches or triggering events.
- **Relational Operator Block** 

: This block compares two signals (e.g., < , > , = = ). It is useful in creating conditional branches, setting thresholds, and comparing values in control systems.
- **Compare To Constant Block** 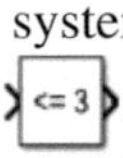

: This block compares an input to a constant value. It simplifies thresholding operations, especially in control systems or signal processing applications.
- **Bitwise Operator Block** 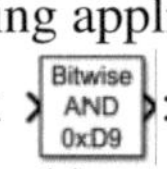

: This block performs bitwise operations (AND, OR, NOT, XOR) and is used in embedded systems, digital logic, and low-level data manipulation.
- **Hit Crossing Block** 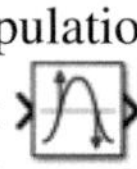: This block detects when a signal crosses a specified threshold and is useful for event-based simulations, triggering actions when certain conditions are met.
- **Detect Increase Decrease and Change Blocks** 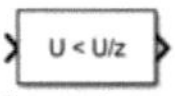

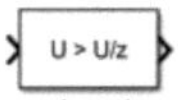

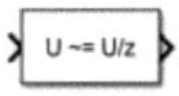

: These blocks detect when a signal value has increased, decreased, or changed from their previous values and are useful for event-based simulations, triggering actions when certain variations are met.

iv. **User-Defined Function Blocks**

- **MATLAB Function Block** 

: This block allows custom MATLAB code to be embedded directly in Simulink. It is useful for implementing complex algorithms, custom mathematical functions, and any logic that requires flexibility beyond standard Simulink blocks. As an example, the MATLAB function shown in Fig. 8.10 returns the inverse and determinant of a matrix with arbitrary dimensions.
- **Interpreted MATLAB Function Block** 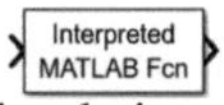

: This block executes MATLAB commands in real time during simulation. It is suitable for incorporating custom MATLAB functions that do not need to be compiled and is often used for quick prototyping.

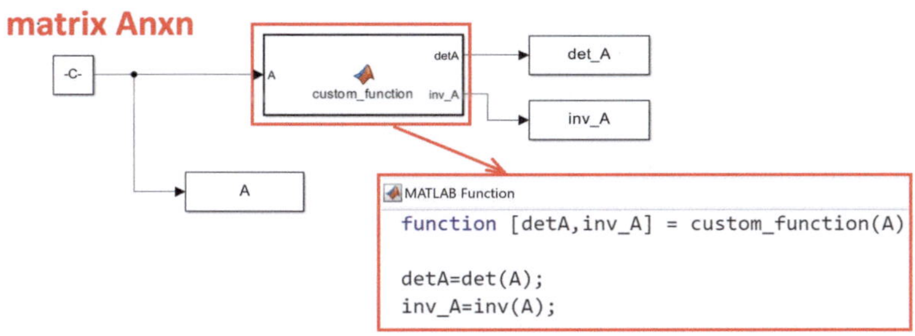


**Fig. 8.10** MATLAB function block that returns the inverse and determinant of a matrix with arbitrary dimensions

- **Simulink Function Block** 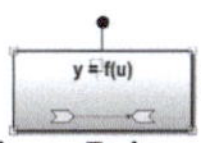

: This block allows you to define and encapsulate a reusable function. It is particularly useful for modeling modular systems, where functions need to be called and executed multiple times or from different parts of a model. They will be discussed further in Sect. 8.3.
- **S-Function Block** 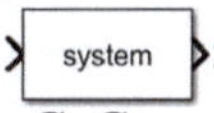

: This block allows for custom block creation using S-functions in C, C++, or MATLAB. It is ideal for highly customized operations, complex algorithms, and extending Simulink with new functionality. S-Functions are discussed in detail in Sect. 8.5.
- **MATLAB System Block** 

: This block integrates MATLAB System objects into Simulink. It is useful for object-oriented designs and for deploying MATLAB System objects in a Simulink model, often used in communication and DSP applications.
- **Initialize, Reset, and Terminate Function Blocks** 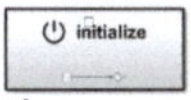

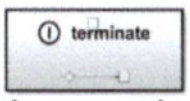

: These blocks provide custom initialization, reset, or termination actions in a model. They are used to set up, reset, or clean up resources or variables for simulations with complex starting or stopping behaviors.

## *8.3.6 Signal Routing Blocks*

- **Mux** : This block combines multiple signals into a single vector signal. It can merge scalar or vector signals.
- **Bus Creator** : This block combines multiple signals into a bus signal. It is used in grouping related signals into a single bus for organized signal routing. In a Bus, signals have dedicated names; their orders can be modified; and new signals can easily be added or removed. To create a bus from selected signals,

simply highlight the signals, select the three dots that appear at the top of these signals, and select the "Create Bus" option .

Since Mux and Bus Creator blocks have similar definitions, we discuss their different applications below:

**Use Mux When:**

1. You need to combine signals of the same type into a vector for processing. Note that you can combine signals with different dimensions or sample times together with a Mux block, but it is not preferred.
2. The combined signals are to be used by blocks like: Matrix Gain, Selector, or Sum.
3. You prefer lightweight and simple grouping for numeric computations.

**Use Bus Creator When:**

1. You want to make your model more **modular and readable**, especially for large systems.
2. You plan to use **Bus Selector** to extract specific signals later.
3. You are designing a model that requires structured, named data (e.g., for code generation or organizational clarity).

**Virtual Bus Versus Non-virtual Bus:**

A virtual bus, the output of a "Bus Creator" block, is essentially a logical grouping of signals that exists only for visual organization and ease of modeling within Simulink. A non-virtual bus represents a real data structure (treated as a single entity during simulation and in generated code), making it more rigid and structured compared to a virtual bus. It is useful in embedded systems. In a virtual bus, data can have different dimensions or sample times, but they cannot in a non-virtual bus. A virtual bus does not require additional memory or computational resources, but a non-virtual bus requires memory allocation to hold the structured data. To convert a virtual bus to a non-virtual bus, you need a "Signal Conversion" block that is shown on the next page.

- **Demux** : This block breaks a vector signal into individual scalar signals. It is used in separating combined signals for individual processing. Note that in complex models, accessing a specific constituent signal can be cumbersome, since signals are stored and accessed based on the orders they are added in Mux and Demux blocks. For example, to access a specific signal out of the Demux block, you need to trace it back to when the signals were put together in the Mux block and find their orders.
- **Bus Selector** : This block extracts specific signals from a bus via their names, which is useful in complex systems. It is used in accessing specific components of a grouped signal. To add any constituent signal to the output, highlight its name and use the option "Add to Output" .

- **Selector** 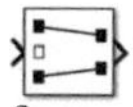: This block allows the selection of specific elements, rows, or columns from a vector or matrix signal. It is used in extracting specific data elements, such as a specific sensor reading from a sensor array.
- **Switch** 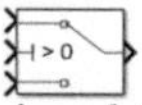: This block passes one of two inputs to the output based on a control signal (logical or conditional). For instance, for the block shown here, if the input to the middle port is bigger than zero, the top signal passes through; otherwise, the bottom signal passes through. This condition can be altered by changing the comparison operator or the threshold for comparison. It can be used in implementing decision-making logic (i.e., **if/elseif/else** or **switch** commands), such as choosing between manual and automatic control modes.
- **Manual Switch** 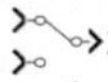: This block allows manual switching between two input signals during simulation. This is particularly useful for testing and debugging models or comparing different system configurations interactively.
- **Multiport Switch**: This block passes one of several inputs to the output based on an index or selector input and is used when implementing more complex decision-making systems.
- **Merge** 

: This block combines multiple input signals into a single output signal, one at a time. It is used in control systems to merge signals from different subsystems operating under different conditions. The initial value of the output signal is specified by the "Initial output" parameter.
- **From** 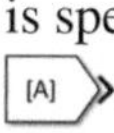

: This block receives a signal from a corresponding "Goto" block.
- **Goto** 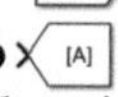

: This block sends a signal to a "From" block. "From" and "Goto" blocks together simplify wiring in large models by allowing signal routing without direct lines.
- **Signal Conversion** 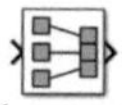: This block converts signal types or dimensions without changing its values (e.g., scalar to vector or fixed point to floating point). It converts a virtual bus to a non-virtual bus and vice versa. It ensures compatibility between blocks requiring different signal formats.
- **Signal Specification** 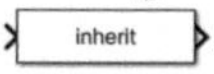

: This block specifies properties of a signal, such as dimensions, data types, or sample time, and is used in enforcing or defining signal constraints.
- **Rate Transition** 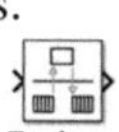: This block handles transitions between different sample rates in a model. It is used in multi-rate systems to ensure proper signal transfer between subsystems with different sample times. This block has an input window for an "Initial Condition". This parameter applies only to slow-to-fast transitions. It specifies the initial output of the "Rate Transition" block at the beginning of a transition, when there is no output signal due to the absence of input signal coming from the slow block connected to the input of the "Rate Transition" block.

### 8.3.7 Ports/Subsystem Blocks

- **Subsystem** : This block encapsulates a group of Simulink blocks into a single reusable block. It improves model organization, readability, and modularity by grouping related components into one block (e.g., grouping control logic into a single subsystem for clarity). You can add as many "In" and "Out" ports as needed to a subsystem and even convert it to a "Mask", which will be discussed in Sect. 8.8.7.
- **"In" port** 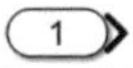: This block represents an input port for a subsystem or a top-level model and is used to pass external signals into a subsystem or the top-level model. For example, feeding sensor data into a control system model.
- **"Out" port** 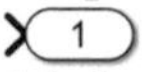: This block represents an output port for a subsystem or a top-level model. It sends signals out of a subsystem or top-level model to other parts of the model (e.g., outputting computed control signals to an actuator).
- **Enabled Subsystem** 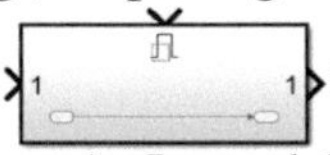: This block executes only when an enable signal is active (nonzero). It models conditional execution systems, such as enabling a controller when a machine is operational or a signal crosses a threshold.
- **Enable Port** : This block enables or disables the execution of a subsystem based on an "enable" signal. It is used to model conditional execution, such as activating a controller when a system is operational or ready.
- **Triggered Subsystem** 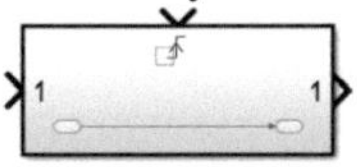: This block executes when a trigger signal changes, typically on a rising edge, falling edge, or both. It models event-driven systems, such as generating a pulse when a specific condition is met or responding to an external event.
- **Trigger Port** : This block executes a subsystem based on an event-driven trigger (e.g., rising edge, falling edge, or both). It is useful in event-driven systems, like initiating a sequence when a signal crosses a specific threshold.
- **Function-Call Port** 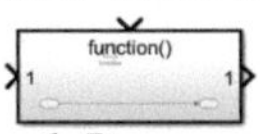

: This block allows a subsystem to be executed by a function-call signal. It supports modular design in systems where subsystems are triggered programmatically, such as tasks triggered by a scheduler.
- **If Action Subsystem** 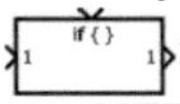: This block executes conditionally based on the output from an "If Block". 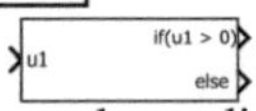

: This block models conditional logic, such as executing different actions depending on whether a signal is positive, zero, or negative.
- **Switch Case Action Subsystem** 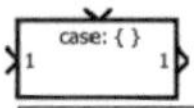

: This block executes based on the output of a "Switch Case Block" 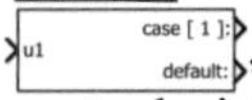

, which evaluates an input signal against predefined cases. It implements decision-making logic where multiple distinct actions depend on the value of an input (e.g., switching between operational modes).

- **Action Port** 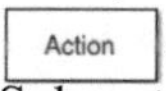

: This block connects an "If Action Subsystem" or "Switch Case Action Subsystem" to decision-making logic.
- **While Iterator Subsystem** 

: This block repeats execution of a subsystem as long as a condition remains true. It simulates iterative processes, such as iterative algorithms or loops, that converge based on a condition (e.g., solving equations or finding roots).
- **For Iterator Subsystem** 

: This block executes the subsystem for a predetermined number of iterations specified by an input signal. It models iterative processes with a fixed iteration count, such as summing a series or processing elements of an array.
- **Model Block** 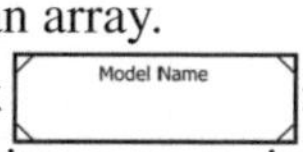

: This block represents a reference to another Simulink model, allowing you to include an entire model within another. It facilitates modular design and reuse by allowing a single model to be included in multiple parent models. Useful for large-scale systems or shared components.

### *8.3.8 Dashboard Blocks*

Dashboard blocks in Simulink allow users to create interactive controls and displays for monitoring and tuning model parameters during simulations. They are found in the "Dashboard" library. Below are some of the most common dashboard blocks:

- **Knob** : This block provides a rotary control for adjusting scalar parameter values interactively. It is used for fine-tuning model parameters (e.g., adjusting gain or input voltage) during simulation.
- **Slider** 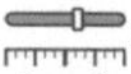: This block allows interactive adjustment of scalar parameter values using a horizontal slider. It is useful for varying a parameter continuously over a range, such as changing a setpoint in a control system.
- **Slider Switch** : This block provides an interactive toggle for binary (on/off) parameter values. It is used for switching between modes (e.g., manual/automatic control) or enabling/disabling features during simulation.
- **Rocker Switch** : This block provides a two-state toggle with a visual rocker-style switch and is used for toggling between binary states (e.g., power on/off or enable/disable).
- **Toggle Switch** : This block provides an on/off toggle control and simplifies binary parameter tuning during simulation.
- **Rotary Switch** : This block allows selection among multiple discrete values or options in a rotary format. It is used for selecting between multiple operational modes or configurations during simulation.

- **Combo Box** : This block provides a dropdown menu for selecting one value from a predefined set of options. It is used to switch between multiple discrete settings or configurations, such as selecting a controller type (e.g., PID, fuzzy logic, or adaptive control) and it is also useful for models with multiple modes of operation or testing various parameter sets.
- **Checkbox** : This block provides a toggle control for a binary (true/false or on/off) parameter. It enables or disables specific features during simulation, such as toggling a system's fault detection or enabling/disabling logging. It also allows for quick control over binary settings without requiring additional switch blocks.
- **Radio Button** : This block allows the selection of one option from a group of mutually exclusive choices. It is ideal for choosing between multiple discrete states or options, such as operational modes (e.g., "Manual", "Automatic", "Standby"). It simplifies the interface for models where only one setting can be active at a time.
- **Lamp** : This block displays a visual indicator that changes color based on signal values and can be used as a status indicator (e.g., showing "green" for safe operation, "red" for fault conditions, and maybe gray for default).
- **Gauge** : This block displays scalar values in a circular gauge format, similar to an analog meter. It monitors real-time values like speed, pressure, or temperature during simulation.
- **Linear Gauge** : This block displays a signal as a linear bar indicator, similar to a progress bar. It is ideal for visualizing percentage values, like fuel levels or progress completion.
- **Display** 42 : This block shows numeric values of a signal in real time during simulation and is used for monitoring specific output values or states numerically.
- **Dashboard Scope** : This block displays signal values in a compact time-domain scope directly on the dashboard. It combines compact visualization with other dashboard elements for real-time monitoring.
- **Push Button** : This block triggers an event or sets a signal to a predefined value when clicked. It simulates manual triggers like resetting a system or activating a specific mode.
- **Edit Field (Numeric/Text)** 42 : This block allows users to enter numeric or text values interactively during simulation. It is used for manually updating parameters or providing inputs during run time.

### *8.3.9 Masks*

In Simulink, **masks** are a feature that allows user to customize the appearance and behavior of blocks, enhancing usability, abstraction, and reusability. Below, you can find a brief list of their applications:

i. **Simplify Complex Blocks**: Masks enable you to hide the internal details of a block or subsystem by encapsulating its logic behind a simplified interface.

They allow you to present a clean and intuitive user interface to the user, making the model more user-friendly.

ii. **Parameterization**: You can define mask parameters that allow users to configure the block's behavior without delving into the block's internal structure. Parameters are connected to variables or settings inside the masked subsystem or block.

iii. **Custom User Interface**: Masks allow you to create a dialog box for a block with custom fields, dropdowns, checkboxes, and other UI controls for setting parameters. This customization improves model readability and user interaction.

iv. **Reusability**: By masking subsystems, you can create reusable components with a standardized interface. This is particularly useful in large models with repetitive functionality.

v. **Hiding Implementation Details**: Masking helps in maintaining model abstraction by hiding the underlying implementation (e.g., detailed blocks in a subsystem). This is useful for protecting intellectual property or when sharing models with users who don't need to know the details.

vi. **Enhancing Visual Appearance**: You can customize the block's icon, making the model more visually intuitive by adding text, shapes, or dynamic visual elements that respond to block parameters.

vii. **Dynamic Behavior**: Masks can control dynamic behavior, such as enabling/disabling certain parameters or controls based on the selection in a dropdown menu. This can guide users to provide valid configurations, reducing errors.

To illustrate using subsystems, signal routing, and dashboard blocks as well as creating masks for subsystems, we will look at an engineering example. In this example, the speed of a car is determined based on engine, drive shaft, and tire parameters as well as the gearshift selection (1 for Park, 2 for Reverse, 3 for Neutral, 4 for Drive, 5 for Manual) and throttle position (between 0 and 100%). Gearshift selection and manual gear selections are done using a rotary switch, since can only take a limited number of discrete values. Throttle position is simulated using a slider as it can take on a continuous range of values. The engine RPM, gear ratio, and car speed are displayed on "Display" blocks. Logic for the engine RPM based on the throttle position, automatic and manual transmission gear ratios based on the gearshift selection, engine RPM, and car speed, and the car speed based on the engine RPM and the gear ratio is all incorporated into separate subsystems as can be seen in Fig. 8.11.

The engine RPM is given by the sigmoid function in Eq. 8.4:

$$RPM = \frac{RPM_{max}}{1 + e^{-k(tp - tp_0)}} \tag{8.4}$$

where **tp** is the throttle position read from the slider block, which is between 0 and 1; **k** is the rate of RPM increase (a typical value is 10); and $\mathbf{tp_0}$ is the throttle position when RPM starts to increase (a typical value is 0.3). This equation is implemented

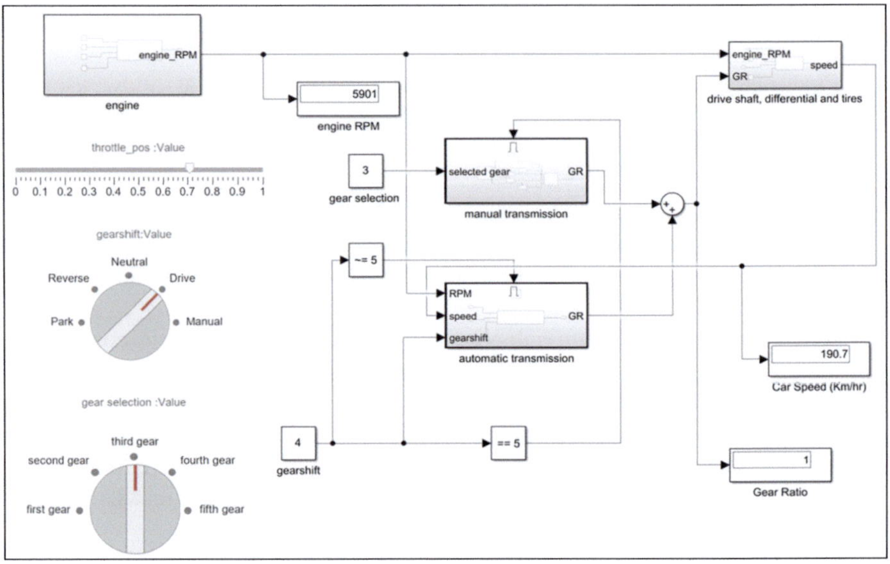


**Fig. 8.11** Car speed simulation in a car with automatic plus manual transmission system

in the "engine" subsystem using a "MATLAB Function" block as can be seen in Fig. 8.12. The three parameters of this subsystem are $\mathbf{RPM_{max}}$, $\mathbf{tp_0}$, and **k**, which must either be directly typed into the constant blocks or be provided in MATLAB workspace prior to running the Simulink model.

The red halo around these three constant blocks in Fig. 8.12 indicates that the values for them are unknown.

To avoid this issue and hide the details of this subsystem, we can convert it to a **masked subsystem**. To do so, right-click on the subsystem and select "Create Mask" under the "Mask" item. An editor window similar to Fig. 8.13 will appear.

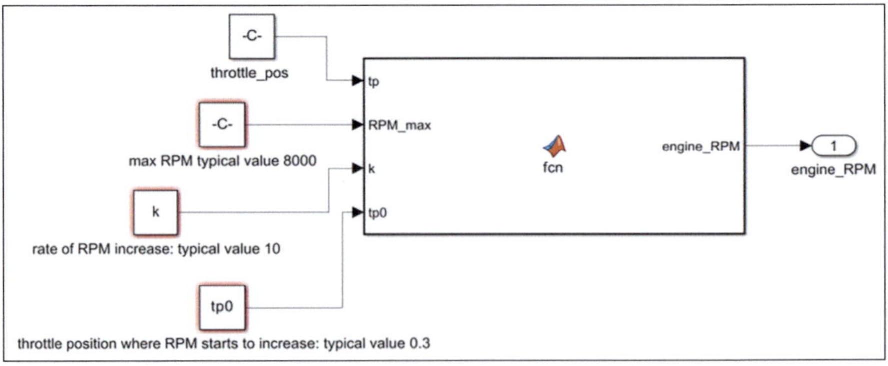


**Fig. 8.12** The "engine" subsystem in Fig. 8.11

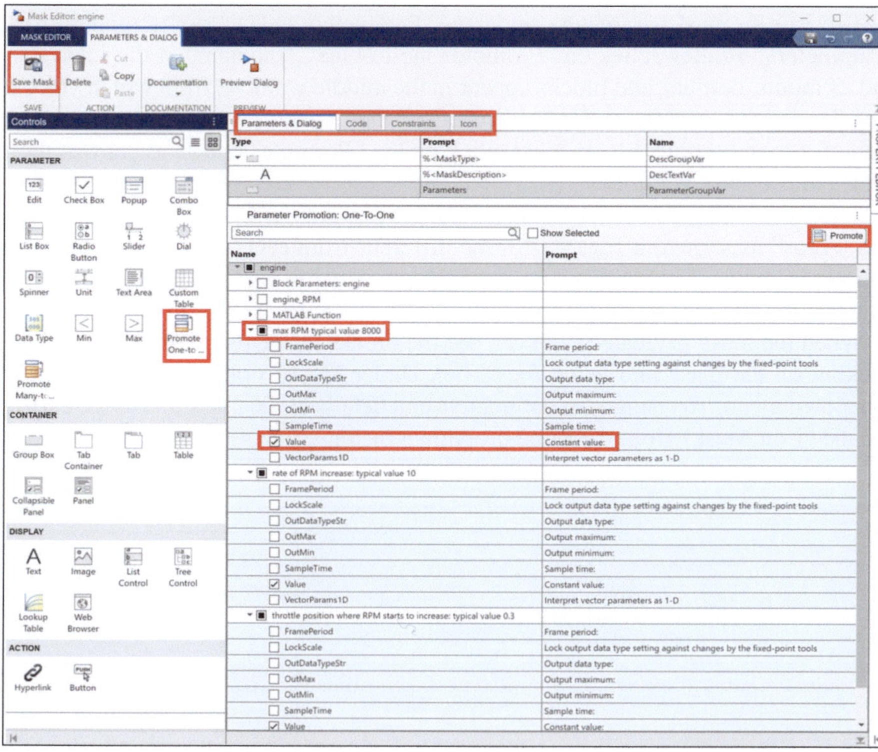


**Fig. 8.13** Mask editor window for the "engine" subsystem

Here, we can create the mask parameters using a variety of methods, such as using edit windows, dials, slider, popup menu, or using "Promote One-to-One" (which is how we are going to do it here); we can add constraints for the mask; we can create pushbuttons and hyperlinks; we can write separate code for the subsystem (e.g., to initialize the parameters as we will see); and we can even add a picture, plot, or text to this mask that represents an engine under the "Icon" tab. To add an image for the block, first download or create an image for an engine. Next, click on the "Icon" tab and click on the "**image()**" button from the left panel. Then, change the name of the image to the name of the image file you just created/downloaded:

```
image('engine.png')
```

To create the mask parameters, under the "Parameters and Dialog" tab and from the "Parameters" window, click on "Promote One-to-One". The name of the subsystem and its inputs, outputs, and blocks appear in the middle window, from which you can pick the desired parameters. Here, we checked the "Value" box under the dropdown menu for each constant block. Next, click on the "Promote" button to convert these values into a mask parameter.

Once the mask parameters are promoted, we can change their "Prompt", "Name", and "Alias" as shown in Fig. 8.14 to be different from each other (otherwise, you will get an error). Note that, you must click on "Save Mask" on the top left corner to make any changes effective.

When the mask is created, you can see the blocks inside the mask only by right-clicking on the mask and using the "Look Under Mask" option from the "Mask" dropdown menu. To change the parameters and details of the mask discussed earlier, use the "Edit Mask" from the same dropdown menu.

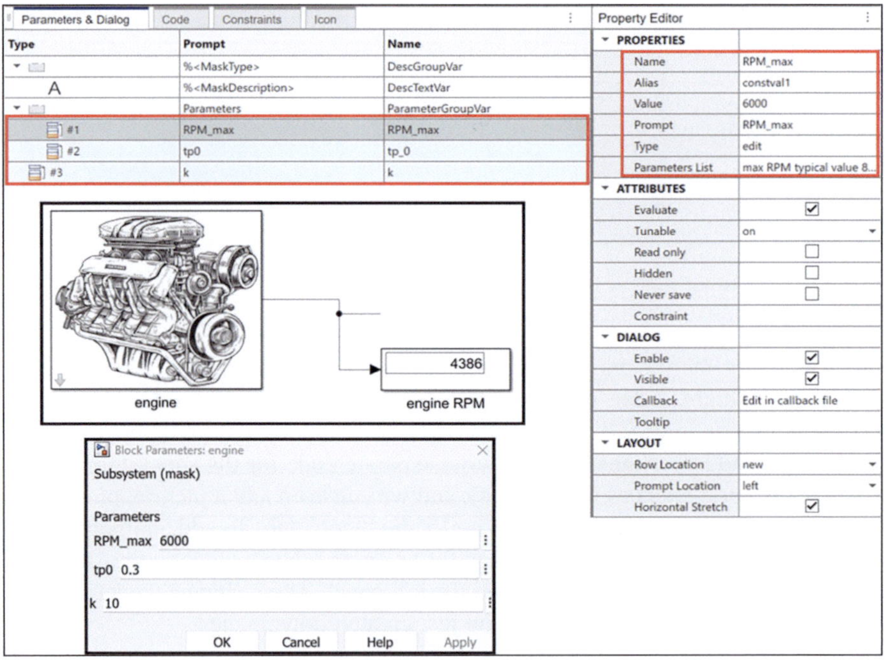


**Fig. 8.14** Mask parameters, appearance, and parameter input window for the "engine" block mask

The logic for the “manual transmission” subsystem/mask is created using “Switch Case” and “Switch Case Action” blocks as can be seen in Fig. 8.15, where gear ratios of 4.2, 2.5, 1.7, 1.2, and 1 are considered for the first to the fifth gears, respectively (using the constant blocks inside the “Switch Case Action” blocks for each case). The “Merge” block combines the outputs of different “cases” of the switch block and sends out the selected gear ratio. Here, we created a mask for this block and added a picture for the mask. We also added an “Initialization” function for the “InitialOutput” value of the “Merge” block using the “Code” tab under the mask with the following code, which utilizes the function **set_param**:

```
classdef MaskInitilizatiob
methods(Static)
    function MaskInitialization(maskInitContext)
        set_param('mask_demo/manual transmission/Merge','InitialOutput','0');
    end
end
end
```

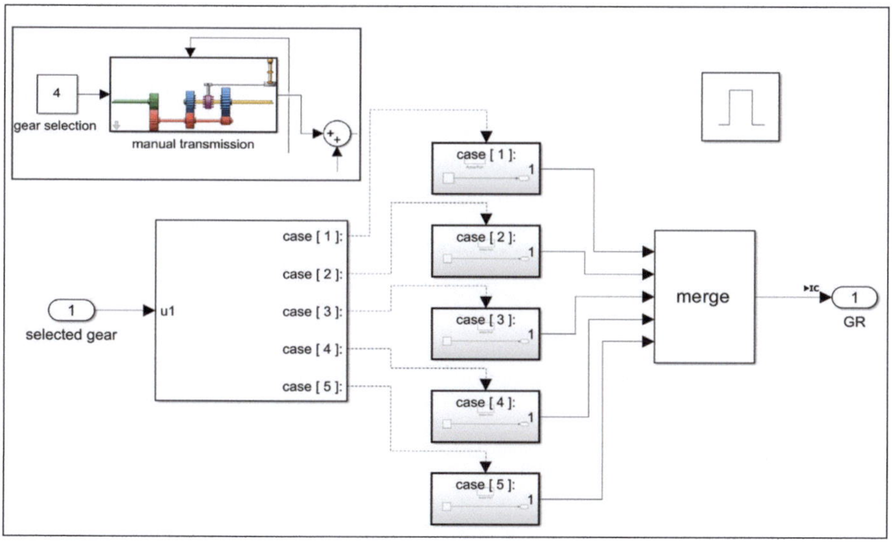


**Fig. 8.15** Logic and appearance of the “manual transmission” mask

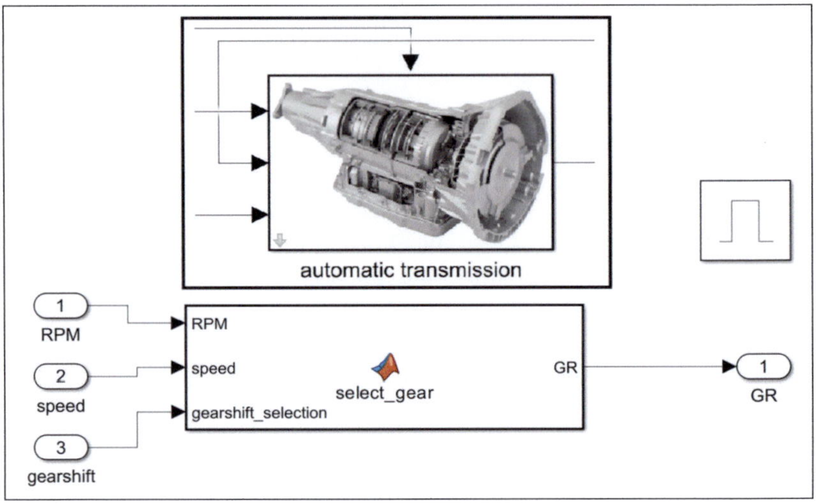


**Fig. 8.16** Logic and appearance of the "automatic transmission" mask

Note that this function needs three inputs: the first one is the full name/address of the block, the second one is the parameter to set/change, and the third one is the value that has to be provided **as a string/character and not as a numerical value**. Also, it can be seen that Object-Oriented Programming (OOP) is used inside this tab for object initialization.

For the "automatic transmission" subsystem/mask, the logic is created using a "MATLAB Function" block as can be seen in Fig. 8.16 and a simple mask with an image icon is added to change its appearance.

MATLAB code for this function is provided below:

```
function GR = select_gear(RPM, speed, gearshift_selection)

GR=0;

% Define gear ratios for each gear

reverse_ratio = 3.0; % Reverse gear ratio

first_ratio = 4.2; % 1st gear ratio

second_ratio = 2.5; % 2nd gear ratio

third_ratio = 1.7; % 3rd gear ratio

fourth_ratio = 1.2; % 4th gear ratio

fifth_ratio = 1.0; % 5th gear ratio

park_ratio = 0; % Park gear ratio (no movement)

neutral_ratio = 0; % Neutral gear ratio (no movement)
```

```
% Check the gearshift selection

switch gearshift_selection
    case 1 % Park

    GR = park_ratio;

    case 2 % Reverse

    GR = reverse_ratio;

    case 3 % Neutral

    GR = neutral_ratio;

    case 4 % Drive

        % Logic for selecting gear in drive mode

        if RPM < 1500

            if speed < 20

                GR = first_ratio; % 1st gear

            elseif speed < 40

                GR = second_ratio; % 2nd gear

            elseif speed < 60

                GR = third_ratio; % 3rd gear

            else

                GR = fourth_ratio; % 4th gear

            end

        elseif RPM >= 1500 && RPM < 3000

            if speed < 20

                GR = first_ratio; % 1st gear

            elseif speed < 40

                GR = second_ratio; % 2nd gear

            elseif speed < 60

                GR = third_ratio; % 3rd gear
```

```
            elseif speed < 100
                GR = fourth_ratio; % 4th gear
            else
                GR = fifth_ratio; % 5th gear
            end
        else % (RPM > 3000)
            if speed < 60
                GR = third_ratio; % 3rd gear
            elseif speed < 100
                GR = fourth_ratio; % 4th gear
            else
                GR = fifth_ratio; % 5th gear
            end
        end
    end
end
```

The “drive shaft, differential and tires” subsystem requires two parameters: tire radius (a typical value of R for a sedan car is 30 cm) and Final Drive Ratio (a typical value for FDR is 3.5). The car speed is calculated using Eq. 8.5 in this subsystem, which is implemented using a “MATLAB Function” block:

$$speed = \frac{3.6 \times RPM \times (2\pi R)}{60 \times GR \times FDR} \tag{8.5}$$

where RPM is the engine RPM determined by the “engine” mask, and GR is the gear ratio determined by user’s selection of the gearshift and manual gear (for the “Manual” drive option).

Note that both “automatic transmission” and “manual transmission” subsystems are **Enabled Subsystem**s as can be seen in Figs. 8.11, 8.15, and 8.16. The reason for this selection is to allow the user to either select the “Drive” option with automatic transmission (**gearshift ~= 5** will enable this system) or with manual transmission (**gearshift == 5** will enable this system) and find the gear ratio based on the transmission selected (using the “Sum” block).

In Fig. 8.17, a mask is created for this subsystem with a relevant image, two parameters, and one constraint for the tire radius (**R**). To create a constraint for a parameter, click on the “Constraints” tab, and then click on the “Parameter” button 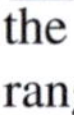 from the top menu. A constraint window will appear in which you can specify the data type, dimension, sign, finiteness, having fractions or being integers, and the range for that parameter as can be seen in the bottom part of Fig. 8.17. A custom

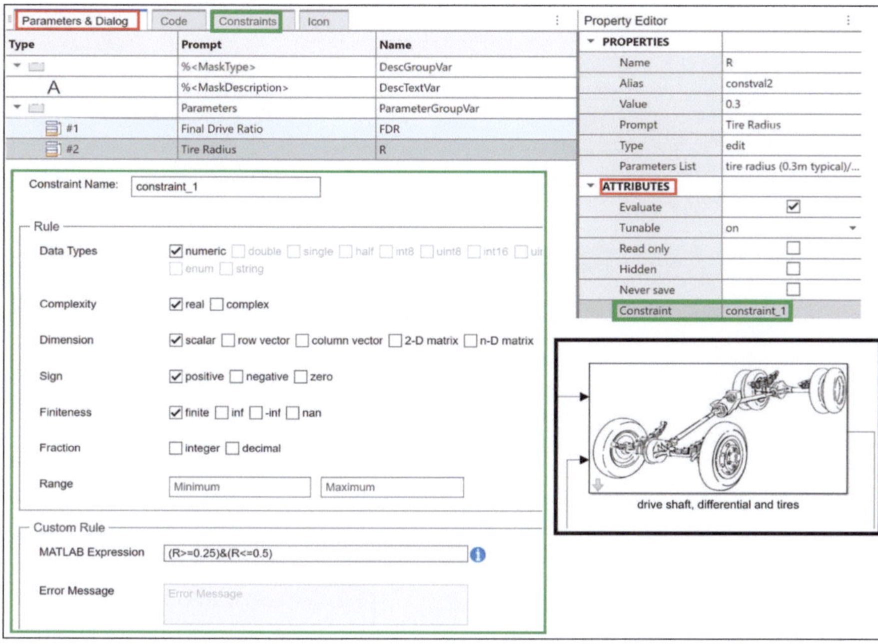


**Fig. 8.17** Parameters, appearance, and constraint of the "drive shaft, differential and tires" mask

rule may also be defined using MATLAB logical expressions. Once a name for this constraint is provided, you need to go back to the "Parameters and Dialog" tab, click on the desired parameter, and then from the "Property Editor" window on the right click on the "Constraint" dropdown menu under "ATTRIBUTES", select the constraint you just created, and apply it to this parameter.

Figure 8.18a shows the final appearance of this Simulink model with the masks added. Note that if you encounter an error indicating an "Algebraic Loop" when running this model, you can click on the "Model Settings" under the "Modeling" tab and change the "Algebraic loop" option from **warning** to **none** from its dropdown menu under the "Diagnostic" branch.

Note that all dashboard blocks need to be properly connected to their corresponding blocks in the model. For instance, to use the slider block that simulates the throttle position, after adding a "Slider" block from under "Simulink/Dashboard", you need to double-click on this block, set its **Minimum** and **Maximum** values, and other optional items, such as the location of the label or the **Ticks**. Then, click on the blank "Connect" window and click on the constant "throttle_pos" block inside the "engine" mask (first, you must open this mask using "Mask/Look Under Mask"). When the "slider" block is connected to the "throttle_pos" block and determines its value using the slider position (user has to move it to the desired position), you can see a checkmark next to the "throttle_pos" block under the "Connect" window as can be seen in Fig. 8.18b. A similar process must be done for the "Rotary Switch" blocks

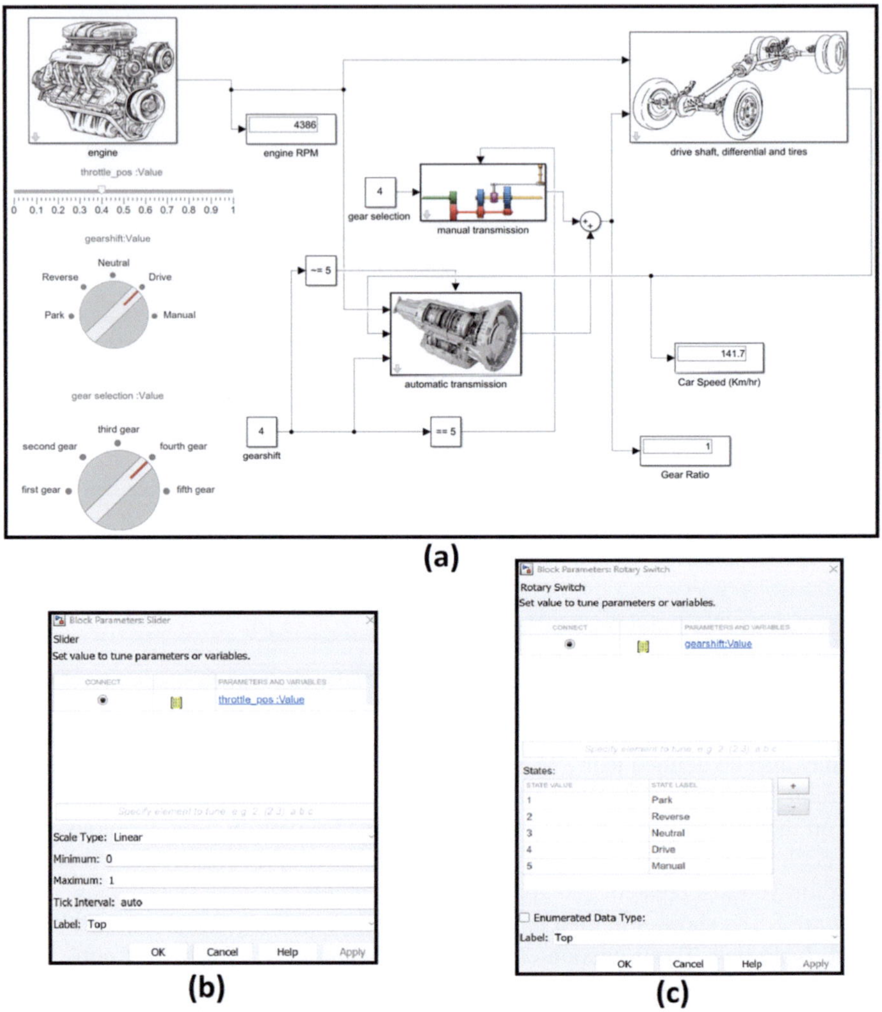


**Fig. 8.18** **a** Final appearance of the Simulink model "mask_demo.slx" with added masks; **b** Editor window for the "Slider" dashboard block which is connected to the "throttle_pos" constant block inside the "engine" mask; **c** Editor window for the "Rotary Switch" dashboard block which is connected to the "gearshift" constant block inside the "mask_demo" model

that determine the gearshift position and the manual gear selection. The window for the "gearshift Rotary Switch" is shown in Fig. 8.18c, where the columns of "STATE VALUE" and "STATE LABEL" have to be manually filled as shown.

## 8.4 Simulink Settings for Simulation of Dynamic Systems

i. **Selecting Solver Type**: Simulink solvers are algorithms that compute the model's behavior over time. The selection depends on the system dynamics (continuous, discrete, or hybrid) and accuracy requirements.

**Types of Solvers:**

- **Fixed-Step Solvers**
  - Use a constant step size for all time steps.
  - Best for:
    - Real-time systems where simulation time must match real-world time.
    - Discrete systems or when using hardware-in-the-loop (HIL) simulations.
  - Common fixed-step solvers:
    - **ode1** (Euler): Fast but less accurate.
    - **ode3**: Medium accuracy.
    - **ode4** (Runge–Kutta): Good accuracy for smooth systems.
- **Variable-Step Solvers**:
  - Adjust the step size dynamically based on system dynamics.
  - Best for:
    - Continuous systems with varying dynamics.
    - Higher accuracy requirements and non-real-time simulation.
  - Common variable-step solvers:
    - **ode45**: Good for non-stiff, smooth systems.
    - **ode15s**: For stiff systems (e.g., systems with fast and slow dynamics).
    - **ode23s**: For stiff systems with moderate accuracy needs.

**How to Choose a Solver:**

- **Continuous Dynamics** (e.g., electrical circuits, mechanical systems): Use variable-step solvers (**ode45**, **ode15s**).
- **Discrete Dynamics** (e.g., digital controllers, sampled systems): Use fixed-step solvers (**ode1**, **ode3**).
- **Hybrid Dynamics** (combination of continuous and discrete): Use variable-step solvers with event detection (**ode45** with zero-crossing detection).

As an example of simulating a discrete system, consider a Low-Pass Filter (LPF) given by $G(z) = \frac{0.2}{1-0.8z^{-1}}$ applied to a sine wave (amplitude 1 and frequency $\pi/2$ rad/s) that is noisy (a uniform random noise with a seed of 500 added to the sine wave), and the autocorrelation of the filtered signal is also calculated using the model shown in Fig. 8.19.

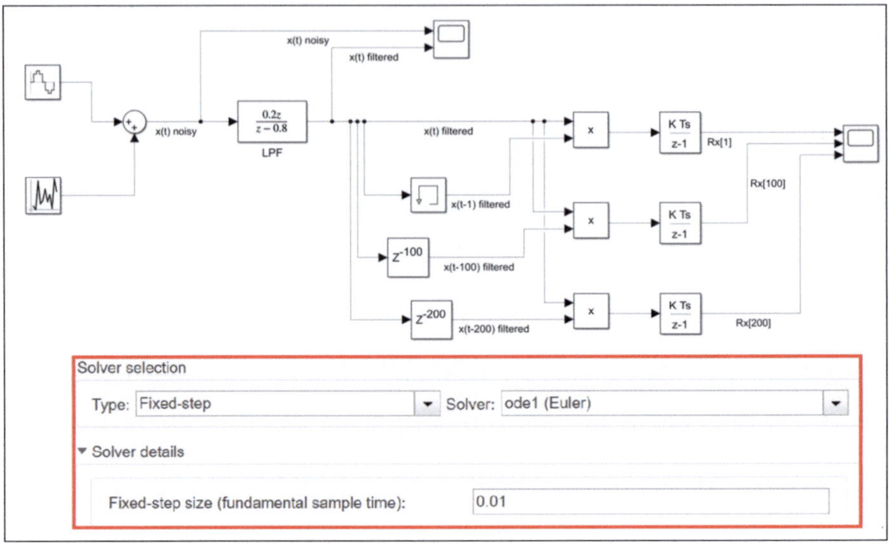


**Fig. 8.19** A discrete system involving a LPF applied to a noisy sine wave and calculating autocorrelation which is solved using ode1 (Euler) fixed-step size solver

To calculate the autocorrelation of the signal $R_x[K] = \sum_{i=1}^{i=N} x[i]x[i-K]$ for different **K** values, we use the "Memory" (**K = 1**) and "Delay" (arbitrary **K**) blocks from Simulink under the "Discrete" tab blocks, as well as the "Product" and "Discrete-Time Integrator" blocks. A sample time of 0.01 s is used in all blocks. To apply a solver with fixed step size, we need to click on the "Model Settings" icon under the "MODELING" tab, go to the "Solver" branch, and select "Fixed-step" from the dropdown menu under "Solver selection" and **ode1 (Euler)** from the drop-down menu under "Solver". We should also use 0.01 for the **Fixed-step size** option. These solver options are also shown in Fig. 8.19.

Figure 8.20a shows the noisy and filtered signals. Figure 8.20b shows the autocorrelation for 3 different delay values of 1, 100, and 200. Since the sine wave period is π/2 (period of 4 s) and the step size is 0.01 s, a delay of 1 sample does not alter the sine wave too much, so the autocorrelation is increasing in the positive direction; a delay of 100 samples corresponds to a phase shift of π/2 which resembles a cosine wave and its correlation with a sine wave is harmonic, and a delay of 200 samples corresponds to a phase shift of π which resembles a negative sine wave and its correlation with a sine wave is increasing in the negative direction.

ii. **Selecting Simulation Times and Step Size**

Regarding the simulation time, consider the following tips:

- **Start Time**: The default value is 0. While the start time cannot be changed under the "Simulation" tab in Simulink, you can go to "Model Settings" under the

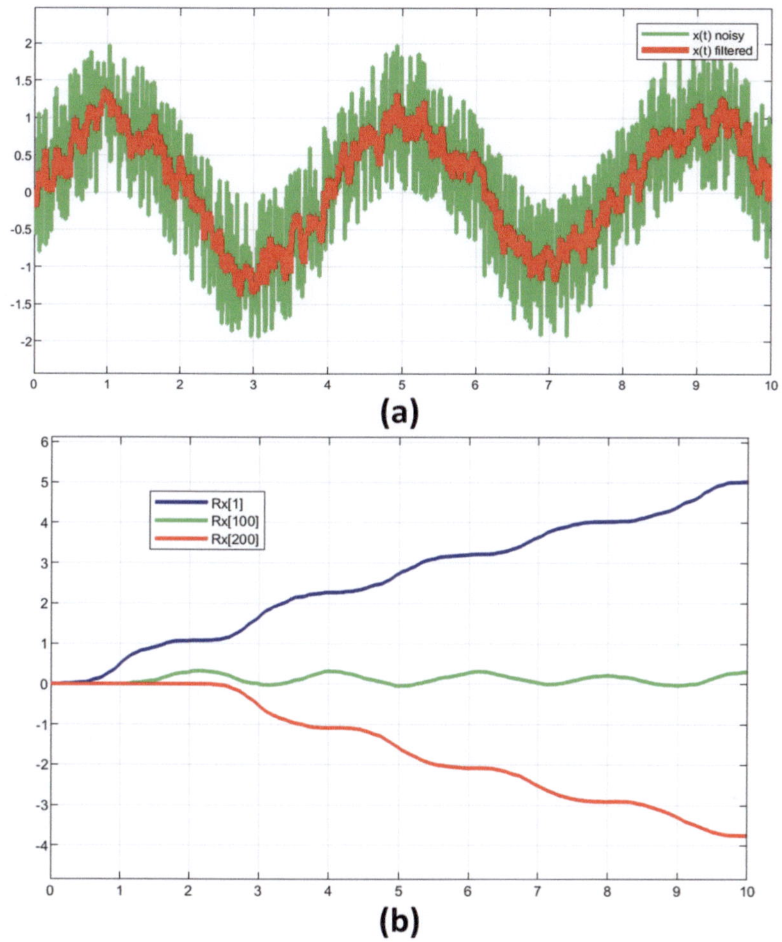


**Fig. 8.20** **a** Noisy versus filtered signal; **b** Autocorrelations for different delays

"MODELING" tab, go to the "Solver" branch, and then change the **Start Time** there.

- **Stop Time:**
  - Should match the time required to observe system behavior.
  - Example:
    - For a system stabilizing at 10 s, set the **Stop Time** to 15 s for a margin.
  - Set larger stop times for steady-state or long-term behavior analysis.
  - If the stop time is not exactly known in advance, you can use the "Stop" block in Simulink under the "Sink" blocks. Consider a projectile subjected to air drag and lift for which no analytical formula for the flight time exists, and

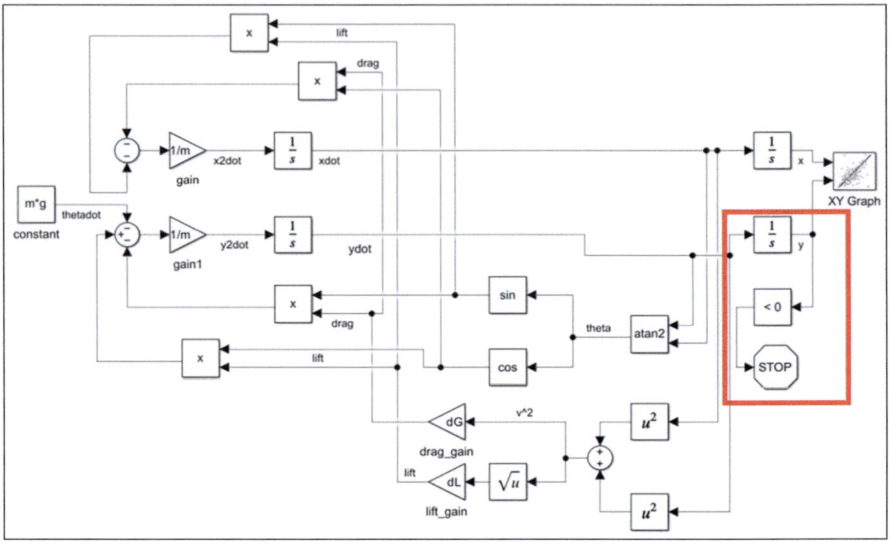


**Fig. 8.21** Using a "Stop" block for a projectile subjected to air drag and lift to terminate the simulations when the projectile hits the ground

you want to ensure that the stop time you entered is large enough to capture the entire projectile motion. However, the simulation must stop right when the projectile hits the ground (since simulation after this point leads to negative y values which corresponds to the projectile moving under the ground). Since the "Stop" block stops the simulation when it receives nonzero values, you can compare the projectile y value against zero using a "Compare to Zero" block and connect the output of this block to the "Stop" block. This is shown in Fig. 8.21. Just pick a large "Stop Time" to begin the simulation.

- **Real-Time Simulations:**
  - For real-time systems, the stop time must match the real-world time you aim to simulate.

Step size determines how often the solver computes the model's state. Regarding the step size, consider the following tips:

- **Fixed-Step Size:**
  - Directly defined by the user.
  - Ensure that the step size is small enough to capture critical system dynamics.
  - Example:
    - For a system with a maximum frequency of 100 Hz, use a step size smaller than **1/(10 * 100) = 0.001** s.

- **Variable-Step Size:**
  - Automatically adjusted by the solver.
  - User specifies:
    - **Max Step Size:**
      - Upper limit for step size.
      - Example: For a simulation of 0–10 s, set max step size to 0.1 s.
    - **Min Step Size:**
      - Lower limit for step size (rarely used).

## 8.5 Simulink "DEBUG", "MODELING", and "Format" Tabs

### *8.5.1 DEBUG Tab*

In this subsection, we will be exploring Simulink debugging tools. Figure 8.22 shows a servo system, in which the input ramp function is the desired angular position which goes up from 0 to 4000 radians in 10 s. DC motor dynamics is modeled using a second-order transfer function given by $(s) = \frac{0.01}{0.005s^2+0.06s+0.1001}$, and a PID controller is used for position control. Desired and actual angular positions are logged and can be monitored using the "Data Inspector" tool under the "Simulation" tab. To log any signal, left-click on it once and use the "Log Selected Signal" option from the "3 dots" that appear above the signal.

First, we perform a PID tuning task by opening the PID block window, importing the DC motor model from the workspace, tuning the step response of the controlled system, and finally adjusting the PID gains (to learn more about this process, check Sect. 10.3).

For debugging purposes, we want to add two breakpoints for the simulation when the actual angle passes values of 100 and 500 radians to check on the file outputs. To do so, first click on the "actual angle" signal in the model. Then, under the Simulink "DEBUG" tab, click on the "Add Breakpoint" icon . A window like the one in Fig. 8.23 would appear, where you can specify a logical condition for the signal to cause the simulation to stop. Use this method to create two breakpoints for the "actual angle" signal using >100 and >500 settings. A red circle must appear above this signal as can be seen. Now you can see the list of breakpoints by clicking on the "Breakpoint List" icon from this tab as shown in Fig. 8.23. You can also "Enable/Disable" and "Clear" any breakpoint that you have created.

As can be seen in Fig. 8.23 under the "Data Inspector" window, once the "actual angle" signal passes 100, the simulation is stopped, and you can see the signal up to the time point before this breakpoint was activated. When a signal is logged, you

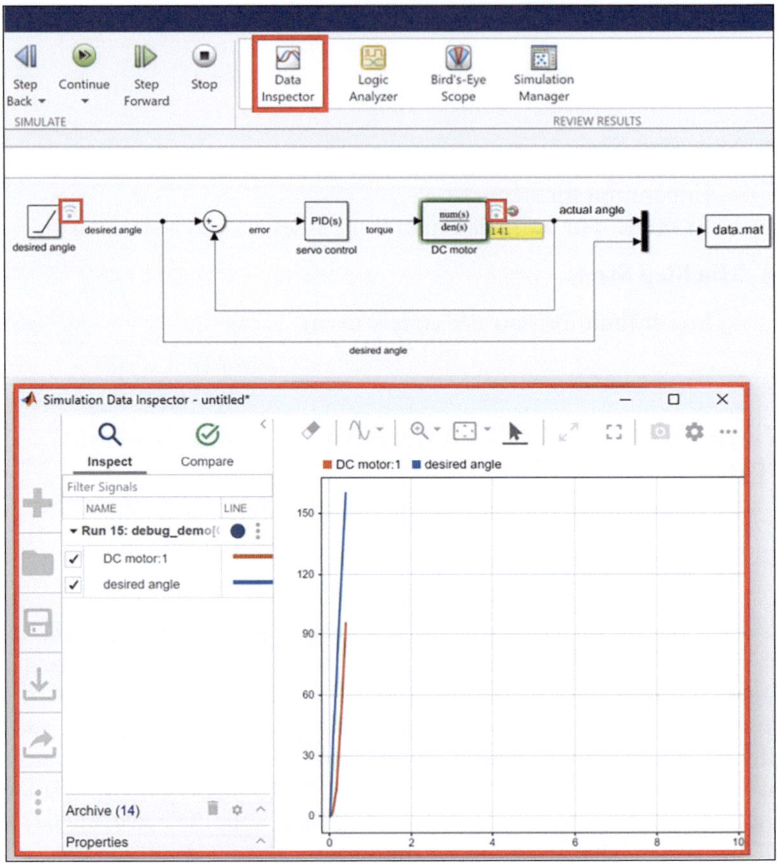


**Fig. 8.22** Simulink model of a servo system

can also see the signal value in yellow boxes next to the signal during simulation, as can be seen in Fig. 8.23.

Note that you can use the "Continue" button to move from one breakpoint to the next. Buttons "Step Forward" and "Step Back" can be used to move the simulation one sample time forward or backward, respectively. The "Stop" button can stop the simulation at this time.

There are three other tools next to these buttons: "Step Over", "Step In", and "Step Out". These buttons are used when subsystems or functions exist in a Simulink model. The "Step Over" button allows you to advance the simulation through the current time step, executing the entire content of subsystems or functions without pausing inside them. The "Step In" button pauses inside subsystems or functions, allowing you to debug their internal operations step by step. If you're

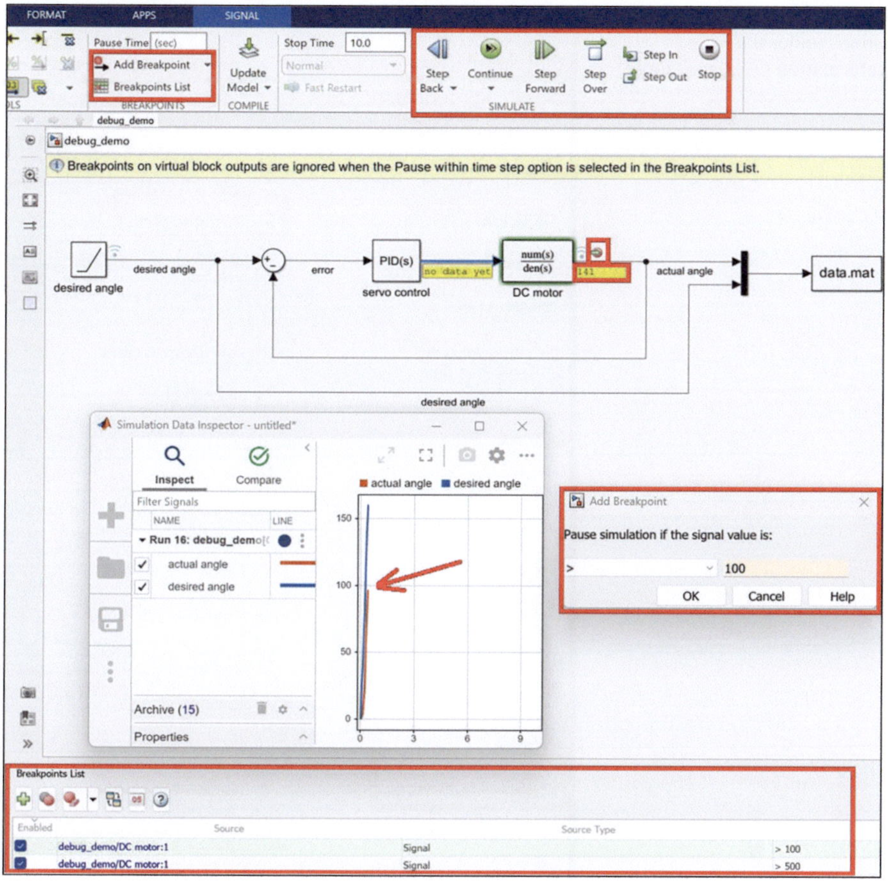


**Fig. 8.23** Adding breakpoints to the servo system in Fig. 8.22

already inside a subsystem or function, the "Step Out" button completes execution and steps out to the parent model.

Another great debugging tool is the "Information Overlays" . Under this button, you can see sample times color-coded, "Execution Order", "Signal Dimension", "Signal Unit", "Block Link", etc., as shown in Fig. 8.24.

Among these options, sample time legend and signal dimension overlays are the most important tools for debugging as can be seen in Fig. 8.25 for the system in Fig. 8.11.

Note that to visualize the sample times for signals, we can also use the "Sample Time" button  from the "palette" on the left side of Simulink window (check Fig. 8.1).

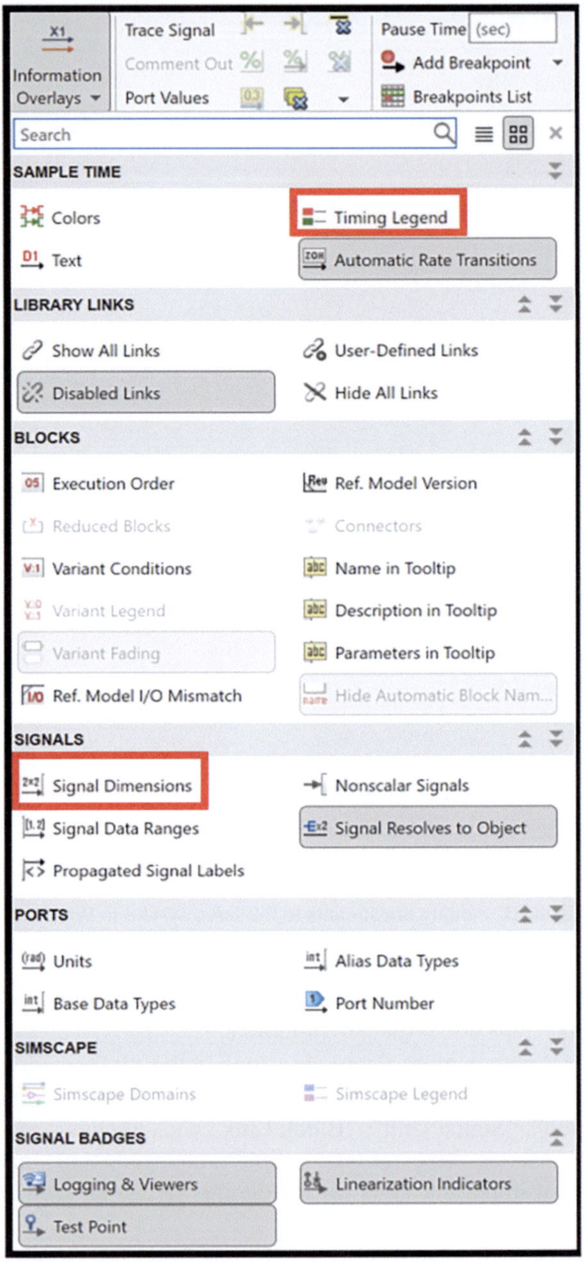


**Fig. 8.24** Debugging options under the "Information Overlays" tool

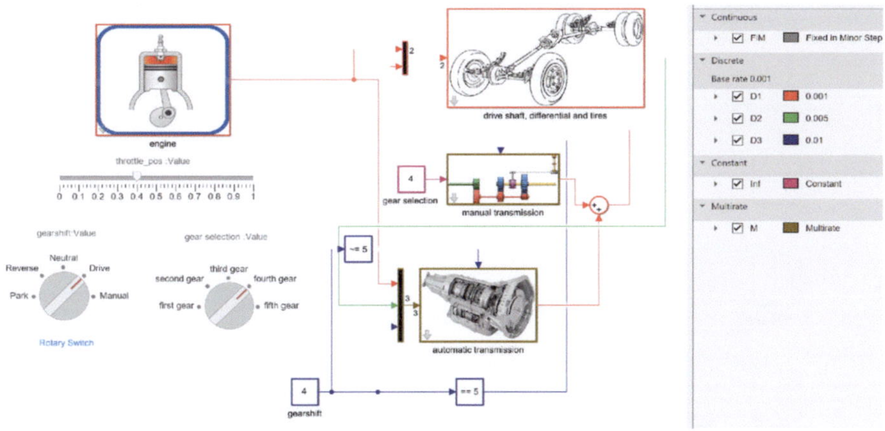


**Fig. 8.25** Sample time color-coding and dimension overlay in a Simulink model

### 8.5.2 *MODELING Tab*

Under the "Modeling" tab in Simulink, you can find tools for building, configuring, and enhancing your Simulink model. There are several important buttons, including "Model Settings", "Model Explorer", "Model Data Editor", "Model Advisor", and "Model Update".

To demonstrate these buttons, we will use the demo shown in Fig. 8.26 that demonstrates "Conditional Signal Assignment" using an "Assignment" block that passes its input signal (middle left port) to one of the output indices based on external provided indices (bottom left port). This block is very useful in resource allocation, such as when a microcontroller has only one free pin for a PWM signal, since the user can select (here, using the toggle switch) to output either to Motor 1 or Motor 2 for speed control. In this model, if the user changes the toggle switch position, it will change the sign of constant input to the middle port of the "Switch" block (5 or −5), which sends index 1 or index 2 to the "Assignment" block corresponding to Motor 1 or Motor 2. As can be seen in this figure, with the toggle switch being "on", Motor 1 receives the PWM signal, but Motor 2 doesn't receive any signal.

"Model Settings" has options for "Solver", "Data Import/Export", "Math and Data Types", "Diagnostics", "Hardware Implementation", "Model Referencing", Simulation Target", "Code Generation", "Coverage", and "HDL Code Generation". We have used some of these options throughout this book. For instance, if we want Simulink to export the simulation time to the workspace as a variable titled **t**, we click on "Model Settings", and then under "Data Import/Export" and "Save to workspace or file", we check the box for "Time" and type "t" for the variable's name in front of it.

"Model Explorer" has options for "Model Hierarchy" as well as details about the model blocks.

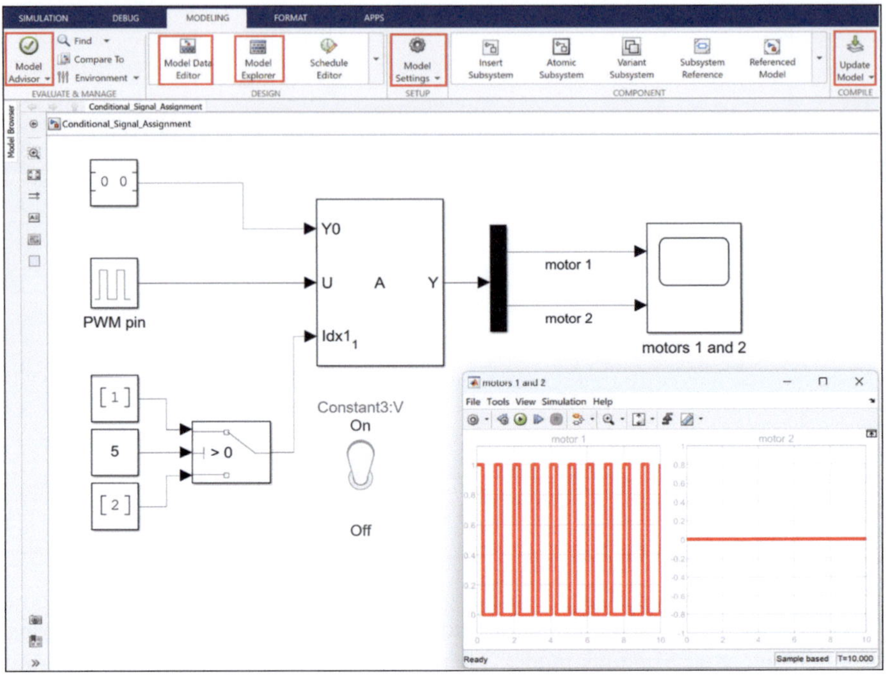


**Fig. 8.26** Simulink "MODELING" tab in a "Signal Assignment" demo file

Figure 8.27 shows the data available under "Model Explorer" for the system mentioned above (variables under Base Workspace) and for the block details under the model name.

If more details about the signals are required, we can click on the "Model Data Editor" as can be seen in Fig. 8.28 for this model.

Note that the green checkmark for the "Model Advisor" indicates that the model we made is compliant with the modeling guidelines and standards. If there was any error, "Model Advisor" would indicate that.

Keep in mind that after applying some modifications to a model, it is always advised to use the "Update Model" button (or Ctrl + D) to ensure that no "static error" is present during model compilation. A "static error" in a Simulink model compilation refers to an error detected during the process of converting your Simulink model into executable code, where the issue is identified based on the model structure itself, without requiring any simulation to run. This usually indicates a problem with the model's design, such as missing connections, incompatible data types, or incorrect block configurations, that would prevent the code from compiling successfully.

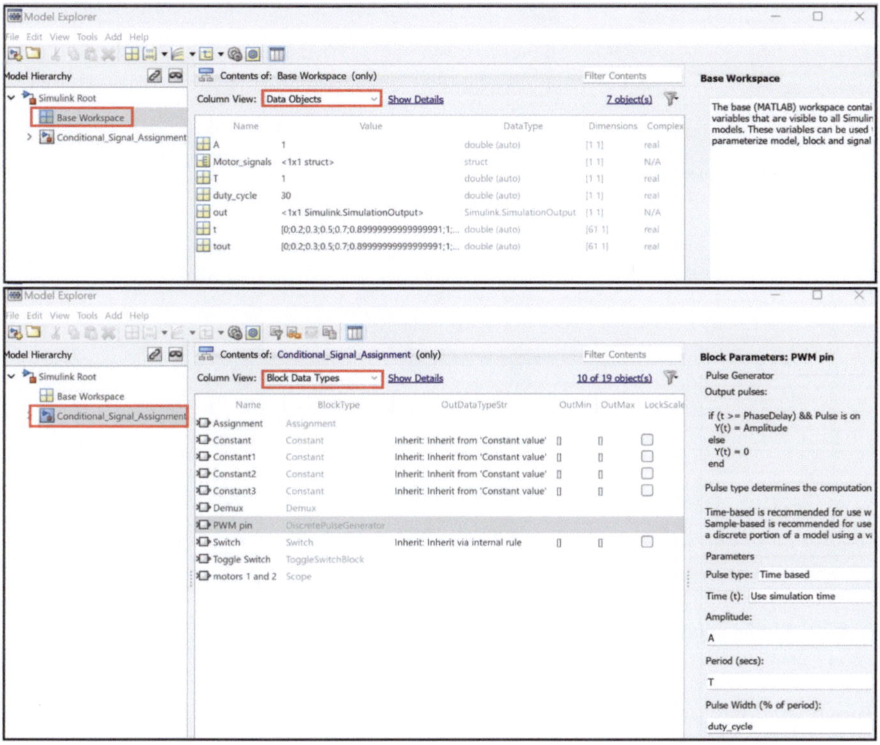

Fig. 8.27 Model Explorer under Simulink "MODELING" tab

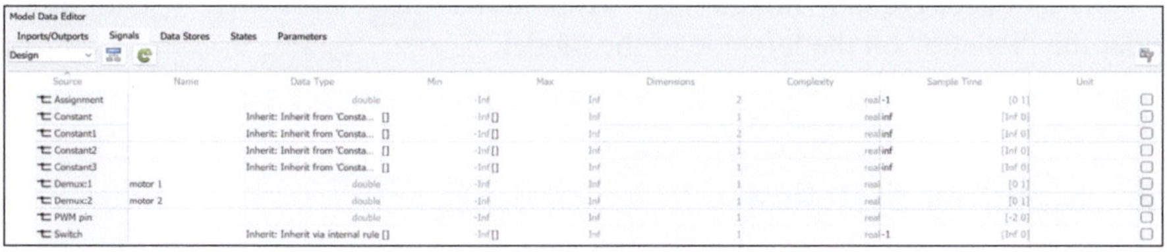

Fig. 8.28 Model data editor under Simulink "MODELING" tab

### 8.5.3 *FORMAT Tab*

The "Format" tab in Simulink provides tools to improve the visual organization and appearance of your model. These options help enhance readability, alignment, and consistency in the layout. Figure 8.29 shows the model in Fig. 8.27 with several formatting modifications.

Under this tab, we can flip or rotate blocks (keyboard shortcuts are Ctrl + I and Ctrl + R), show or hide their names, auto-arrange them, add pictures to the blocks or the background, add annotations to the model, change the color of the blocks or add/

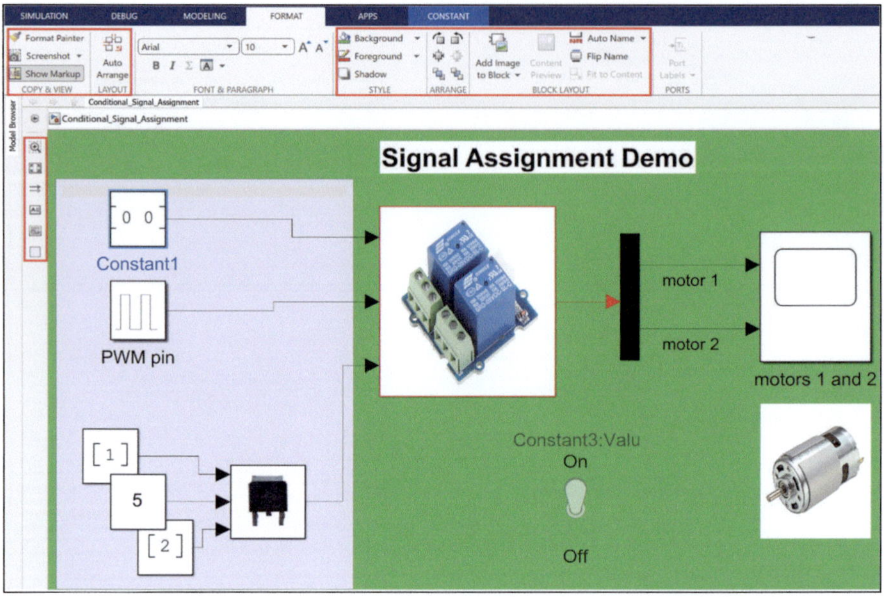


Fig. 8.29 Model appearance modification using the tools under Simulink "FORMAT" tab

remove shadows to/from them, bring blocks to the front or send them behind other blocks, zoom in/out of the model or fit the model window to the content (keyboard shortcut is the space key), and add an "Area" with a specific color to a part of the model (the gray box on the left side of Fig. 8.29).

Note that background "Image" , "Annotation" , and "Area" buttons are under the "Palette" on the left side of the model.

## 8.6 Simulink Stateflow

Stateflow in Simulink is a powerful tool for modeling and simulating **event-driven** systems using state machines and flow charts. It is particularly useful for systems that exhibit reactive behavior, meaning they respond to events or changes in conditions.

Some applications of Stateflow are as follows:

- **Control System Design**: Modeling supervisory logic, mode switching, and fault management in control systems. Example: Switching between different control modes in an HVAC system or automotive cruise control.
- **Embedded Systems**: Developing control logic for real-time embedded systems. Example: Automotive systems such as Anti-lock Braking Systems (ABS), transmission control, and driver assistance systems.

- **Robotics**: Designing logic for robotic navigation, task scheduling, and behavioral control. Example: Switching between obstacle avoidance and target-following behaviors in a robot.
- **Aerospace and Defense**: Modeling mission management systems, fault detection, and recovery strategies. Example: Autonomous drones transitioning between takeoff, flight, and landing states.
- **Event-Driven Systems**: Modeling systems driven by asynchronous events. Example: Industrial process control, such as event-based triggering in manufacturing systems.
- **Workflow Automation**: Designing workflows with complex decision-making processes. Example: Customer service chatbots or automated ticketing systems.

Some important terminologies in Stateflow are **states**, **transitions**, **events**, and **actions**, which allow the **representation of system logic visually**, thus making the system more convenient to understand and debug.

Stateflow blocks are under the "Stateflow" tab in the Simulink library. To start, let's create a blank model and add a "Chart" block  from the Stateflow library to the blank model. A Stateflow chart is a graphical representation of a finite state machine consisting of states, transitions, and data. You can create a Stateflow chart to define how a system reacts to external input signals, events, and time-based conditions. Inside a "Chart", when you double-click on it, there are lots of tools on the left side, called the "object palette", for creating logic inside the empty space in the middle, called the "chart canvas", as shown in Fig. 8.30.

A **state** is a fundamental component of a Stateflow chart that represents a specific mode or condition of a system. A state is considered **active** when the system is operating in that state. States can be **atomic** (with no internal sub-states) or **composite** (contains nested sub-states).

Each state can have three types of **actions**: Upon **entry**, **during** (being in the state), and upon **exit** as can be seen in Fig. 8.31a.

The **default transition** in Stateflow defines the initial state or starting point when a Stateflow chart or a composite state becomes active. It specifies which state the system should enter first in the absence of any other triggering conditions. In Fig. 8.31a, the default state is **state2** and variables **x** and **y** are set to 0 (**actions**).

Moving from one state to another is called a **transition** and can have a name, it requires some **conditions** to happen, and during which some **actions** can be carried out. To apply a transition, simply click on one state and drag the generated wire to another state. Next, click again on the top of this wire to see a question mark. When you click on this question mark, you will see the options for creating the **Name**, **Condition** (that must be inside a bracket), and **Action** (that must be inside curly braces) for this transition as shown in Fig. 8.31b–d.

Not only can a condition for a transition come from some change in the variables, inputs, or outputs, but it can also come from timing conditions using commands **after**, **before**, **at**, **every**, and **timer**:

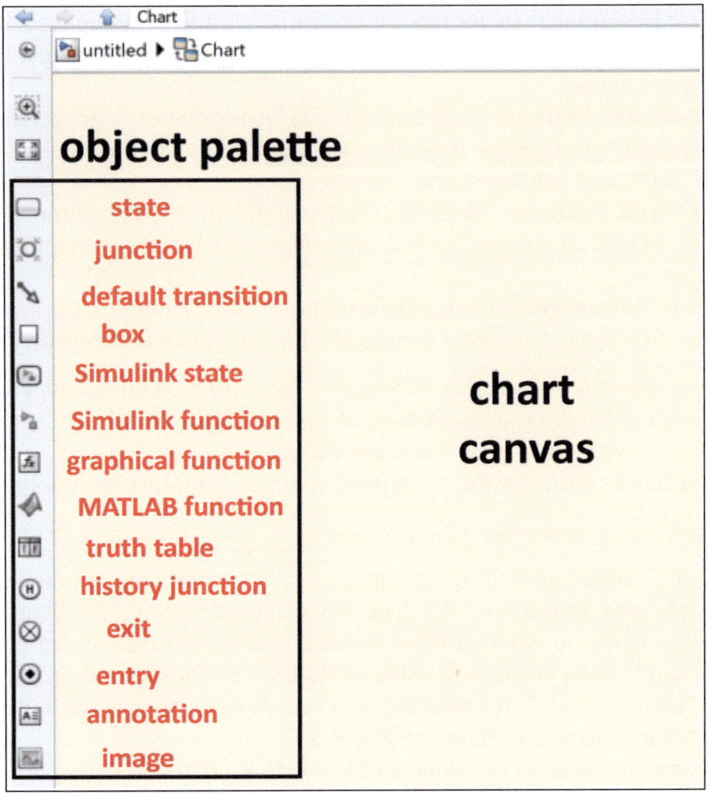


**Fig. 8.30** Stateflow chart and the commands inside it

```
[(motor1==1)&(motor2==1)]
[after(3,sec)] % after 3 seconds
```

A chart by default does not have any input from or output to Simulink when you create them, but you can add inputs and outputs by right-clicking in the chart canvas and using the option "Add Inputs and Outputs". This is shown in Fig. 8.32. Not only can inputs and outputs be created this way, but we can also create local variables, constants, parameters, and data store memory for a chart.

If the logic for transitions and events is complicated, we can use Simulink functions, MATLAB functions, or graphical functions as shown in Fig. 8.33.

Creating MATLAB functions is straightforward and has been discussed in Sect. 2.1. Simulink functions are basically subsystems and creating them has been discussed in this chapter. To create a graphical function, first we type the name and syntax of the function (function declaration line) in the top space of the graphical function. Next, we right-click on the function and use the option "Add Pattern in Function" and select the logic control that we need (**if-elseif-else**, **switch-case-otherwise**, **for** loop, **while** loop, etc.). In Fig. 8.33c, **if-else** is selected, which leads to

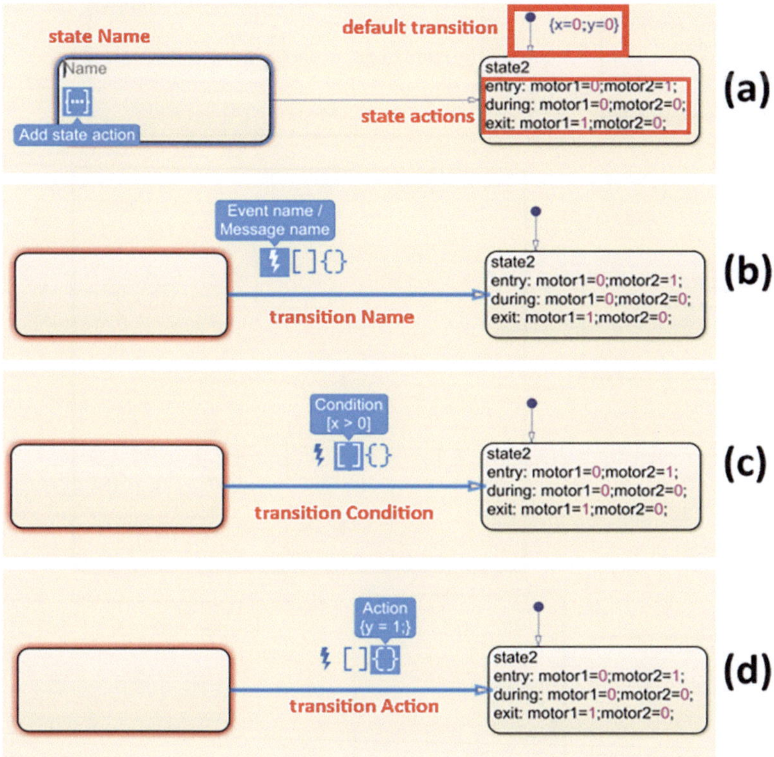


**Fig. 8.31** **a** Name and actions of a state and default transition of a chart; **b** Name of a transition; **c** Condition(s) of a transition; **d** Action(s) of a transition

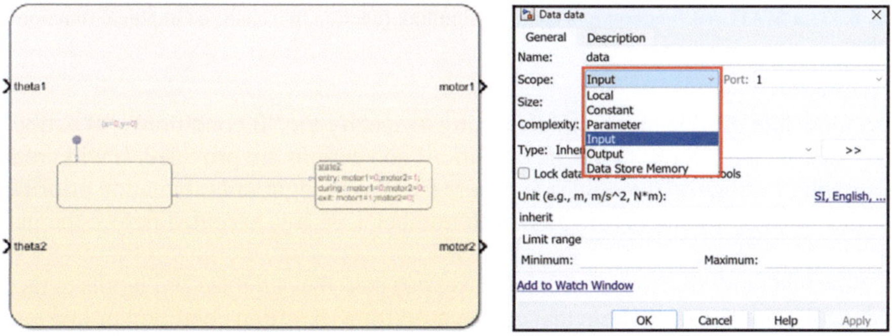


**Fig. 8.32** Creating inputs and outputs for a chart

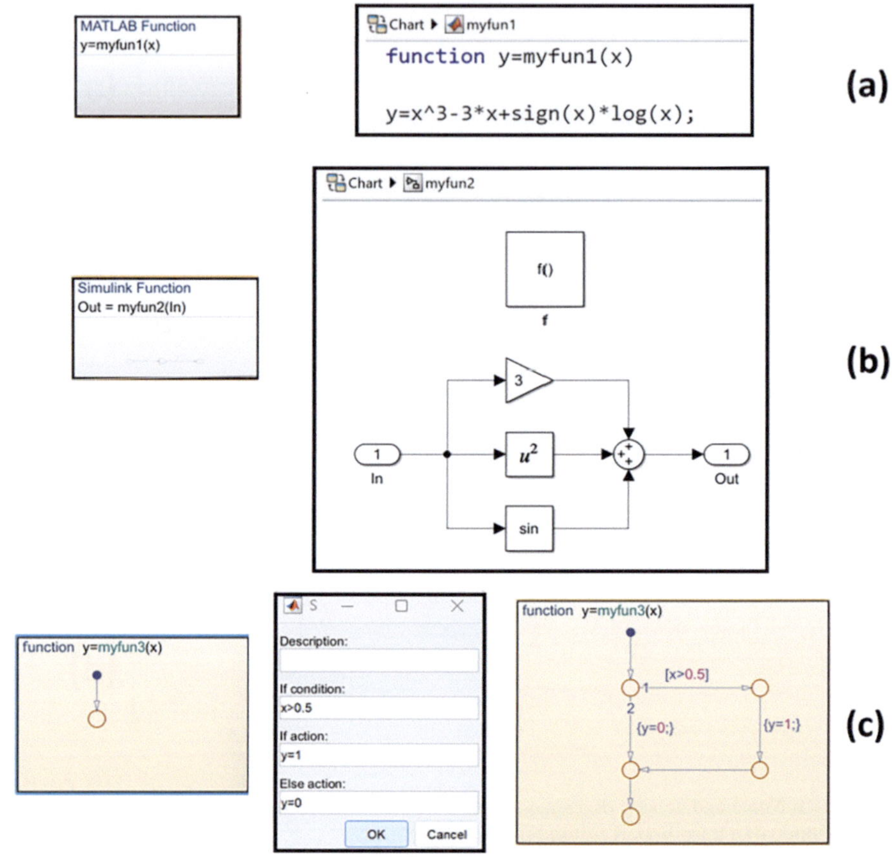


**Fig. 8.33** **a** MATLAB function in a chart; **b** Simulink function in a chart; **c** Graphical function in a chart

a window like the one in the middle asking to specify the "If condition", "If action", and "Else action" items. Once the conditions and actions are provided, it will create the graphics shown on the right side, where the flow of logic and calculation priorities are indicated by numbers (the path with number 1 on it is executed before the path with number 2).

Now that these functions are created, we can pass the variables and inputs to them and get the outputs from them that can be used for a transition condition or inside an action.

Sometimes transition from a state can have a **junction**, which means depending on one set of conditions, transition happens to one state and under another set of conditions, transition happens to another state. This is shown in Fig. 8.34a using the **junction** command .

Another element inside a chart is a **box**, which is used for organizing and grouping the contents of a chart. A box is "non-executable" and is merely a visual element.

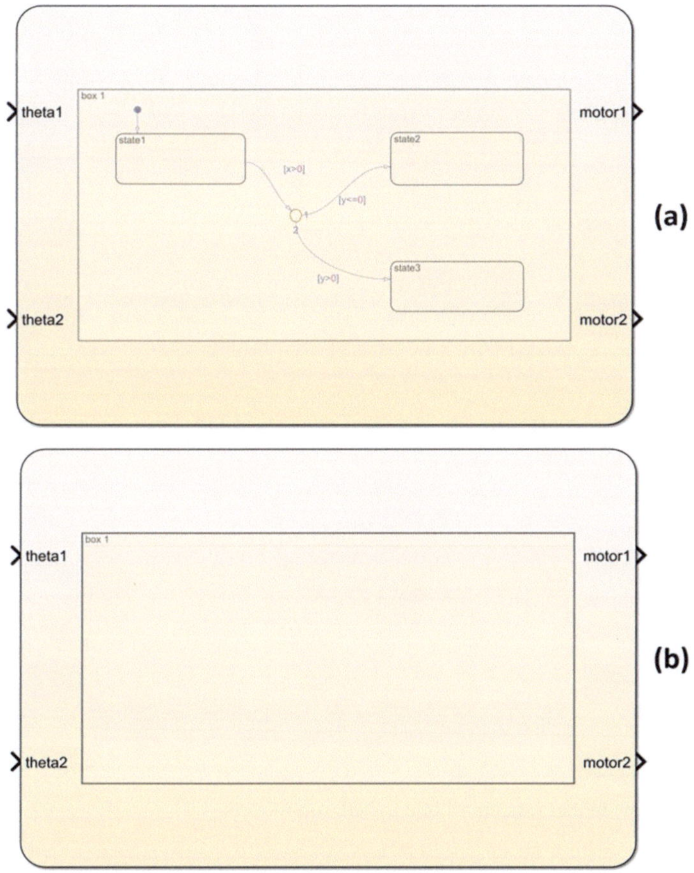


**Fig. 8.34** A junction used in a transition and organization of chart contents using a box; **b** Hiding a box content as a subchart

One advantage of using a box is the ability to hide its contents from outside of the chart by right-clicking on the box and selecting the "Subchart" from the "Group and Subchart" dropdown menu as shown in Fig. 8.34b.

To demonstrate using Stateflow for logic control with what we have learned so far, our demo shown in Fig. 8.35 performs motor angular velocity control of a 2-DOF planar robotic arm (no kinematic or kinetic equations are used for the motion of robot links, and their angles called **theta1** and **theta2** are simply the integral of the angular velocity provided by the logic control system).

The logic for turning on and off the two motors are stated as follows:

- Starting from both motors off,
  - If the robot switch is on and **q1(theta1) <1.2** rad, motor 1 must be turned on.
  - If **theta1 > 0.8**, turn motor 2 on.

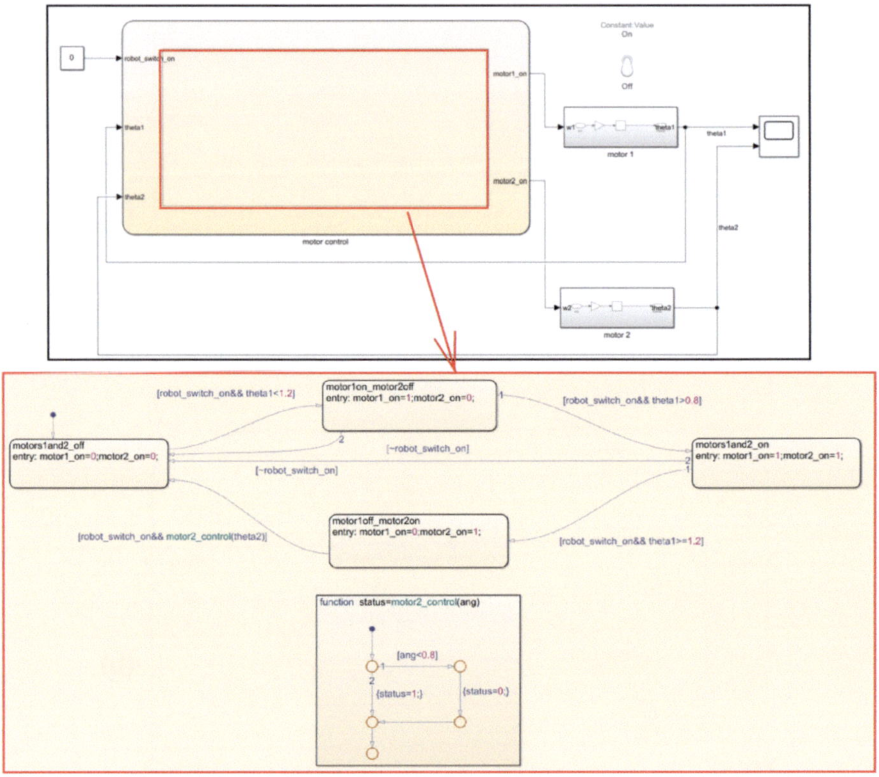


**Fig. 8.35** Stateflow model for control of a 2-DOF robotic arm

  - When **theta1 >= 1.2** rad, turn motor 1 off.
  - When **theta2 >= 0.8** rad, turn motor 2 off.
  - Whenever during operation, robot switch is turned off, turn both motors off.

In Fig. 8.36, we can see the graphical representation of the logic flow via blue highlights around the active states for different scenarios in this system.

Note that the transitions in this system might happen very fast, so you might need to slow down the animation of the logic flow to verify it. To do so, under the "DEBUG" tab, change the "Animation Speed" to "Slow" or "Medium" instead of the default option "none".

Let's look at a few other important items in the stateflow.

- To "comment out" a state and the transition lines leading to it, we can use the keyboard shortcut "Ctrl + Shift + X" or right-click on it and use the option "Comment Out", which turns the state and its transition lines gray as shown in Fig. 8.37a.
- To pause a transition and check the status of a chart, you can create a "Breakpoint" anywhere along the"transition" line by clicking on it and using the option "Add

**Fig. 8.36** Visualization of logic flow via blue highlights around the active states for different scenarios in the 2-DOF robotic arm

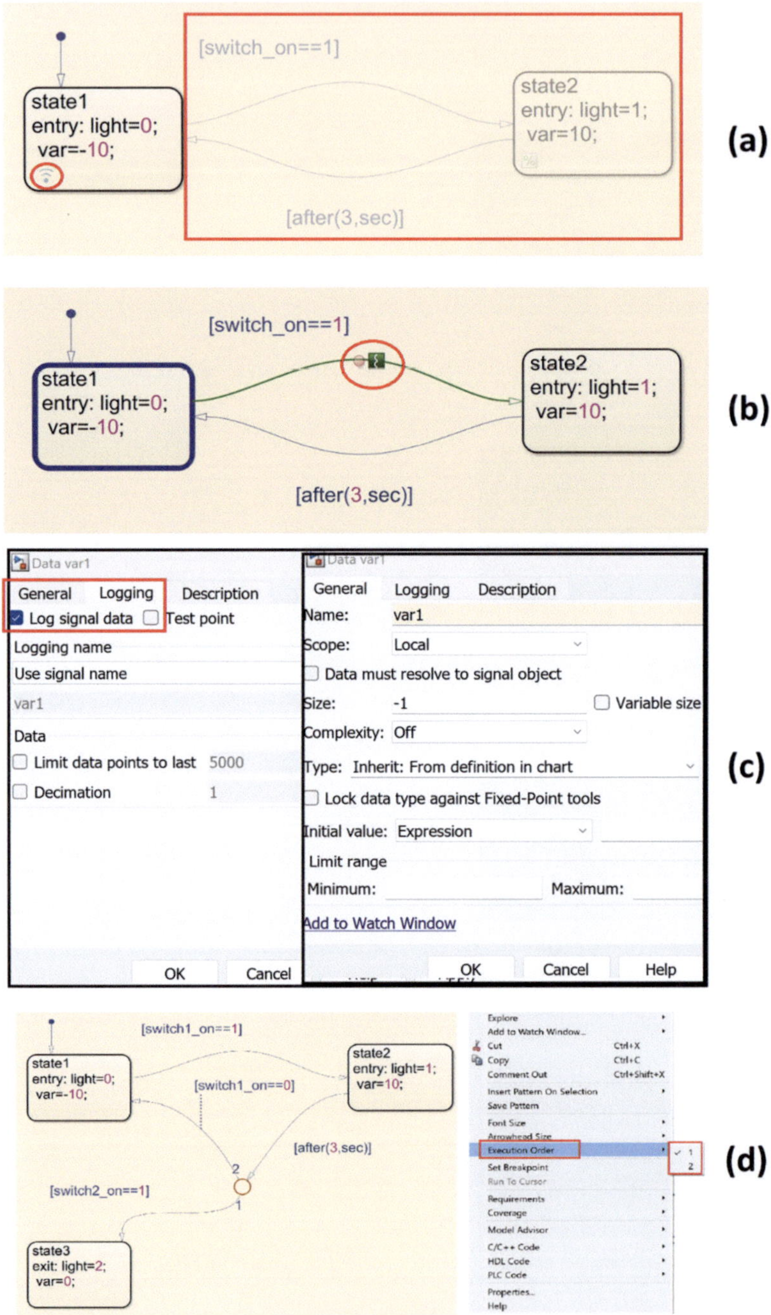


**Fig. 8.37** **a** Commenting out a state and its transitions; **b** Adding a breakpoint to a transition; **c** Creating a local variable for a chart and logging it; **d** Changing the execution order in a junction

Breakpoint". This will add a red circle on the transition line as can be seen in Fig. 8.37b, and when running the model, a green box would appear on the transition line to show that the simulation is paused. Breakpoints can be removed by right-clicking on them and using the "Clear Breakpoint" option.

- Not only can a chart have inputs and outputs, but it can also have local variables. To create a local variable for a chart, right-click anywhere in the canvas and use the option "Local Data" from the dropdown menu under "Add Other Elements". These local variables are only accessible within that chart (they can be changed via states and transitions of the chart) and are not shared with other charts. We can also log them for monitoring purposes, if we check the box "Log signal data" under the "Logging" tab in the "Data" window when creating this variable as shown in Fig. 8.37c. When it is logged, we can see the history of the signal using the "Data Inspector" tool under the "Simulation" tab.
- As you have seen so far, any junction has an "Execution Order" which is indicated using the numbers on its transition lines (1, 2, etc.). It means that first the transition with number 1 will be checked, and if it cannot happen, then other transitions will be checked based on their numbers. We can change this execution order by selecting any transition line, right-clicking on it and using the option "Execution Order". Then, click on the execution priority number we desire for this transition line as shown in Fig. 8.37d.

Under the "MODELING" tab, we can use the "Model Explorer" button to access all the states, variables, inputs, outputs, external data, and other items of a stateflow model as shown in Fig. 8.38 to modify, delete, rename, and copy them. If we click on a state, under the "Logging" tab on the right side, we can check the box for "Log Self Activity" to monitor the activity of that state over time. A logging symbol would appear inside the state block as shown in Fig. 8.37a for **state1** inside the red circle on the left.

## 8.7 Simscape for Physical System Modeling

Simscape allows users to model and simulate physical systems, such as mechanical, electrical, hydraulic, and thermal systems, using a physical network approach.

Key features of Simscape are as follows:

i. **Physical Modeling**: Simscape enables the creation of models using physical components like resistors, motors, gears, and pumps. These components represent real-world physical elements and are connected in a way that mirrors the actual physical system.
ii. **Multidomain Simulation**: Simscape is designed to handle simulations that involve multiple physical domains. For example, you can simulate a system that combines mechanical, electrical, and hydraulic elements within a single model.

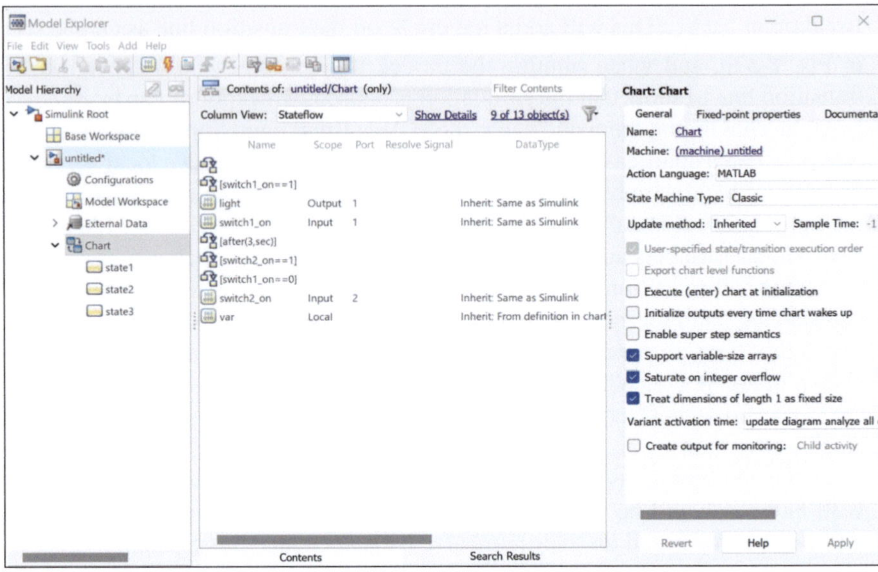


**Fig. 8.38** Model explorer of a chart in stateflow

iii. **Pre-built Component Libraries**: Simscape comes with extensive libraries of predefined components for various physical domains, which can be used to build models quickly without needing to define equations manually.

iv. **Equation-based Modeling**: Users can also define custom components by specifying the underlying equations, providing flexibility to simulate unique systems or behaviors that are not covered by the pre-built components.

v. **Integration with Simulink**: Simscape integrates seamlessly with Simulink, allowing for the inclusion of control systems, signal processing, and other dynamic system elements within the same simulation environment.

vi. **Real-Time Simulation and Code Generation**: Simscape models can be used for real-time simulation and can generate C code for deployment in hardware-in-the-loop (HIL) simulations or embedded systems.

To illustrate the capabilities of Simscape, we compare simulating a series RC circuit with a DC voltage source (a periodic pulse function) by monitoring the voltage across the capacitor and the current in the circuit using both Simulink and Simscape.

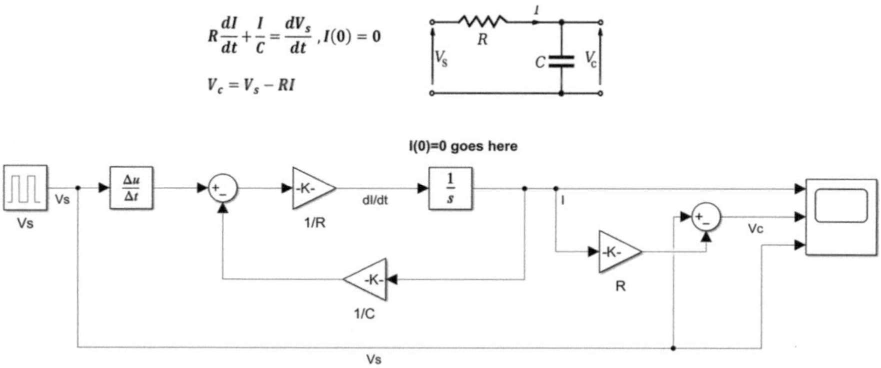


**Fig. 8.39** Modeling of a series RC circuit with a DC voltage source and measurements of the loop current and the voltage across the capacitor in Simulink

To use Simulink, we need to form the governing ODE of the system, which is shown on the top of Fig. 8.39. The physical representation of this system along with the Simulink model to solve this ODE can also be seen in this figure.

Not only do you need to know the voltage-current formulas for a resistor and a capacitor and be able to apply the KVL law to derive the governing ODE, but you also need to mathematically create that ODE in Simulink and form a separate signal for voltage across the capacitor to simulate the system's behavior.

Simscape allows you to achieve this goal without doing any of the tasks above by utilizing its built-in blocks for physical systems, which in this care are "controlled voltage source", "electrical reference", "resistor", "capacitor", "voltage sensor", and "current sensor" as shown in Fig. 8.40. These blocks can be accessed under "Simscape/Electrical" or "Simscape/Utility" and can be easily wired together to form a series RC circuit. Note that you need to place the voltmeter in parallel to the capacitor and the ammeter in series with the resistor.

It is important to note that Simscape blocks (called physical systems or PS) and Simulink blocks cannot connect to each other. Under the hood, vast amounts of optimizations, non-linear equations, and all sorts of other parameters are being tweaked to get Simscape to work as indented. So, to communicate to and from Simscape to Simulink, we must use specific blocks and connectors, which will handle the communications. For this case, we will use a "Simulink-PS converter" after the "pulse generator" block to connect it to the "controlled DC voltage source" block and two "PS-Simulink convertor" blocks to connect the voltmeter and ammeter measurements and pass them to the scope block.

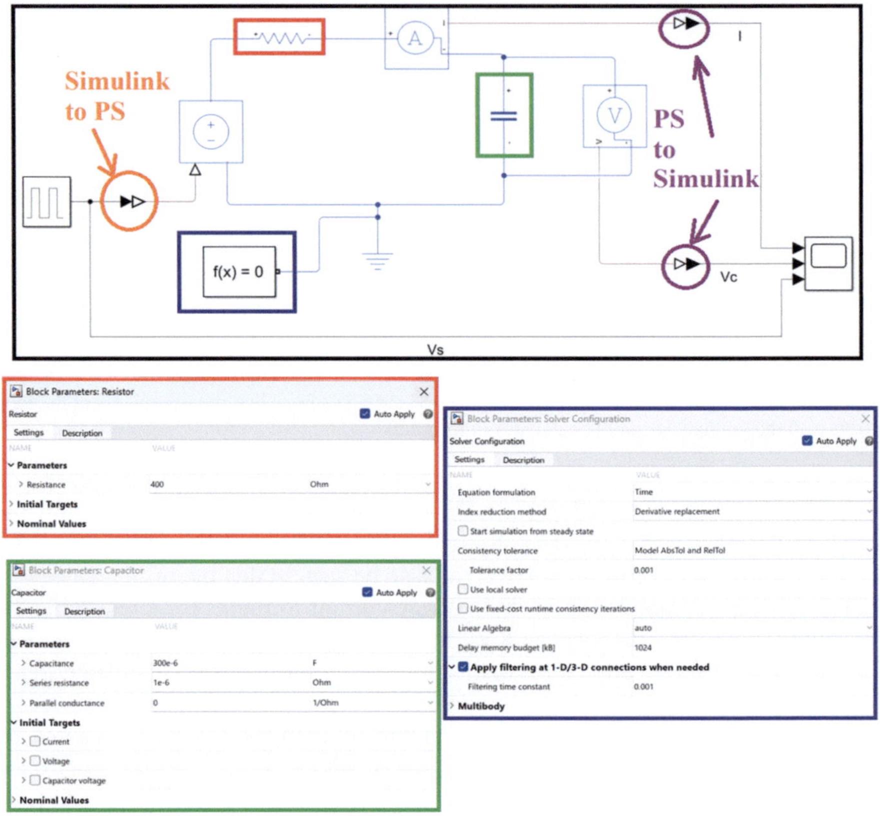


**Fig. 8.40** Modeling of a series RC circuit with a DC voltage source and measurements of the loop current and the voltage across the capacitor in Simscape

To set the values of capacitance and resistance, we can simply double-click on these blocks and provide the numbers as shown in Fig. 8.40. Initial conditions are also provided via these blocks.

The last block that must always be used when using Simscape is the "solver configuration" block, which must be connected to a point on the circuit. Unless necessary, we typically do not change the values and parameters inside this block.

Simulation results of the system in Fig. 8.40 are provided in Fig. 8.41, where charging and discharging of the capacitor can be observed.

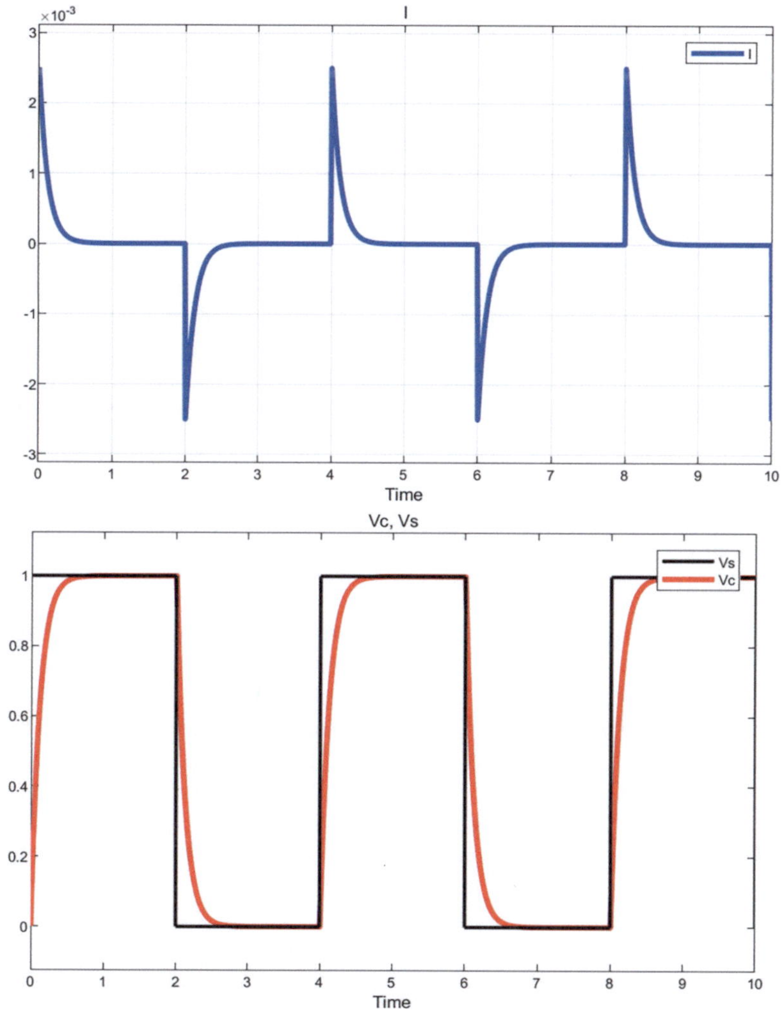


**Fig. 8.41** Simulation results for the series RC circuit with a pulse as DC voltage source: current versus time (top); capacitor and source voltages versus time (bottom)

For our second example, we will explore the multi-domain simulation capabilities of Simscape. Here, we will use Simscape to simulate the response (speed) of a DC motor to a Pulse Width Modulation (PWM) signal for different duty cycles. We also allow the armature resistance to vary with the operating temperature.

To make the system shown in Fig. 8.42, we need the following blocks: "DC motor", "Pulse Generator", "Controlled Voltage Source", "Temperature Source", "Solver Configuration", "Ideal Rotational Motion Sensor", "PS-Simulink Converter", "Electrical Reference", "MRRef" (mechanical reference), "Terminator", "Simulink-PS Convertor", "Gain", and "Scope".

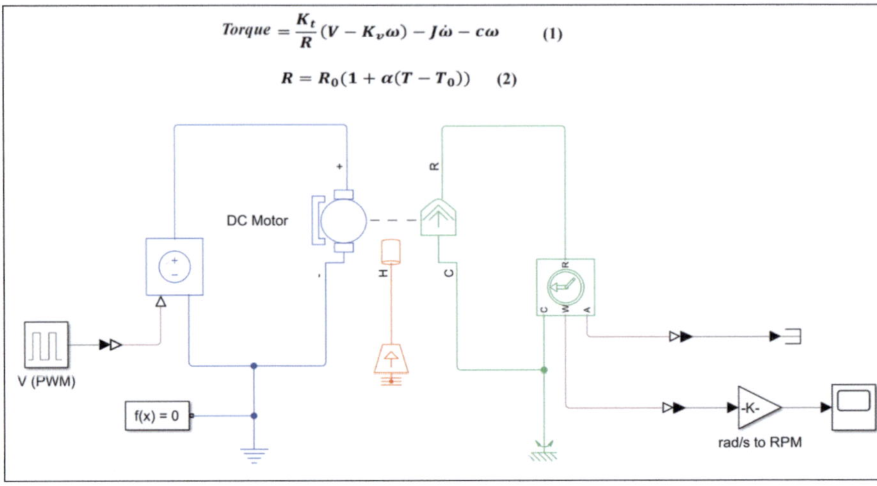


**Fig. 8.42** Simulation of RPM control for a DC motor with temperature-depending armature resistance using PWM voltage source

Although governing equations are not required for building this model, they are provided at the top of Fig. 8.42 so you know the effect of each parameter for DC motor block shown in Fig. 8.43 on the system's performance. In the first equation of Fig. 8.42, **V** is the source voltage, $\omega$ is the angular velocity (here converted to RPM using the gain block), **J** is the rotor inertia, **c** is the rotor damping factor, $\mathbf{K_v}$ is the back-emf constant, $\mathbf{K_t}$ is the torque constant, and **R** is the armature resistance. The second equation of Fig. 8.42 shows how **R** depends on the temperature, where $\mathbf{R_0}$ is the initial armature resistance at temperature $\mathbf{T_0}$, $\alpha$ is the rate of change of resistance per temperature change, and **T** is the operating temperature.

Note that the color of each domain is different in Fig. 8.42: blue for electrical, green for mechanical, and orange for thermal systems.

Also note that the "Ideal Rotational Motion Sensor" block can measure angular position, velocity, and acceleration of an inertia, where in this case only velocity is needed, so acceleration is terminated.

Now we can change the duty cycle in the "pulse generator" block (pulse width % parameter) and the operating temperature and investigate their effects on motor steady-state RPM. Figure 8.44 shows steady-state RPM ($\omega$) versus duty cycle (**pw**) as it goes from 15 to 90% and operating temperature as it goes from 25 to 225 degrees Celsius. This surf plot is generated using the following MATLAB code (we need to add a "to workspace" block for the RPM to the diagram in Fig. 8.42 and add two parameters **pw** and **T_op** in the "Pulse Generator" and "Temperature Source" blocks, respectively):

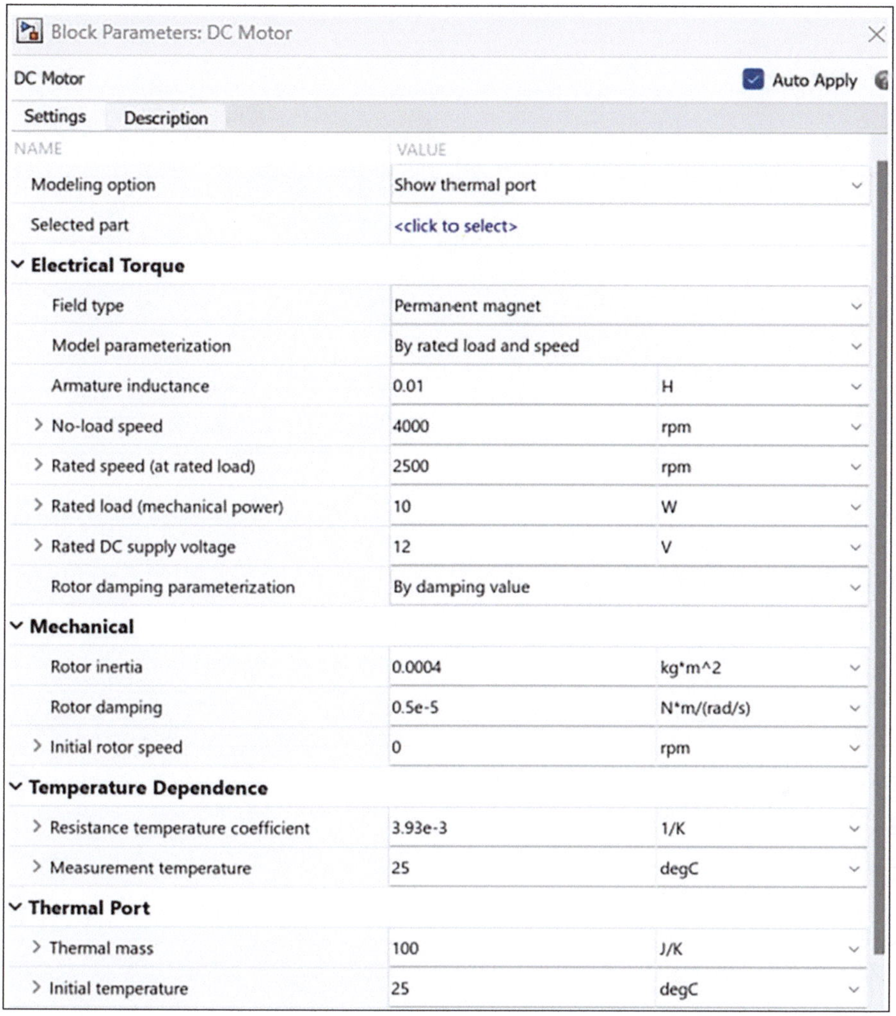


**Fig. 8.43** Block parameters of the DC motor block

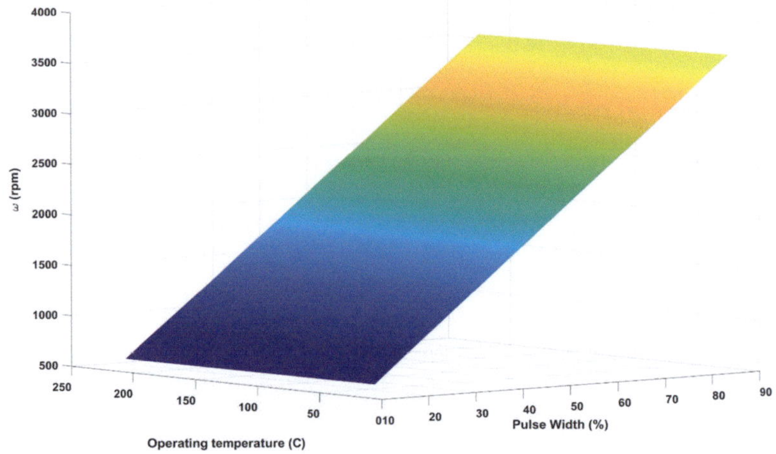


**Fig. 8.44** Steady-state RPM of DC motor shaft versus pulse width of the source voltage and the operating temperature

```
[PW,T]=meshgrid(15:15:90,25:40:225);
for ii=1:size(PW,1)
    for jj=1:size(PW,2)
        pw=PW(ii,jj);
        T_op=T(ii,jj);
        sim('DC_motor_PWM_RPM_monitor.slx');
        RPM(ii,jj)=w(end);
        clear w;
    end
end
surf(PW,T,RPM);
shading interp;
xlabel('Pulse Width (%)');
ylabel('Operating temperature (C)');
zlabel('\omega (rpm)');
view(-40,7)
```

As can be seen, steady-state RPM increases with the duty cycle, but it slightly decreases due to the rise in operating temperature.

For our last example, we will look at a **Simscape Multibody**. We will construct a pendulum with its support and animate the motion for a constant applied torque (here, 2 Nm). The signals to be monitored are the pendulum angular position and the pin reaction forces.

Figure 8.45 shows the diagram for this system, where the pendulum and support subsystems are made up of two "Solid" blocks connected by a "Rigid Transform" block (depicted in the red and green boxes).

To make this system, we need the following blocks: "Solid", "Cylindrical Solid", "World Frame", "Rigid Transform", "Revolute Joint", "Mechanism Configuration", "Subsystem", "Constant", "Solver Configuration", "Scope", and "Terminator".

The red color around the blocks in the pendulum subsystem means they require external parameters that are not provided numerically. This can be seen in Fig. 8.46, which shows the parameter windows for all of the blocks involved in this diagram (parameters inside red boxes).

The solid block could be an extrusion of a profile (Extrusion Type: General) that can be provided by a variable defining the vertices of the profile, such as the **csMain**

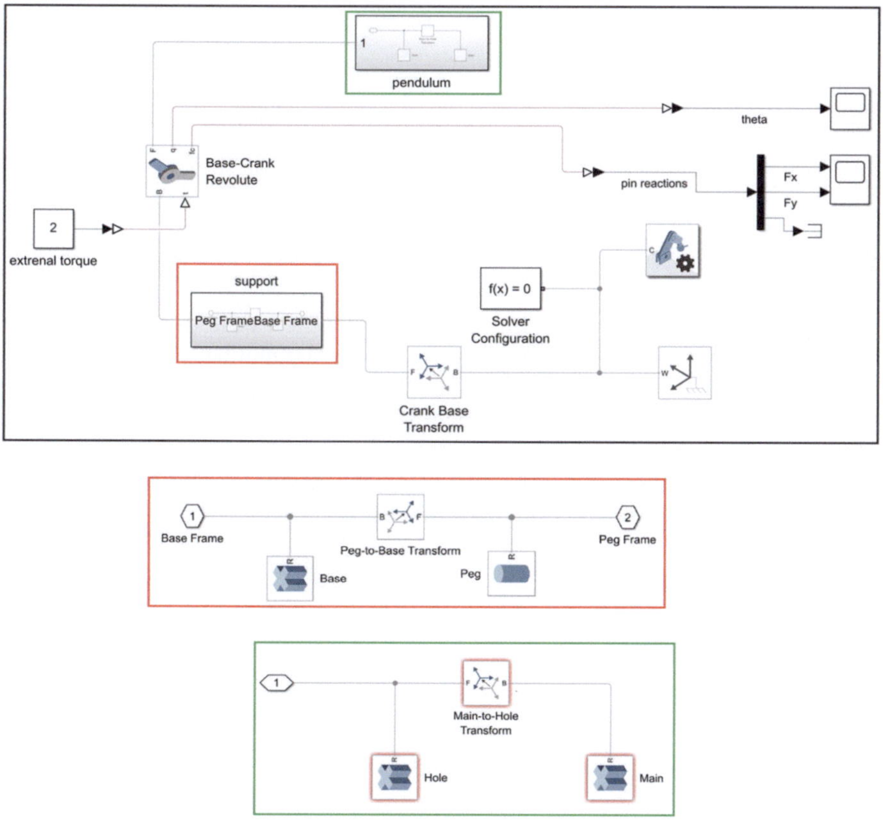


**Fig. 8.45** Simscape multibody system for a pendulum

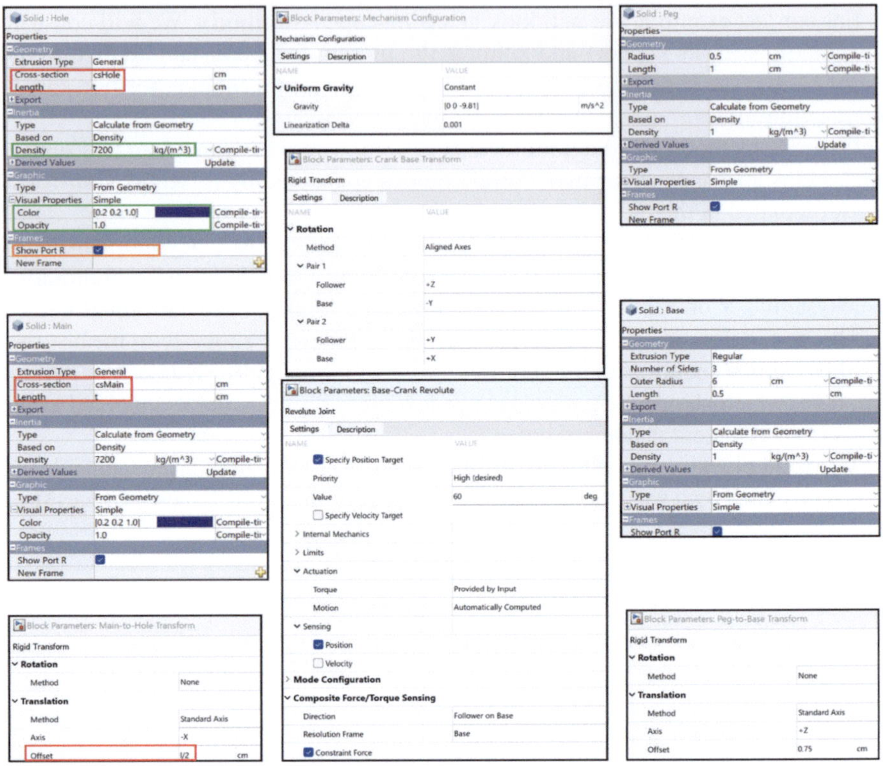


**Fig. 8.46** Parameters for blocks in the pendulum multibody system

variable in the "pendulum/main" block, or extrusion of a well-known profile such as a polygon (Extrusion Type: Regular) defined by the number and dimension of the sides, such as the "support/Base" block. There are other **Body Elements** under "Simscape/Multibody", such as **Cylindrical Solid** like the **pendulum/peg**, **Revolved Solid**, and**Spherical Solid**.

For this diagram, we need to define the parameters needed for the "Solid" blocks' extrusions, which can be done using the code below:

```
% Body Geometry Parameters
l = 20;    % Hole-to-hole distance
w = 2;     % Link width
d = 1.2;   % Hole diameter
t = 1;     % Link thickness

% Main Solid Cross-section:
A = linspace(-pi/2,pi/2)';
B = linspace(pi/2,-pi/2)';
csRight = [l/2+w/2*cos(A) w/2*sin(A)];
csLeft = [-l/2 w/2; -l/2 + d/2*cos(B) d/2*sin(B); -l/2 -w/2];
csMain = [csRight; csLeft];

% Hole Solid Cross-section:
C = linspace(pi/2,3*pi/2)';
D = linspace(3*pi/2,pi/2)';
csHole = [w/2*cos(C) w/2*sin(C);
d/2*cos(D) d/2*sin(D)];
```

To access these parameters, we can click on "Modeling/Model Explorer" and then double-click on the "Base Workspace" from under the "Simulink Root" in the left window.

Gravity must be applied in the "Mechanism Configuration" block.

The two "Rigid Transform" blocks inside the subsystems are provided with rotations and translation offsets to assemble the **Peg** on the **Base** for the "Support" subsystem and the **Hole** with the **Main** for the "Pendulum" subsystem. The "Rigid Transform" block between the **World** frame and the "Support" subsystem defined the orientation and location of the support with respect to the world frame.

The "Revolute Joint" block defines the initial conditions (here, initial angle of 60 degrees) as well as the measurements (here, pendulum angular position and pin reaction forces).

Figure 8.47 shows a snapshot of this system during animation as well as the graphics for individual solid blocks. Note that to make the pendulum part, we had to use a main "Solid" block and a hole "Solid" block shown at the bottom of Fig. 8.47 as there is no negative solid object in Simscape.

One major advantage of Simscape multibody is animation of the physical system from which a video could also be made. After running the model, go to "Tools/ Video Creator" and set up the video parameters, such as **Frame Rate** and **Video Format**, and click on "Create". Figure 8.48 shows three frames of this animation.

While it is possible to make more complicated physical models using the "Multi-Body" blocks, it can be a daunting task if the system to be modeled is very complicated. In such cases, we can design the physical system in **SolidWorks** or **Pro-Engineer** CAD software and then export them to Simscape using an add-in called "Simscape Linker".

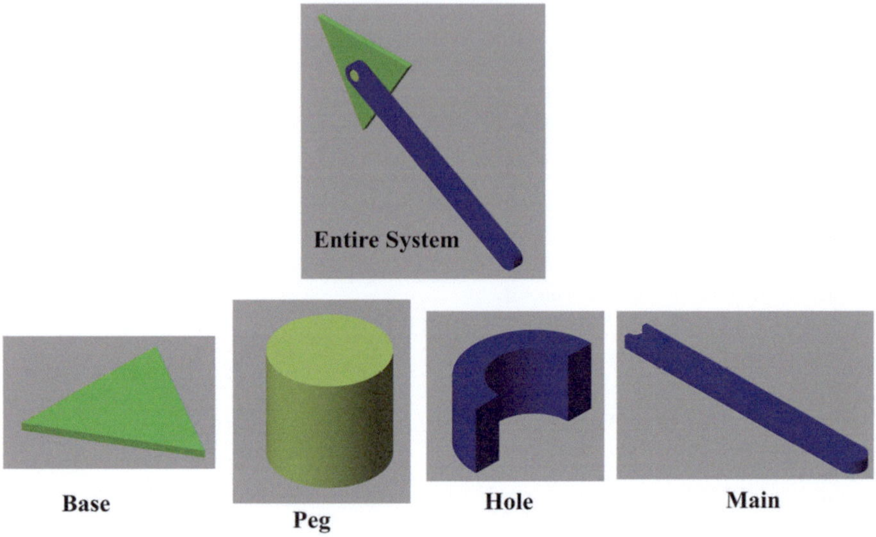


**Fig. 8.47** Graphics of the entire system and individual blocks for the pendulum and support

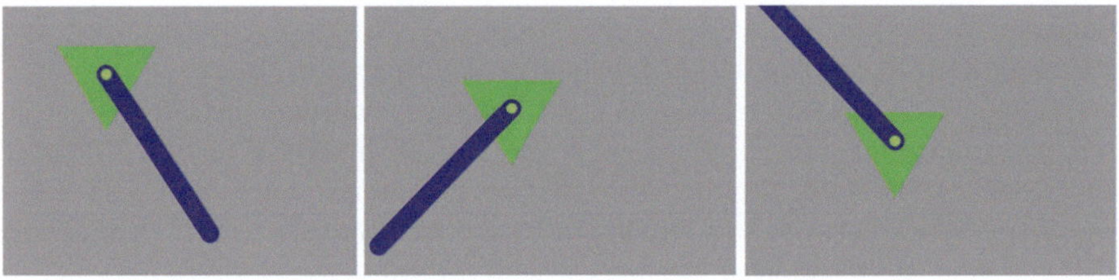

**Fig. 8.48** Three frames of animation for the pendulum

To install this linker and use it, please refer to the following link:
https://www.mathworks.com/help/smlink/ug/installing-and-linking-simmechanics-link-software.html.

## 8.8 S-Functions

**S-functions (System Functions)** are a powerful feature in MATLAB Simulink that allow users to create custom blocks for simulation models. They are **user-defined functions** that can be written in MATLAB, C, C++, or Fortran, and they provide a mechanism to extend Simulink's built-in capabilities. S-functions allow for the implementation of custom algorithms and the integration of code that may not be directly supported by Simulink's standard block libraries.

Applications of S-Functions are as follows:

i. **Custom Algorithms**: S-functions are used to implement complex algorithms that are not available in Simulink's built-in blockset. This includes advanced mathematical computations, control algorithms, or signal processing techniques.
ii. **Legacy Code Integration**: S-functions allow the integration of existing code written in C, C ++, or Fortran into a Simulink model. This is particularly useful in cases where existing code needs to be reused in a simulation environment.
iii. **Interfacing with External Hardware**: S-functions can be used to interface Simulink models with external hardware, such as sensors or actuators. This is often used in real-time simulation or hardware-in-the-loop (HIL) testing.
iv. **Custom Block Development**: Developers can create custom blocks that behave like built-in Simulink blocks but are tailored to specific needs, such as implementing custom signal routing or data management logic.
v. **Optimization and Control**: S-functions can be employed in optimization tasks where specific custom optimization routines are needed, or in control systems where custom controllers are implemented.
vi. **Simulation of Complex Systems**: For simulating systems with dynamics that are difficult to model using standard blocks, S-functions provide the flexibility to define the exact behavior required.

There are different methods for creating S-Functions. Here, we will look at the Legacy Code Tool (LCT) and the "S-Function Builder" block in Simulink.

To use LCT, first you need to create a **source file** and a **header file**. Let's say we want to add an S-Function that calculates the ratio of sum of two numbers and the difference between them.

The main file includes the header file and the computational operations:

```
#include "ratio.h"
double myratio(double a, double b)
{
double c;
c=(a+b)/(a-b);
return(c);
}
```

The header file will include the **function signature**:

```
#ifndef RATIO_H_INCLUDED
#define RATIO_H_INCLUDED
double myratio(double a,double b);
#endif // RATIO_H_INCLUDED
```

Set up the mex compiler using the following command:

```
mex -setup
```

Then create a structure for the S-Function:

```
def=legacy_code('initialize')
```

The resulting structure will be shown in the command window:

```
def = struct with fields:
                 SFunctionName: ''
InitializeConditionsFcnSpec: ''
                 OutputFcnSpec: ''
                  StartFcnSpec: ''
              TerminateFcnSpec: ''
```

```
                   HeaderFiles: {}
                   SourceFiles: {}
                  HostLibFiles: {}
                TargetLibFiles: {}
                      IncPaths: {}
                      SrcPaths: {}
                      LibPaths: {}
                    SampleTime: 'inherited'
                       Options: [1×1 struct]
```

Now, it is time to provide the function name, inputs and outputs, the name of the source file, and the name of the header file:

```
% define the S function name
def.SFunctionName='my_s_fun';

% define the S function specs (inputs and outputs)
def.OutputFcnSpec='double y1 = myratio(double u1, double u2)';

% define the source file
def.SourceFiles={'main.c'};

% define the header file
def.HeaderFiles={'ratio.h'};
```

Finally, we generate the C code for the S-Function and compile it to a mex file:

```
% generate the c code for S function
legacy_code('sfcn_cmex_generate',def)
```

```
% compile s function from c file into a binary
legacy_code('compile',def)
```

Now, we are ready to create a Simulink block for this operation. We add an "S-Function" block and enter the name of the S-Function that we just created. Once we click on the "Apply" button, it will create two input ports for the Simulink block as specified by the function signature. Figure 8.49 shows the S-Function block and the Simulink testing of this block.

The second method is to use the "S-Function Builder" block in Simulink. Once you insert and double-click on the block, you will see that it uses the same structure of **.cpp** file as we used in LCT, but all you need to provide are the types of inputs, types of outputs, and the computations.

You can right-click on the bottom window titled "Ports and Parameters", add inputs and outputs and provide their scopes (input or output), types, dimensions, and complexities (e.g., real versus complex). You also can change their names.

Figure 8.50 shows a block in a test Simulink file that calculates the radii of four circles in a spur gear (pitch circle, base circle, addendum circle, and dedendum circle) given its modulus, number of teeth, and pressure angle.

Figure 8.51 shows the code for this block, where the custom code is depicted in red boxes. Note that the inputs and outputs are all written as **pointers** (* behind the name of variables).

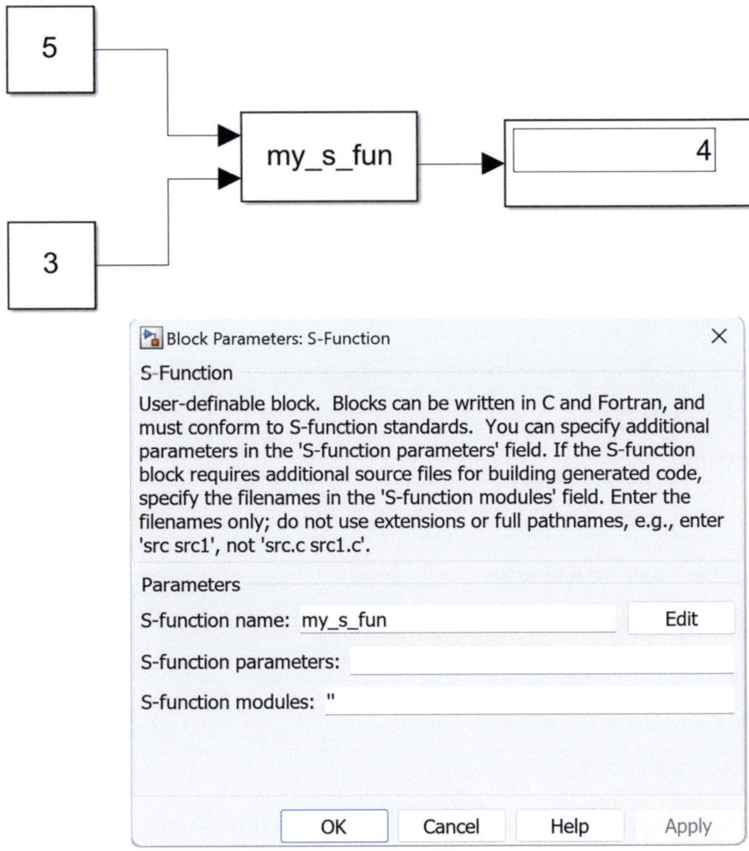


**Fig. 8.49** S-Function block created in Simulink using LCT

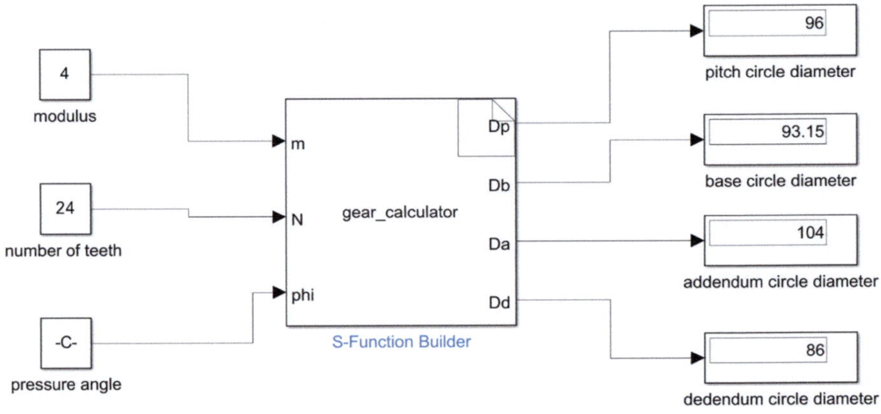


**Fig. 8.50** S-Function builder block used to calculate radii of four circles in a spur gear

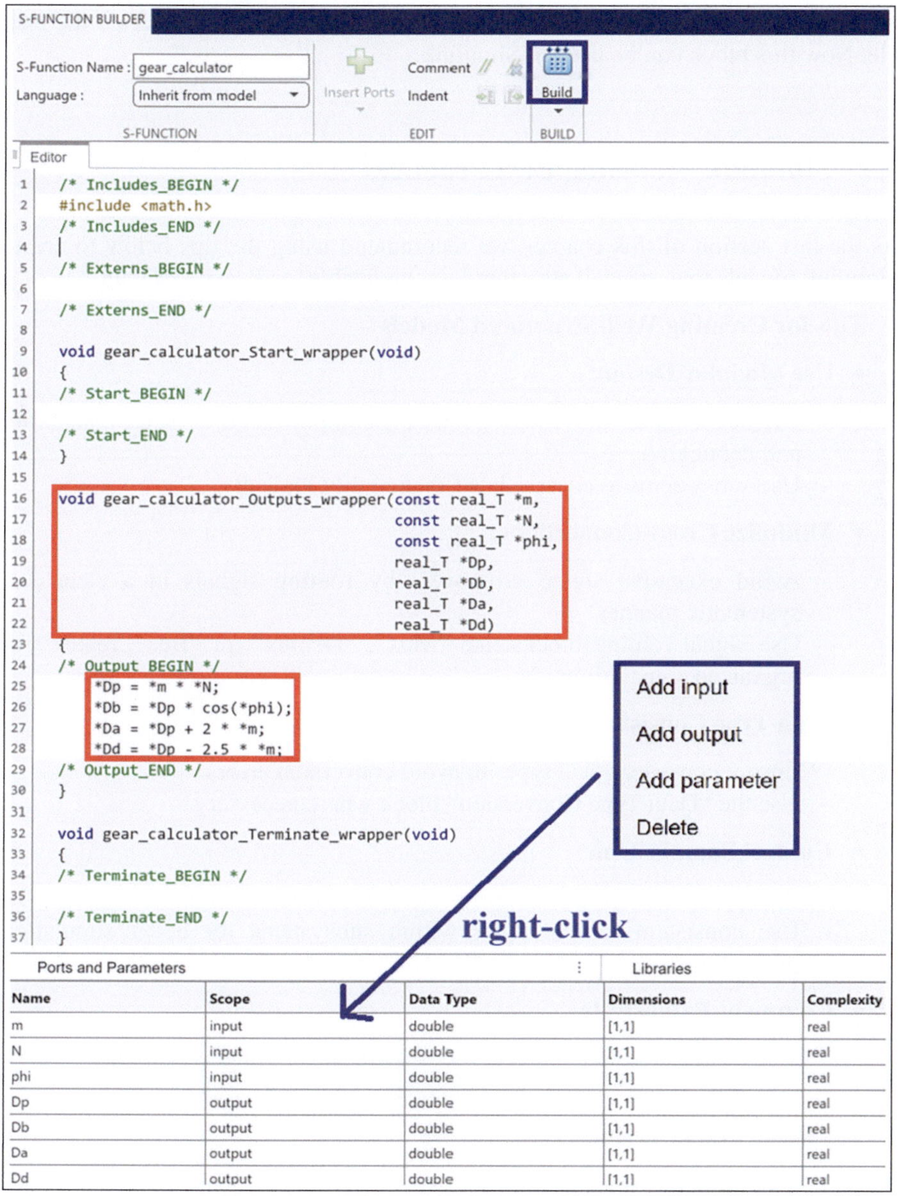


**Fig. 8.51** Code for the S-Function builder block used in Fig. 8.50 with added parts depicted in red boxes

When the custom code is added, click on the "Build" button to create the mex file. Now this block can be used in Simulink.

## 8.9 Simulink Modeling Best Practices

As the last section of this chapter, we recommend using the tips below to create Simulink models that are well structured, maintainable, and easy to debug.

i. **Tips for Creating Well-Structured Models**

- **Use Modular Design:**
  - Divide the model into smaller, reusable components to enhance readability and debugging.
  - Use subsystems to encapsulate functionality logically.
- **Minimize Cross-Connections:**
  - Avoid excessive signal crossovers by routing signals in a clear and systematic manner.
  - Use signal routing blocks like "Mux", "Demux", or "Bus Creator" for organized signal grouping.
- **Data Type Consistency:**
  - Ensure consistent data types to avoid conversion errors.
  - Use the "Data Type Conversion" block where necessary.
- **Control Sample Time:**
  - Set appropriate sample times for discrete blocks and signals.
  - Use consistent sample times within subsystems for better simulation performance.
- **Comment Extensively:**
  - Add descriptive comments to blocks, subsystems, and signal lines.
  - Use annotations to describe the model's high-level logic.
- **Enable Model Verification:**
  - Use Model Verification blocks like "Assertion" to validate signal ranges and conditions during simulation.
- **Test Thoroughly:**
  - Simulate under various conditions and edge cases to ensure robustness.
  - Use "Test Harnesses" to test subsystems independently.

ii. **Naming Conventions and Model Hierarchy**

- **Consistent Naming:**
  - Use meaningful, descriptive names for signals, blocks, and subsystems.
  - Use camelCase or snake_case for variable names (e.g., **engineRPM**, **vehicle_speed**).
- **Prefix Signals:**
  - Prefix signal names with their purpose or type (e.g., **input_**, **output_**).
  - Example: **input_throttle**, **output_torque**.
- **Subsystem Naming:**
  - Name subsystems based on their functionality (e.g., **SpeedController**, **GearLogic**).
- **Avoid Generic Names:**
  - Avoid names like **Subsystem1** or **Signal1**.
- **Organized Model Hierarchy:**
  - Group related blocks into logical subsystems.
  - Create a hierarchical structure for large models to reduce clutter and enhance clarity.

iii. **Grouping and Using Subsystems**

- Encapsulation:
  - Use subsystems to encapsulate functionality logically. This makes the model easier to understand and debug.
  - Examples:

    A Throttle Control subsystem for throttle-related logic.
    A Transmission Control subsystem for gear selection.
- **Parameterization:**
  - Use subsystem parameters that can be easily modified. Implement masks for subsystems to expose key parameters.
- **Avoid Deep Nesting:**
  - Limit the depth of subsystems. Deeply nested subsystems can make debugging challenging.
- **Reusable Subsystems:**
  - Convert frequently used subsystems into libraries or reference models to ensure consistency.
- **Types of Subsystems:**
  - Use appropriate types based on functionality:

**Enabled Subsystems:** Execute when a condition is met.
**Triggered Subsystems**: Execute on specific events.
**For Iterator Subsystems**: Loop through input data.
**Switch-Case Subsystems**: Implement decision-making logic.

- **Clarity in Signal Grouping:**
  - Use "Inport" and "Outport" blocks for clear signal entry and exit points in subsystems.
  - Label input and output signals inside subsystems.

**Practice Problems**

1. Create a Simulink model to simulate the dynamics of a nonlinear oscillatory system governed by the Duffing equation using basic Simulink blocks and show the response in a "Scope" block. The Duffing equation is a second-order nonlinear differential equation that describes the motion of a damped oscillator (mass) with a nonlinear stiffness term (nonlinear spring) exposed to an external force. More specifically, use the following governing ODE with the parameters given below:

$$m\ddot{x} + \delta\dot{x} + \alpha x + \beta x^3 = \gamma \cos(\omega t) \; s.t. \; x(0) = 0.1m, \dot{x}(0) = 0\frac{m}{s} \tag{8.6}$$

   Mass (**m**): 1 kg.

   Damping Coefficient (**δ**): 0.1 Ns/m.

   Linear Stiffness (**α**): −1 N/m.

   Nonlinear Stiffness (**β**): 1 N/m$^3$

   Forcing Amplitude (**γ**): 0.5 N.

   Forcing Frequency (**ω**): 1.2 rad/s.

   Final Simulation Time ($\mathbf{t_f}$): 100 s.

2. Redo problem 1 but this time use a script to change the parameter **ω** from 1 to 1.5 to 10, run the Simulink model that exports the x values to workspace, and plot the *x* values versus time for all **ω**'s in the same plot with a legend. Place the blocks for the oscillating system inside a "Subsystem" block.
3. Redo problem 2 but using a "MATLAB Function" block instead of the subsystem that implements the governing ODE. Compared to the code in problem 2, which code runs faster? Why?
4. Create a virtual reality environment for the system in problem 1 and animate it in **Simulink 3D Animation**. Change the initial displacement to 20.
5. Create the system in problem 1 with $\alpha = 1, \beta = 0$ using Simscape blocks under the "Foundation Library/Mechanical" and animate it.
6. Create the system in problem 1 with $\alpha = 1, \beta = 0$ using Simscape Multibody blocks and animate it.

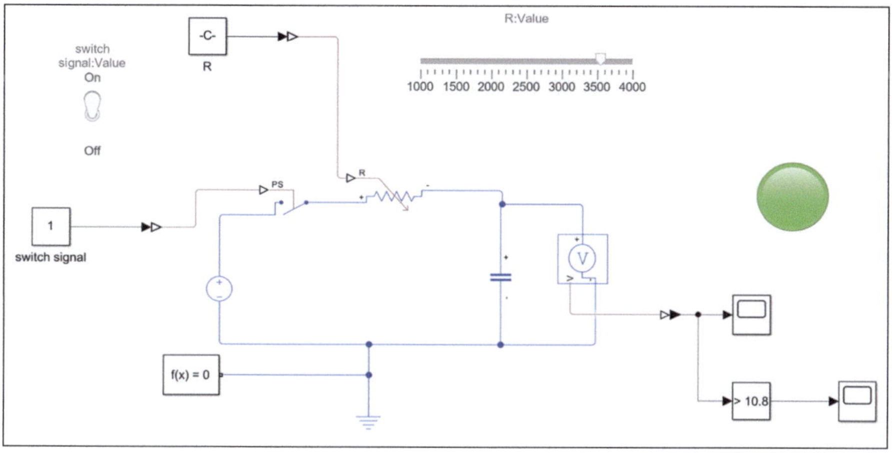


**Fig. 8.52** The series RC circuit with dashboard blocks

7. Change the subsystem in problem 2 to a **mask** that receives and sets the values of $\delta$, $\alpha$, $and \beta$. Add an appropriate image for a mass-spring-damper with external force to the mask. Run the Simulink file directly and without the need for a script and plot x on a "Scope" block.
8. Create a series RC circuit with a switch, a variable resistor (between 100 and 400 ohms), a fixed capacitor of 5000 $\mu F$, and a fixed DC voltage source of 12 V similar to Fig. 8.52 using the Simscape blocks. The switching is controlled by a "toggle switch" block from "Dashboard" and the value of **R** is controlled by "variable slider" block from "Dashboard". Include a "lamp" block that turns on with a green color when voltage across the capacitor reaches 90% of its max steady-state value and turns off with a red color otherwise. Use gray color for an undefined state in the "lamp" block.
9. Create a **Stateflow** diagram for controlling the motors of a 2-DOF robotic arm similar to Fig. 8.53. You don't need to include any kinetic or kinematic equations for the robot in this diagram, and the signals coming out of the stateflow chart will only have values of 1 or 0 to indicate whether the motor is on or off. The subsystems "motor1" and "motor 2" multiply this signal by a gain that represents the steady-state angular velocity of the motor and then integrate the result to get the angular position for the motor.

The motor logic control is summarized below:

- If the robot switch is on and **theta1 < 1.2** rad, motor 1 must be turned on.
- If **theta1 > 0.8** rad, turn motor 2 on.
- When **theta1 >= 1.2** rad, turn motor 1 off.
- When **theta2 >= 0.8** rad, turn motor 2 off.
- Whenever the robot is switched off during operation, turn both motors off.

Use a graphical function that receives **theta2** and returns a value of 0 or 1 to help with the motor logic control.

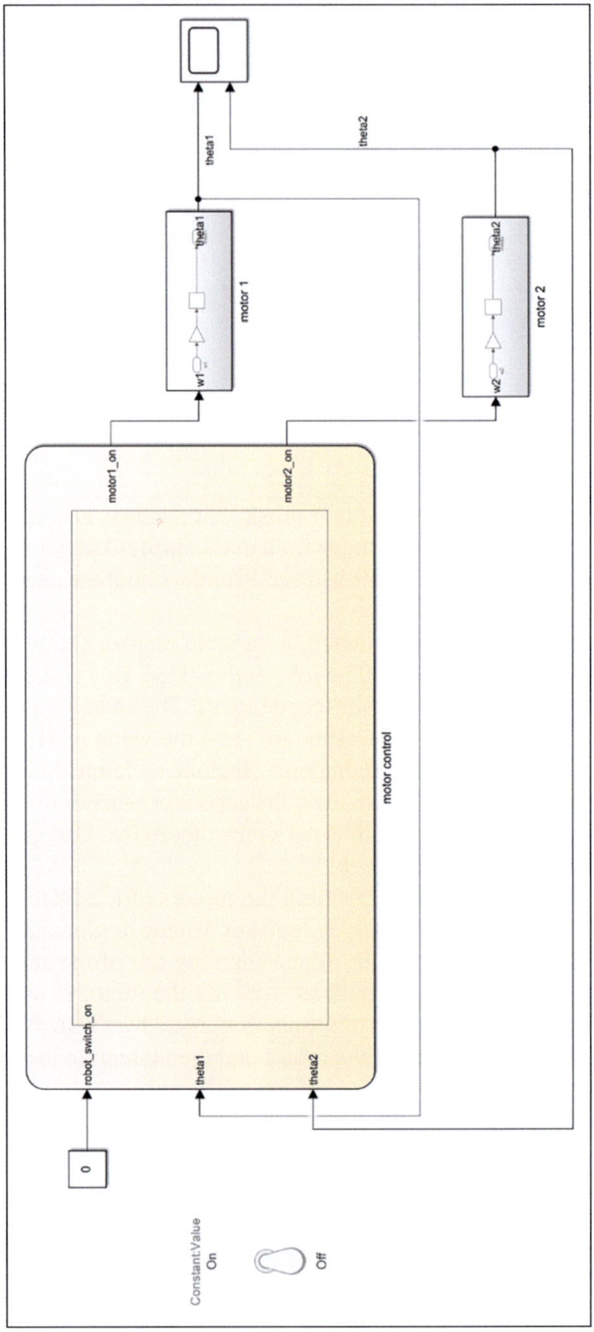


**Fig. 8.53** Motor logic control for a 2-DOF robot using Stateflow

10. You are using **Stateflow** to design a logic control for a system that has four states {1, 2, 3, 4}. The inputs to the logic control are an on/off signal from a toggle switch and a variable value from a "slider" dashboard block. The outputs of the stateflow chart are the **system_state** value connected to a display block, a related value called **LED_light** (**LED_light = system_state - 1**) that is connected to a "Scope" block, and a "lamp" block. Colors of the lamp corresponding to the **LED_light** values of 0, 1, 2, and 3 are red, orange, blue, and green, respectively. The color gray corresponds to an undefined value.

The logic and details of the system is summarized below:

- The default state is state 4.
- If the toggle switch is turned on, the system will go to state 3.
- Three seconds after switching to state 3, we will make a decision using a **junction** with two possible outcomes: if the output of a **Simulink Function** given the input from the slider is bigger than 15 (**output = 20*input + 10**), then switch to state 1; otherwise, switch to state 2.
- If the toggle switch is turned off, move from either state 1 or state 2 to state 4.
- When moving to each state, set the value of **LED_light** to the appropriate value mentioned above upon entry to that state.

The general form for this diagram is shown in Fig. 8.54.

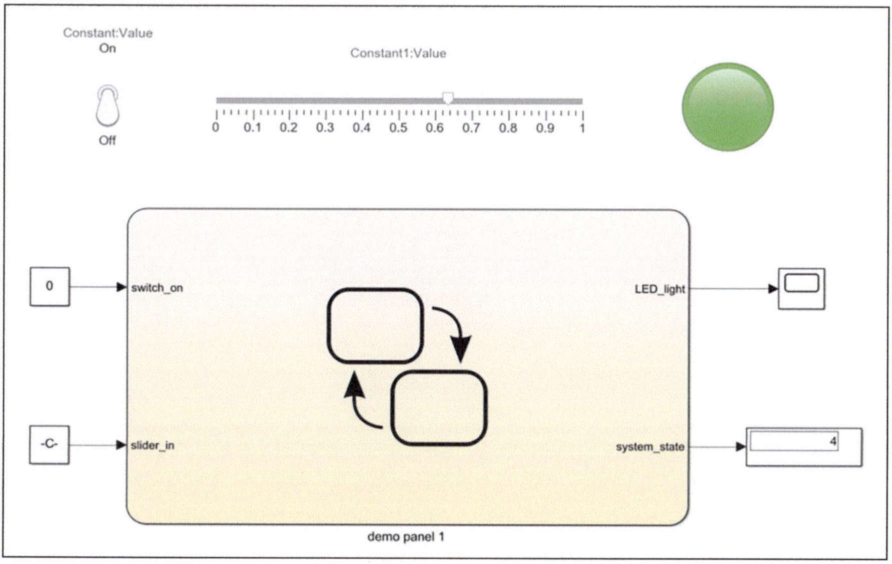


**Fig. 8.54** Logic control for problem 8.10

# Chapter 9
# Interfacing with Hardware and Other Software

## 9.1 Hardware Control and Real-Time Data Acquisition

### *9.1.1 Hardware Setup for MATLAB Interfacing*

MATLAB has many libraries and add-ons that allow it to interface with a wide variety of hardware, including Arduinos, Raspberry Pis, and Lego EV3s. This chapter will cover how to set up MATLAB to interface with some of these systems, how to use Simulink to handle real-time data acquisition, and how to link MATLAB with online databases.

The first step in interfacing with MATLAB is installing the correct add-ons. This process is straightforward and can be achieved by navigating to the Home tab, selecting the Add-Ons dropdown, and launching the Add-On Explorer by clicking Get Hardware Support Packages (Fig. 9.1).

Once the Add-On Explorer is open, scroll or search for the add-on for the desired hardware and click install. Keep in mind that many of the hardware options have two separate add-ons, one for base MATLAB and one for Simulink capability. For some devices, such as Arduinos, the installation of add-ons in MATLAB will require additional hardware setup. Typically, additional explanations will be provided in the installation window.

### *9.1.2 Arduino Interfacing*

One of the systems MATLAB is prepared to interface with is Arduino boards. When installing the Arduino add-ons, MATLAB will prompt the user to connect the Arduino device so that programs can be installed directly on it. During this process, MATLAB will identify the port the Arduino is connected to, but the port can also be found later using the Ports tab of Device Manager (if Windows is being used).



M. Andalibi and S. A. Cherkauer, *Advanced MATLAB for Engineers*,
https://doi.org/10.1007/978-3-032-10677-3_9

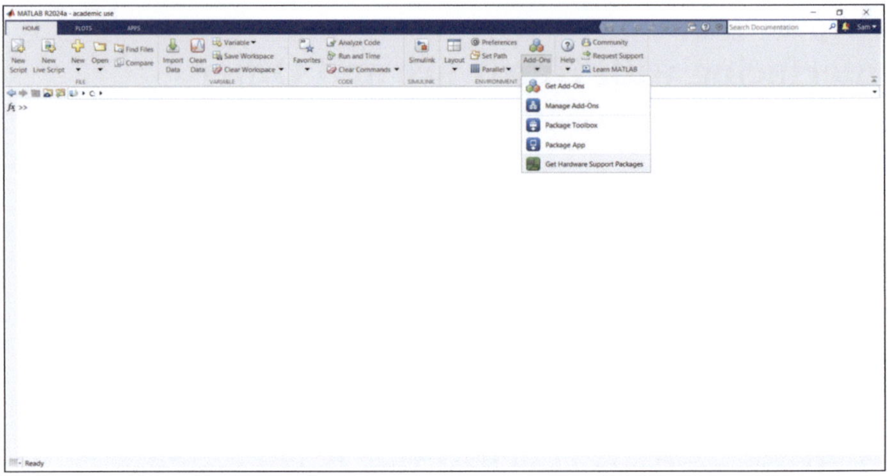


**Fig. 9.1** Installing MATLAB's hardware add-ons

Once the Arduino add-ons are installed and the device is set up, an **Arduino** object must be defined and given inputs of the port and the device type as strings. In this case, an Arduino Uno was used and attached to the port named COM3. The **AvailablePins**, **AvailableDigitalPins**, **AvailablePWMPins**, and **AvailableAnalogPins** properties of the object are all matrices of pin numbers that can be useful later in the interfacing process.

```
a = arduino("COM3", "Uno")
a = arduino with properties:
Port: 'COM3'
Board: 'Uno'
AvailablePins: {'D2-D13', 'A0-A5'}
AvailableDigitalPins: {'D2-D13', 'A0-A5'}
AvailablePWMPins: {'D3', 'D5-D6', 'D9-D11'}
AvailableAnalogPins: {'A0-A5'}
AvailableI2CBusIDs: [0]
Libraries: {'I2C', 'SPI', 'Servo'}
```

Next, pin modes need to be set to correctly read sensor data and provide outputs. MATLAB will automatically set initial pin modes, but further mode changes must be made manually using the **configurePin** function. This function takes the **arduino** object, a string containing the pin name, and a string containing the desired pin mode as inputs and sets the mode of the specified pin. If only the first two inputs are given, MATLAB returns the current mode of the specified pin. All the available pin modes are shown in Table 9.1. The function **configureDigitalPin** can be used similarly to set a digital pin to either **"input"** or **"output"**.

```
configurePin(a, "A0", "Pullup")
configurePin(a, "A0")
ans = 'Pullup'
```

To send or receive a simple 0 or 1 logical signal to or from a pin, the **writeDigitalPin** and **readDigitalPin** functions can be used. Similar to the previous commands, these functions take the **arduino** object and the pin number as inputs. The **writeDigitalPin** takes one additional input, the value that is used to set the pin, in the form of a logical 1 (on) or 0 (off).

The following code reads the initial output from pin **D2**, meaning the initial mode of the pin was automatically set to **'DigitalInput'**. Next, the code sets the mode to **'DigitalOutput'** and outputs a logical 1 out of pin **D2** for one second before once again reading the digital pin input. Since no voltage is going into the pin from the circuit, the digital pin reads 0 both times. When an LED is connected between pin **D2** and ground, as shown in Fig. 9.2, the LED turns on for one second and then shuts off once the code is executed.

```
readDigitalPin(a, "D2")
configurePin(a, "D2", 'DigitalOutput')
writeDigitalPin(a, "D2", 1)
pause(1)
configurePin(a, "D2", 'DigitalInput')
readDigitalPin(a, "D2")
                                        ans = 0
                                        ans = 0
```

For more refined control over pin outputs, **writePWMDutyCycle** and **writePWMVoltage** can be used. These two functions take similar inputs to **writeDigitalPin**, except their third inputs are the pin duty cycle and output voltage, respectively. The **readVoltage** command can be used like **readDigitalPin** to get the output voltage of the specified pin. When the same simple circuit from before (Fig. 9.2) is instead hooked up to the PWM pin **D3**, the following code steadily increases the length of

**Table 9.1** Arduino pin modes

| Pin mode | Description |
|---|---|
| "AnalogInput" | Specify a pin to acquire analog signals |
| "DigitalInput" | Specify a pin to acquire digital signals |
| "DigitalOutput" | Specify a pin to generate digital signals |
| "I2C" | Specify a pin to use with I2C protocol |
| "Interrupt" | Specify a pin in interrupt mode for rotaryEncoder |
| "Pullup" | Specify a pin to use a pullup switch |
| "PWM" | Specify a pin to use a pulse width modulator |
| "CAN" | Specify a pin to use a Controller Area Network bus |
| "Servo" | Specify a pin to use with a servo |
| "SPI" | Specify a pin to use with SPI protocol |
| "Tone" | Specify a pin to use for playTone |
| "Ultrasonic" | Specify a pin to use with an ultrasonic sensor |
| "Unset" | Specify a pin to remove its reserved mode |

Fig. 9.2 Wiring diagram for a digital pin LED

the duty cycle of the pin and then steadily decreases the output voltage of the pin to progressively increase and decrease the brightness of the LED.

```
for i = 1:1:20
   writePWMDutyCycle(a, "D3", i/20)
   pause(0.1)
end
for i = 1:1:20
   writePWMVoltage(a, "D3", 5 - 5*i/20)
   pause(0.1)
end
```

MATLAB also adds a class specifically for controlling servos. This **servo** class is initialized with the **arduino** object and the pin number for the PWM input of the servo. Once a servo object is created, it can be controlled using the **readPosition** and **writePosition** functions. The read function takes the **servo** object as an input and returns the current position of the servo between 0 and 1, where 1 is the maximum deflection angle of the servo. The write function takes the **servo** object and the desired position between 0 and 1 as inputs and moves the servo.

The code below controls a servo connected to the **D5** pin using a potentiometer connected to the **A0** pin (Fig. 9.3). The voltage returned from the potentiometer (between 0 and 5 V) is used to calculate a position between 0 and 1 that is then used to reposition the servo every 0.1 s.

```
s = servo(a, "D5");
while true
   V = readVoltage(a, "A0");
   writePosition(s, V/5)
   pause(0.1)
end
```

Due to the varying nature of real-time input from the potentiometer and other sensors, the servo in the previous example moves in inconsistent jerks. These jerks can be lessoned, and sensor inputs can be improved by controlling the Arduino with a real-time Simulink system. The Simulink tools for Arduino can be found in the Library under the name Simulink Support Package for Arduino Hardware once the appropriate add-on has been installed. Before running the program, the Simulink model must be set up to run on the correct hardware. Navigate to the MODELING

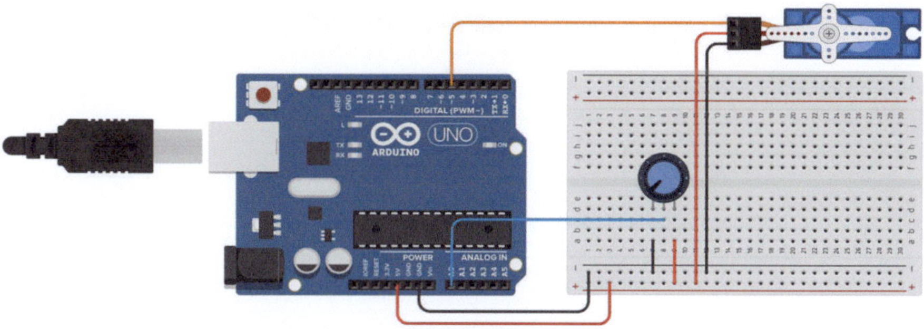


**Fig. 9.3** Wiring diagram for a potentiometer and servo

tab, click Model Settings, and then choose the correct hardware from the dropdown on the Hardware Implementation tab of the Configuration Parameters window (Fig. 9.4).

Applying this change will cause a new Hardware tab to appear at the top of the screen. Once the Simulink model is completed, set the stop time to **inf** and click Build, Deploy, and Start. This will install the model onto the Arduino board, which will then begin running the program. The Arduino can then be disconnected from the computer at this time. An example Simulink program for an Arduino Uno is shown in Fig. 9.5.

In the Simulink model shown in Fig. 9.5, a 5-V potentiometer connected to analog pin **A0** is controlling two servos. The Arduino Simulink analog input block returns the pin voltage as a **uint16** between 0 and 1023 for the Arduino Uno board, so this

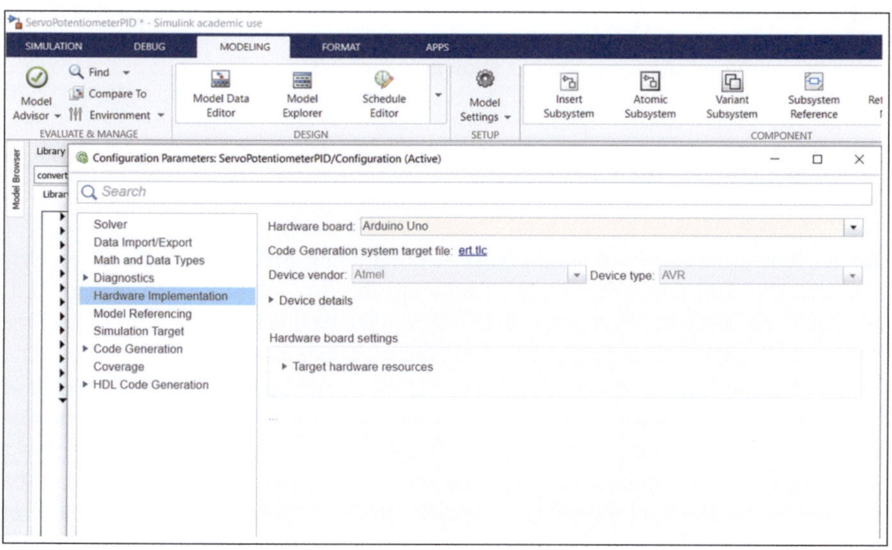


**Fig. 9.4** Simulink real-time hardware setup

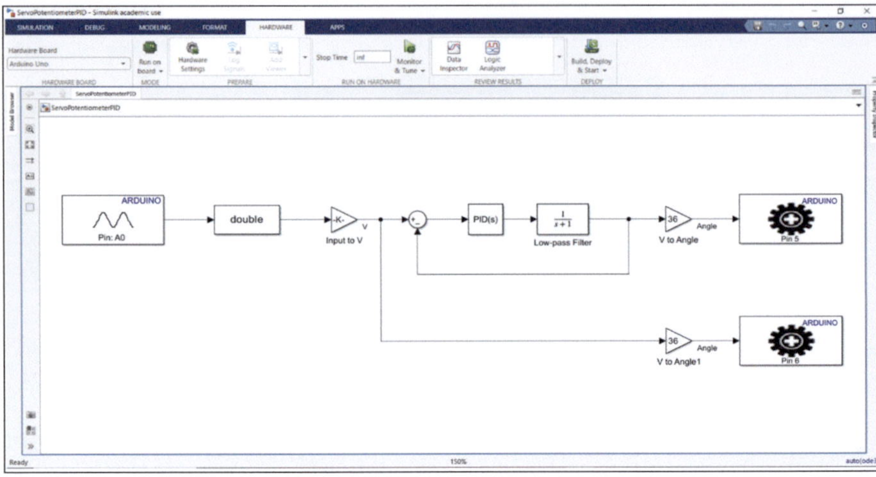


**Fig. 9.5** Arduino simulink model

model first converts the input to a double and then converts the value to a number between 0 and 5 by dividing by 1023 and multiplying by 5. Next, the voltage is converted to an angle in degrees between 0 and 180 for the servos by multiplying by 36. As shown, the servo connected to PWM pin **D6** is given a direct input by the potentiometer, but the servo connected to **D5** has a low-pass filter applied to the voltage. The PID block is set up as a PI controller, where $K_P$ is 5, $K_I$ is 5, and $K_D$ is 0. Notice that when running, the low-pass filter removes some of the input noise, so the servo connected to **D6** responds faster to potentiometer changes but also jitters more than the servo at **D5** when the potentiometer remains unchanged. Each of the servos are wired up the same way as shown in Fig. 9.3.

### *9.1.3 Raspberry Pi Interfacing*

Another hardware system that can be controlled using MATLAB is the Raspberry Pi, where the Add-Ons are installed following the process discussed in Sect. 1.1. MATLAB can connect to a Raspberry Pi through the **raspi** command using either the IP address or the hostname of the desired Pi board. The hostname of a Pi can be found and modified through Rasbian (Raspberry Pi OS) by navigating to the System tab of the Raspberry Pi Configuration menu and viewing or editing the Hostname textbox. This configuration menu can be opened through Application Menu > Preferences > Raspberry Pi Configuration. The Ethernet and WI-FI IP addresses can be found by hovering the mouse cursor over the network icon at the top of the screen. Both of these methods are shown in Fig. 9.6.

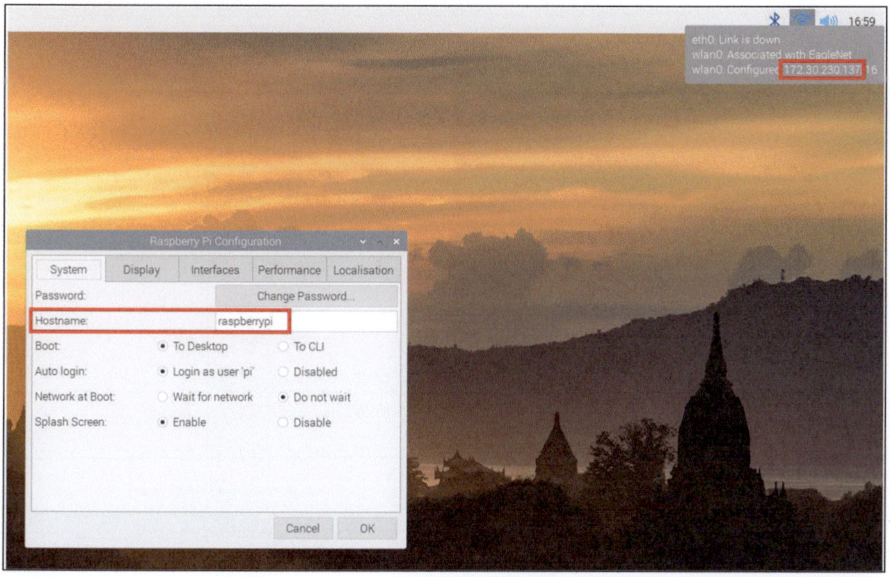


**Fig. 9.6** Raspberry Pi IP address and hostname

Once the IP address or the hostname is found, MATLAB can connect to the Pi using the **raspi** function with three inputs. These inputs are the IP address or the hostname as a string, the username of the Raspberry Pi user as a string (this is **"pi"** by default), and the password of the Raspberry Pi user as a string (this is **"raspberry"** by default). When connecting using the IP address, MATLAB overrides the previous IP connection, so this method should be used when the previously connected board's settings have been changed. Connect to the boards using the hostname when more than one board is being connected. Examples of these two connection methods are shown below:

```
rpi1 = raspi("172.30.230.137", "pi", "raspberry")
rpi2 = raspi("raspberrypi", "pi", "raspberry");
rpi1 = raspi with properties:
DeviceAddress: '172.30.230.137'
Port: 18734
BoardName: 'Raspberry Pi 2 Model B'
AvailableLEDs: {'led0'}
AvailableDigitalPins: [4,5,6,10,12,13,14,15,16,17,18,19,20,21,22,23,24,25,26,27]
AvailableSPIChannels: {'CE1'}
AvailableI2CBuses: {'i2c-1'}
AvailableWebcams: {}
I2CBusSpeed: 100000
AvailableCANInterfaces: {}
```

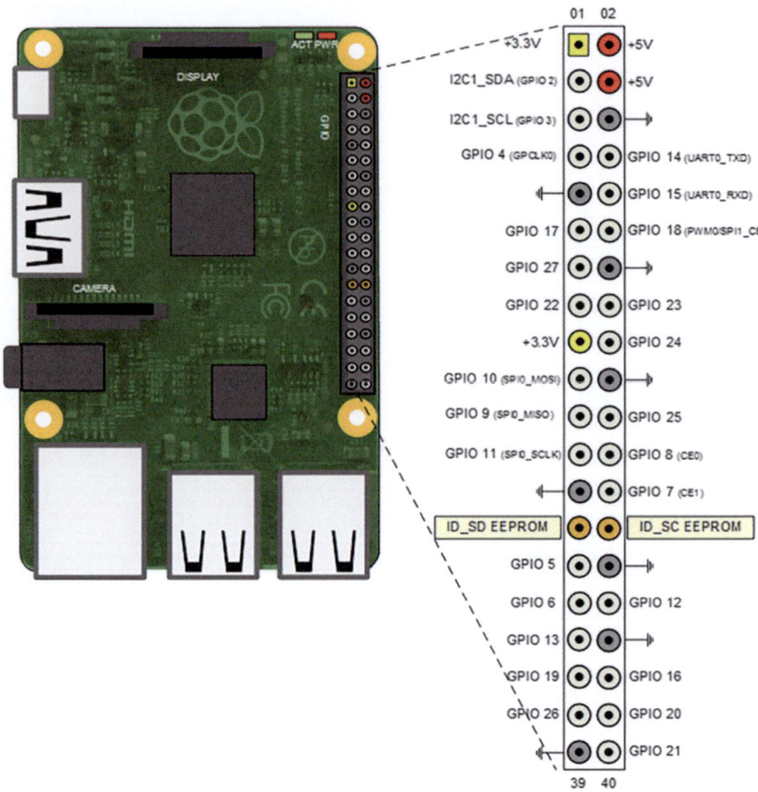


**Fig. 9.7** Pinout chart for Raspberry Pi 2 model B

Once a Raspberry Pi object is created, a pinout diagram of the connected board can be shown by giving the Pi object as an input to the **showPins** function. Figure 9.7 shows an example output from this command.

MATLAB treats Raspberry Pis similar to Arduinos, where they both can use functions and classes such as **configurePin**, **readDigitalPin**, **writeDigitalPin**, **writePWMDutyCycle**, and **servo**. These functions take the same inputs for Pis as for Arduinos, but pins are specified as an integer rather than a string. These pin integers are the numbers following GPIO in the pinout diagrams (Fig. 9.7). Additionally, Raspberry Pis have several different pin modes, which are listed in Table 9.2.

Raspberry Pis only have digital pins, so any functions or pin modes available to the Arduino that make use of analog inputs (such as that from a potentiometer) are not available for the Pi. The following code example uses functions and objects discussed in Sect. 1.2 to move a servo from 0 to 180 degrees and then back to 0 degrees, where the servo is connected to pin **GPIO 4**. Notice that the **writePosition** function takes inputs from 0 to 180 for the Raspberry Pi rather than 0 to 1 for the Arduino.

**Table 9.2** Raspberry Pi pin modes

| Pin mode | Description |
|---|---|
| "DigitalInput" | Specify a pin to acquire digital signals |
| "DigitalOutput" | Specify a pin to generate digital signals |
| "PWM" | Specify a pin to use a pulse width modulator |
| "Servo" | Specify a pin to use with a servo |
| "ArduCam" | Specify a pin to use with an ArduCam |
| "Unset" | Specify a pin to remove its reserved mode |

```
configurePin(rpi, 4, 'Servo')
s = servo(rpi, 4);
writePosition(s, 0)

for i = linspace(0, 180, 50)
  writePosition(s, i)
  pause(0.01)
end
for i = linspace(180, 0, 50)
  writePosition(s, i)
  pause(0.01)
end
```

In addition to these shared features, MATLAB has special functions specifically for Raspberry Pis. These include but are not limited to the **getFile**, **putFile**, and **deleteFile** functions, where **getFile** transfers a file from the Pi to the host computer, **putFile** transfers a file from the host computer to the Pi, and **deleteFile** deletes a file from the Pi. All three of these functions take the Pi object as the first input, the path to the desired file as a character vector for the second input, and the first two functions take the path to the save location as a character vector for a third input. The following code demonstrates the use of these three functions by saving a file named **test.txt** on the Pi to the computer as **test2.txt**, saving **test2.txt** back to the Pi as **test3.txt**, and then deleting the original **test.txt** file from the Pi. Note that these functions cannot create new folders and will overwrite existing files of the same name.

```
getFile(rpi, '/home/pi/test.txt', 'C:\Users\scher\Documents\test2.txt')
putFile(rpi, 'C:\Users\scher\Documents\test2.txt', '/home/pi/test3.txt')
deleteFile(rpi, '/home/pi/test.txt')
```

### 9.1.4 Lego EV3 Interfacing

A third system that MATLAB is able to interface with is the Lego EV3. This system also has standard MATLAB and Simulink support packages that can be installed using the Add-Ons menu similar to the Arduino and Raspberry Pi, but the setup process is simpler since nothing needs to be installed on the EV3 computer. Simulink can also be set up to run on an EV3 using the same process described in Sect. 1.1. Once the support packages are installed, the EV3 can be paired with MATLAB through USB, Bluetooth, or a wireless network connection using the **legoev3** function. This function takes the connection type as a string for its first input, whether that is **"USB"**, **"Bluetooth"**, or **"WiFi"**, and then takes additional inputs depending on the connection type. Connecting through USB is the simplest method, since no other inputs are required, as shown below:

```
lego = legoev3("USB")
                    lego = legoev3 with properties:
                     FirmwareVersion: 'V1.09H'
                          HardwareID: []
                           IPAddress: []
                   CommunicationType: 'USB'
                        BatteryLevel: 100
                    ConnectedSensors: {"  "  "  "}
```

In contrast, connecting through Bluetooth requires the connection port as a second input string and connecting through WiFi requires the IP address and ID of the EV3 computer as two additional input strings. The Bluetooth port can be found in Window's Device Manager once the EV3 is paired with the computer, and the EV3 IP address and ID can be found in the Brick Info menu under Settings on the EV3. Keep in mind that Bluetooth or WiFi connections must first be enabled on the EV3 in Settings or a connection cannot be established. Additionally, the EV3 WiFi dongle must be plugged into the brick before the EV3 can be paired using WiFi. Examples of these two pairing options are shown by the code below:

```
lego = legoev3("Bluetooth", "COM3");
lego = legoev3("WiFi", "192.168.1.3", "00165340e49b");
```

Lego EV3s have four different sensors and two different motors that can be connected to its eight external ports, and MATLAB has built-in functions to handle all five of these systems. The first sensor, the touch sensor, is straightforward and simple to use. First, a **touchSensor** object must be defined with the EV3 object and the number of the connected port (1 through 4) provided as inputs. Then the **readTouch** function can be called with the **touchSensor** object as an input. **readTouch** returns a 1 if the touch sensor is pressed and a 0 if the touch sensor is not pressed. The following example demonstrates reading the input of a pressed touch sensor connected to port 4.

```
touch = touchSensor(lego, 4);
readTouch(touch)
                    ans = 1
```

Similar to the touch sensor, the remaining three sensors and two motors can be controlled using their corresponding objects, named **colorSensor**, **sonicSensor**, **gyroSensor**, and **motor**. All these objects take the same inputs as **touchSensor**, although the **motor** object must be given its port as a string of **'A'**, **'B'**, **'C'**, or **'D'**. The color sensor has two supporting functions that take the **colorSensor** object as their only inputs. The first, called **readColor**, returns the detected color as a string, where the potential colors are **"red"**, **"white"**, **"black"**, **"brown"**, **"yellow"**, **"blue"**, **"green"**, and **"none"**. The second, called **readLightIntensity**, returns the detected light intensity as an integer between 0 and 100, where 0 is dark and 100 is light. The light intensity mode defaults to **'ambient'**, but a second input of **'reflected'** can be provided to determine the intensity of reflected light instead. The sonic sensor only has one supporting function called **readDistance** that returns the relative distance to the detected object in meters. In contrast, the gyro sensor has three supporting functions named **readRotationAngle**, **resetRotationAngle**, and **readRotationRate**. The **readRotationAngle** returns the angle that the sensor has rotated as an integer in degrees, where a positive angle indicates a clockwise rotation. The **resetRotationAngle** function updates the original rotation of the gyro so that **readRotationAngle** once again returns an angle of zero. Finally, the **readRotationRate** function returns the angular velocity of the sensor as an integer value between -440 and 440 degrees per second, where a positive velocity indicates a clockwise rotation.

The **motor** object can be used to control the EV3's large and medium motors. The **motor** object has a **Speed** property that can be read and changed whether the motor is running or stopped. This property is a double datatype between -100 and 100, where -100 is fully reverse and 100 is fully forward. The motor can be started and stopped using the **start** and **stop** functions, respectively, where both functions take only the **motor** object as an input. The **readRotation** and **resetRotation** functions also take this same input and return the rotation of the motor in degrees and reset the rotation reading of the motor to zero, respectively. Note that **resetRotation** only resets the rotation count to zero and does not actually rotate the motor back to zero rotation.

The following code defines a motor connected to port B on the EV3 brick. The program first sets the motor speed to 50% forward and resets the rotation counter to zero. Then the program starts the motor, waits 1 s, reads the rotation of the motor and changes the speed to 50% in reverse, waits 1 more second, reads the rotation of the motor one more time, and then stops the motor.

```
motor = motor(lego, 'B');
motor.Speed = 50;
resetRotation(motor)
start(motor)
pause(1)
readRotation(motor)
motor.Speed = -50;
pause(1)
readRotation(motor)
stop(motor)
                    ans = int32 493
                    ans = int32 64
```

Aside from accessing hardware connected to the EV3's external ports, MATLAB provides functions that can read button inputs and return visual or audio feedback using the brick itself. For example, the functions **writeLCD** and **clearLCD** can be used to write text onto the EV3's screen and to clear the screen, respectively. While **clearLCD** only requires the EV3 object as an input, **writeLCD** requires the EV3 object, a character vector to write to the screen, the row on which to start writing the text, and the column on which to start writing the text. The following code clears the screen and then writes Hello World onto the screen starting at the second row and third column.

```
clearLCD(lego)
writeLCD(lego, 'Hello World', 2, 3)
```

MATLAB can also control the lighting of the input buttons using the **writeStatus-Light** command. This function takes the EV3 object, the light color as a string, and the light mode as a string and changes the light behavior accordingly. The valid color strings are **"green"**, **"red"**, **"orange"**, and **"off"** and the valid mode strings are **"solid"** or **"pulsing"**. If a mode string is not given, MATLAB defaults to **"solid"**. The following code starts flashing the green status light on and off before turning it off completely:

```
writeStatusLight(lego, 'green', 'pulsing')
writeStatusLight(lego, 'off')
```

Additionally, MATLAB can receive inputs from the buttons on the EV3 brick using the **readButton** function. This function takes the EV3 object and button position as a string as inputs and then returns a 1 if the button is pressed or a 0 if the button is not pressed. The five possible button position strings are **'up'**, **'down'**, **'left'**, **'right'**, and **'center'**.

Finally, MATLAB can control sounds played by the EV3 brick. The **beep** function takes the EV3 object as an input and makes the EV3 play a short beep. The **playTone** function takes the EV3 object, the frequency of the sound in Hertz, the play time of the sound in seconds, and the desired volume level and then makes the EV3 play the corresponding sound. The following example has the EV3 play a 500 Hz sound for 3 s at a volume level of 20.

```
playTone(lego, 500, 3, 20)
```

### *9.1.5 Hardware Input Interfacing*

In addition to interfacing with specific hardware systems, such as Arduino, Raspberry Pi, and Lego EV3, MATLAB has the ability to handle various other generic systems. For example, Simulink has a block under Simulink 3D Animation > Virtual Reality

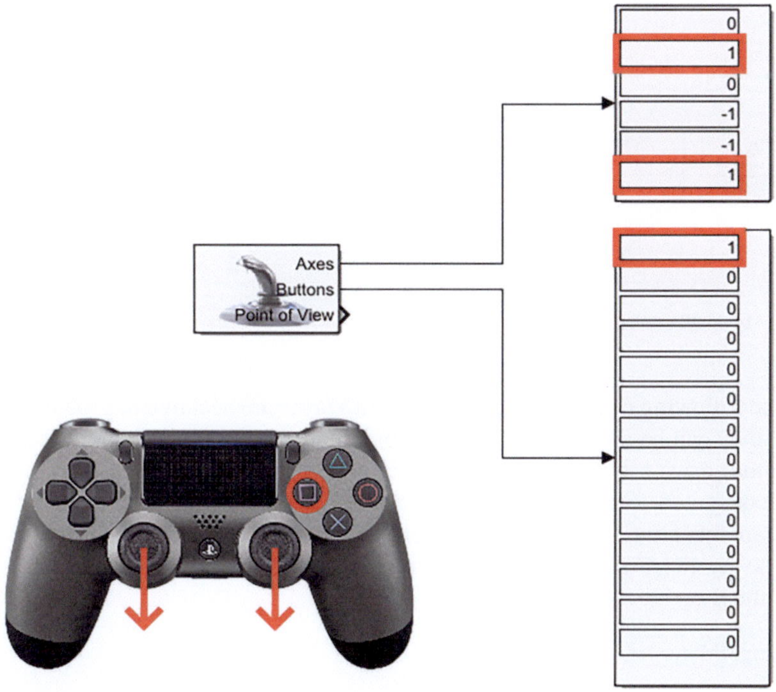


**Fig. 9.8** Joystick input block

> Utilities called the Joystick Input that can automatically interface with a variety of controllers including joysticks and game controllers. Figure 9.8 shows a simple Simulink model that reads the outputs of the Joystick Input block for a game controller connected to a USB port.

In this example, a PS4 game controller was connected to the model where both joysticks were deflected all the way down and the left-most round button was pressed (shown by the positive one values in Fig. 9.8). Since the Joystick Input block can handle a variety of different controllers, the controller inputs and their corresponding outputs must be determined on a case-by-case basis. Note that this Simulink model was made to run forever to get real-time input changes by setting the stop time to **inf**. Unfortunately, this Joystick Input block cannot be compiled, so it cannot be installed onto the Arduino, Raspberry Pi, or EV3 hardware like other Simulink models such as the one demonstrated in Sect. 1.2. Despite this difficulty, the Joystick Input block can be used in conjunction with Arduinos, Raspberry Pis, or EV3s such as the example shown below and in Fig. 9.9.

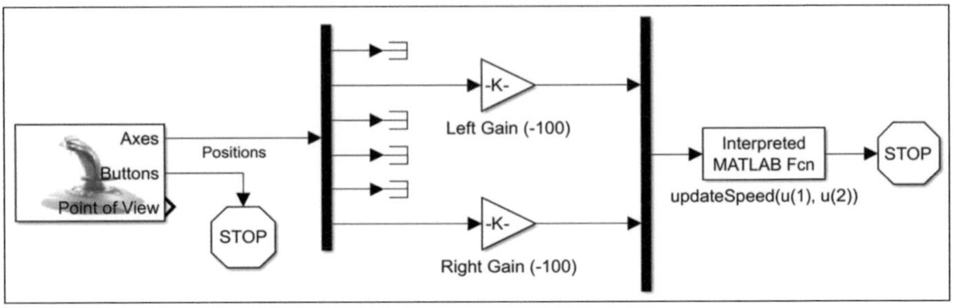


**Fig. 9.9** Joystick control of EV3

In the following code, **leftMotor** and **rightMotor** are declared as global variables so that they can be accessed by the Simulink helper function. The robot is a simple differential-drive robot, with one large motor connected to port B and another large motor connected to port C. The motor speeds are initialized to zero, the rotation measurements are reset to zero, and then the motors are started. Next, the program starts the joystick Simulink model shown in Fig. 9.7, which will run indefinitely until one of the Stop blocks is reached.

```
global rightMotor leftMotor

lego = legoev3("usb");
leftMotor = motor(lego, 'B');
rightMotor = motor(lego, 'C');

leftMotor.Speed = 0;
rightMotor.Speed = 0;
resetRotation(leftMotor)
resetRotation(rightMotor)
start(leftMotor)
start(rightMotor)

sim("EV3Controller.slx");

%% Function to update motor speeds within Simulink
function stop = updateSpeed(VL, VR)

   % Import the global motor objects
   global leftMotor rightMotor

   % Update the two motor speeds
   leftMotor.Speed = VL;
   rightMotor.Speed = VR;

   % Make the function return zero so the Simulink never stops
   stop = 0;
end
```

As shown by Fig. 9.9, the Simulink first reads the inputs given by the PS4 controller. It then takes the vertical axes of the two joysticks, converts them to integers between -100 and 100, and then updates the motor speeds, where the right joystick controls the right motor and the left joystick controls the left motor. Since the Joystick Input block outputs positive one for a downward deflection of the stick as shown in Fig. 9.8, the inputs were multiplied by -100 so that pushing the joysticks fully forward corresponds to full forward throttle and pulling the joysticks fully backward corresponds to full reverse throttle. The Interpreted MATLAB Function block is used to call the **updateSpeed** function from the original MATLAB file, which in turn updates the motor speeds and returns zero to keep the Simulink simulation running. Note that if any button on the controller is pressed, the simulation is stopped.

## 9.2 Interfacing MATLAB with Databases

### *9.2.1 Connecting to a Database*

In addition to interfacing with various hardware, MATLAB is able to connect to and interface with online databases such as MySQL, Oracle, Microsoft SQL Server, and PostgreSQL. MATLAB can also create local SQLite database files that do not require the user to install and connect to an external database. A local SQLite database connection can be established using the **sqlite** function, which takes a database file name and database file mode as inputs. The following code creates and connects to a database file named **test.db**. Once the database exists, the mode can be changed to **"readonly"** to create a read-only connection to the database or the mode can be removed entirely for a regular connection.

```
conn = sqlite("test.db", "create")
                        conn = sqlite with properties:
                            Database: 'test.db'
                          IsReadOnly: 0
                          AutoCommit: 'on'
```

Once the database is done being modified, the connection can be closed using the **close** function with the **sqlite** object as its only input.

```
close(conn)
```

Connections to external databases can be made with the **database** function that takes the database server information as inputs and returns an ODBC or JDBC **connection** object. There are multiple different input configurations for **database**, but the simplest is providing a data source, username, and password all as strings. It is recommended to use the **setSecret** and **getSecret** commands to save confidential information into MATLAB's vault rather than leaving the information unencrypted in the MATLAB code. The **setSecret** function takes a string for the variable name as an input and then opens a textbox for the variable value to be entered and saved. The **getSecret** function takes the same string and returns the saved string. The following code establishes a connection to the ODBC data source **MySQL ODBC** using a username and password saved to MATLAB's vault.

```
datasource = "MySQL ODBC";
setSecret("usernamemysql");
setSecret("passwordmysql");
conn = database(datasource, getSecret("usernamemysql"), getSecret("passwordmysql"))

conn = connection with properties:
DataSource: 'MySQL ODBC'
UserName: ''
Message: ''
Type: 'ODBC Connection Object'
Database Properties:
AutoCommit: 'on'
ReadOnly: 'off'
LoginTimeout: 0
MaxDatabaseConnections: 0
Catalog and Schema Information:
DefaultCatalog: 'toystore_doc'
Catalogs: {'information_schema', 'detsdb', 'mysql' ... and 4 more}
Schemas: {}
Database and Driver Information:
DatabaseProductName: 'MySQL'
DatabaseProductVersion: '5.7.22'
DriverName: 'myodbc5a.dll'
DriverVersion: '05.03.0014'
```

Another way to connect to a database is by using its driver and url. In this method, **database** takes the name of the database, username, password, driver, and url all as string inputs. In the following example, a connection to a database named **dbname** in **PostgreSQL** is established using the host and port numbers **host** and **port**, respectively.

```
databasename = "dbname";
setSecret("usernamepsql");
setSecret("passwordpsql");
driver = "org.postgresql.Driver";
url = "jdbc:postgresql://host:port/dbname";
conn = database(databasename, getSecret("usernamepsql"), getSecret("passwordpsql"), ...
    driver, url)
conn = connection with properties:
DataSource: 'dbname'
UserName: ''
Driver: 'org.postgresql.Driver'
URL: 'jdbc:postgresql://host: ...'
Message: ''
Type: 'JDBC Connection Object'
Database Properties:
AutoCommit: 'on'
ReadOnly: 'off'
LoginTimeout: 0
MaxDatabaseConnections: 8192
Catalog and Schema Information:
DefaultCatalog: 'catalog'
Catalogs: {'catalog'}
Schemas: {'schema1', 'schema2', 'schema3' ... and 1 more}
Database and Driver Information:
DatabaseProductName: 'PostgreSQL'
DatabaseProductVersion: '9.4.5'
DriverName: 'PostgreSQL Native Driver'
DriverVersion: 'PostgreSQL 8.4 JDBC4 (bui ...'
```

### *9.2.2 Editing a Database*

Once a connection to a database has been established, the database can be edited using SQL commands and the **execute** function. This function takes the connection object and a string containing the SQL command as inputs and executes the desired SQL functions. The SQL query given to **execute** cannot be a **SELECT** statement. The example code below creates a new table in **test.db** named **Car** with the categories **Make**, **Model**, **Year**, and **Mileage** and inserts one new row of data.

```
query = "CREATE TABLE Car (Make VARCHAR(250), Model VARCHAR(250), " + ...
    "Year INT, Mileage INT)";
execute(conn, query)
query = "INSERT INTO Car (Make, Model, Year, Mileage) VALUES ('Buick', " + ...
    "'Enclave', '2015', '150000');";
execute(conn, query)
```

Data can also be added to SQL tables using the **sqlwrite** function. This function can be easier to use since it takes the connection object, the table name, and a MATLAB table of data and adds the MATLAB table to the existing SQL table. The following code creates a MATLAB table containing the information of four cars with the same column names as the SQL table and then appends the MATLAB table to the end of the SQL table named **Car**.

```
data = table(["Toyota"; "Chevrolet"; "Ford"; "Toyota"], ...
        ["Rav4"; "Silverado"; "F150"; "Tacoma"], ...
        [2018; 2024; 2016; 2021], ...
        [59000; 120000; 40000; 88000], ...
        'VariableNames', ["Make", "Model", "Year", "Mileage"]);

sqlwrite(conn, "Car", data)
```

## 9.2.3 *Reading a Database*

After a database has been created and a connection has been established, data can be read from the SQL tables using the **fetch** function. This function works in a similar way to **execute**, where the connection object and the SQL query string are given as inputs, but then returns the SQL table as a MATLAB table. An additional **MaxRows** integer input can be given to limit the number of imported rows and a **RowFilter** input can be given to filter the output rows even more. The following code reads all elements of the **Car** table and then reads the table again while filtering out any rows with a **Toyota** make.

```
fetch(conn, "SELECT * FROM Car")
rf = rowfilter("Make");
fetch(conn, "SELECT * FROM Car", "RowFilter", rf.Make ~= "Toyota")
                         ans = 5×4 table
            Make          Model          Year        Mileage
            ______        _______        ____        _______
            "Buick"       "Enclave"      2015        150000
            "Toyota"      "Rav4"         2018        59000
            "Chevrolet"   "Silverado"    2024        120000
            "Ford"        "F150"         2016        40000
            "Toyota"      "Tacoma"       2021        88000
                         ans = 3×4 table
            Make          Model          Year        Mileage
            ______        _______        ____        _______
            "Buick"       "Enclave"      2015        150000
            "Chevrolet"   "Silverado"    2024        120000
            "Ford"        "F150"         2016        40000
```

Another simpler way to read an existing table is through the **sqlread** function. This function takes the connection object and the SQL table name as inputs and returns a MATLAB table with the same format.

```
sqlread(conn, "Car")
                         ans = 5×4 table
           Make           Model          Year        Mileage
          ______         _______         ____        _______
          "Buick"        "Enclave"       2015        150000
          "Toyota"       "Rav4"          2018        59000
          "Chevrolet"    "Silverado"     2024        120000
          "Ford"         "F150"          2016        40000
          "Toyota"       "Tacoma"        2021        88000
```

**Practice Problems.**

(1) Set up a circuit and write the code to control the brightness of an LED using a potentiometer.
(2) Set up a circuit and make a Simulink model to turn an LED on and off using a button.
(3) Program an EV3 to wait until its center button is pressed. Once the button is pressed, move the robot forward until the touch sensor is triggered. Then, have the EV3 play a sound.
(4) Set up a circuit and make a model to control the rotation of two servos using a joystick. The pitch (up and down) deflection should control one servo, and the roll (side to side) deflection should control the other servo. Stop the simulation if any button is pressed on the controller.
(5) Create a new local database with a table named **Person** and columns for **First_Name**, **Last_Name**, and **Age**. Populate the table with five different people and then read the contents of the first three rows.

# Chapter 10
# MATLAB Applications and Toolboxes for Engineering

MATLAB has a variety of applications of toolboxes for engineers. This chapter is not an exhaustive list of these applications and toolboxes, and they are selected based on the background studies of the authors. In each subsection, we introduce the toolbox properties and capabilities along with a few examples.

## 10.1 Curve-Fitting Toolbox

The Curve Fitting Toolbox in MATLAB is a powerful tool that allows you to perform a variety of tasks related to data fitting. Here are some of the key functionalities and tasks you can accomplish with the Curve Fitting Toolbox:

i. **Data Preprocessing**

- Data Import and Visualization: Import data from various sources (e.g., files, workspace, etc.) and visualize it using scatter plots and other types of graphs.
- Data Cleaning: Remove outliers, fill missing values, and smooth noisy data.

ii. **Curve and Surface Fitting**

- Fit Types: Fit various types of curves and surfaces, including polynomials, exponentials, Gaussians, Fourier series, splines, and custom equations to data.
- Automatic Fitting: Automatically fit data using the built-in fit function, which can suggest the best fit type and parameters.
- Interactive Fitting: Use the Curve Fitter app to interactively fit curves and surfaces, adjust fit options, and visualize results.



M. Andalibi and S. A. Cherkauer, *Advanced MATLAB for Engineers*,
https://doi.org/10.1007/978-3-032-10677-3_10

iii. **Model Evaluation and Analysis**

- Goodness-of-Fit Statistics: Evaluate the quality of fits using statistics such as R-squared, adjusted R-squared, RMSE (Root Mean Squared Error), and residuals.
- Confidence Intervals: Calculate confidence intervals for fitted coefficients to assess their reliability.
- Prediction and Extrapolation: Use fitted models to make predictions and extrapolate data beyond the observed range.

iv. **Optimization and Customization**

- Custom Equations: Define and fit custom equations to your data using the fit function or the Curve Fitter app.
- Optimization Options: Customize optimization options, including fitting algorithms, tolerance levels, and starting points.
- Parameter Constraints: Apply constraints to fit parameters to ensure physically meaningful fits.

v. **Visualization and Reporti**ng

- Plotting Options: Visualize fits and residuals with a variety of plotting options, including 3D plots for surface fitting.
- Annotations: Add annotations, legends, and labels to plots for better interpretation and presentation.
- Exporting Results: Export fitting results, including plots and coefficients, to the MATLAB workspace or save them as files for reporting and further analysis.

vi. **Automation and Scripting**

- Batch Processing: Automate the fitting process for multiple datasets using scripts.
- Programmatic Fitting: Use the toolbox functions in MATLAB scripts and functions to integrate curve fitting into larger workflows.

To illustrate these functionalities, we can either use the curve-fitting GUI or perform all tasks using the command window. Here, we pick the former option and load the GUI using the following command:

```
curveFitter
```

Next, we load the x (**years** variable) and y (**pop** variable) data from a file called **curvefittingdata.mat**, which is a modified version of the **.mat** file available in MATLAB called **census.mat** that has the US population versus year from 1790 to 1990. Next, we create a structure from the x and y data and convert it to a table that can be inserted into the Data Cleaner app (**Data Preprocessing**).

This app can be found under the apps tab, under the "MATLAB" category with this icon .

In the loaded file, two data points have been modified: **pop(2)** was set to an outlier value of 25 and **pop(15)** was set to **NaN**.

```
load curvefittingdata.mat
mydata.x=years;
mydata.y=pop;
mydata_table=struct2table(mydata);
```

Next, we import the table created above into this app as shown in Fig. 10.1, select the y data for editing, fill in the missing value (**NaN**), and remove and replace the outlier value. The moving median method with a window length of 5 and a threshold factor of 2.5 is used here to detect the outlier, and a linear interpolation method is used to fill the outlier and the **NaN** values.

After the data is cleaned, the resulting table can be exported to the workspace. This app even allows us to generate a function or script to perform the same tasks on similar data (**Automation and Scripting**).

Note that the Data Cleaner app can also smooth the data and normalize it (common in machine learning tasks as will be seen in Sects. 10.7 and 10.8).

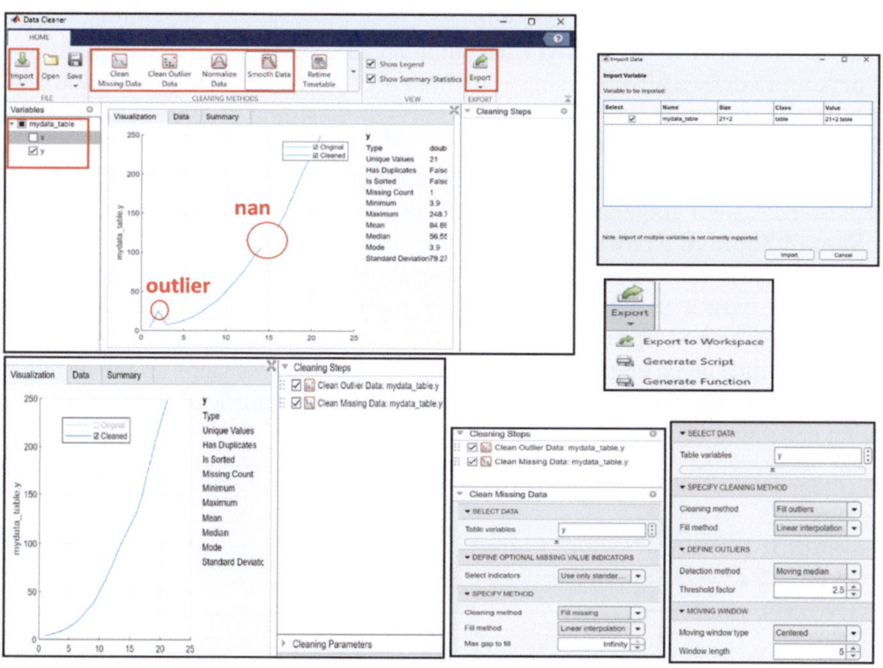


**Fig. 10.1** Using Data Cleaner app to detect and remove outliers and fill in the missing values

**Fig. 10.2** Inserting clean data in the Curve Fitter application generating initial results

Then, we load the cleaned data into the Curve Fitter app by clicking on "Select Data" as shown in Fig. 10.2. MATLAB automatically generates an initial fitted curve (**Curve and Surface Fitting**) using a polynomial of degree one (a line) and **goodness-of-fit** statistics (**Model Evaluation and Analysis**). **Residual plots** can also be added (**Visualization and Reporting**).

In the case of the population data used for this example, an exponential fit generates the best metrics for goodness of fit (**Optimization and Customization**), such as $\mathbf{R^2}$ or **Sum of Squared Errors (SSE)** as can be seen in Fig. 10.3.

Furthermore, the Curve Fitter app also allows you to export a figure, script, function, or even a Simulink lookup table using the tasks performed on the data (**Automation and Scripting**).

As a practice on an engineering problem, try to redo these steps on the results of the Hook's experiment using the code below that creates artificial spring force and spring deflection data:

```
k1=100;
k2=50;
x=0:0.02:1;
F=k1*x+k2*x.^3+15*(rand(1,length(x))-0.5);
F(10)=nan;
F(20)=6;
```

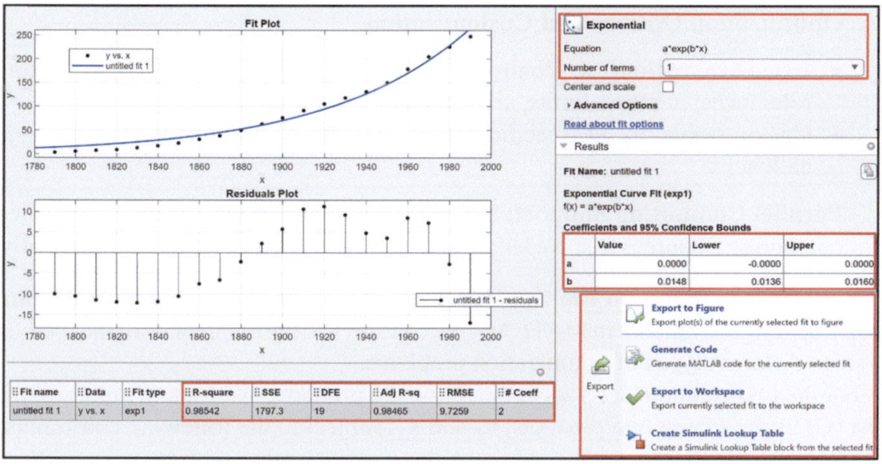


**Fig. 10.3** Curve-Fitting results with an exponential function and exporting the script

## 10.2 Optimization Toolbox

The Optimization Toolbox provides a comprehensive set of functions and tools to solve a wide range of optimization problems, including linear, nonlinear, integer, and multi-objective optimization. It is designed to handle problems with various types of constraints and objectives, making it suitable for a broad array of engineering, scientific, and financial applications.

Key features of the MATLAB Optimization Toolbox are:

i. **Solvers for Different Problem Types**

   - **Linear Programming (LP)**: Solve linear objective functions subject to linear equality and inequality constraints.
   - **Quadratic Programming (QP)**: Optimize quadratic objective functions with linear constraints.
   - **Nonlinear Programming (NLP)**: Minimize nonlinear objective functions subject to nonlinear constraints.
   - **Mixed-Integer Programming (MIP)**: Handle problems involving both continuous and integer variables.
   - **Least-Squares and Curve Fitting**: Solve problems where the objective is to minimize the sum of squares of functions.

ii. **Multi-Objective Optimization**: Solve problems with multiple objectives, providing a set of optimal solutions (Pareto front).

iii. **Global Optimization**: Methods such as genetic algorithms, simulated annealing, and particle swarm optimization for solving complex, multimodal problems.

iv. **Optimization Options and Customization**

- Extensive options to control the behavior of solvers, such as setting tolerances, iteration limits, and choosing algorithms.
- Customizable to fit specific needs through user-defined functions and callbacks.

v. **Parallel Computing Support**: Leverage multi-core processors and distributed computing resources to solve large-scale optimization problems more efficiently.

While the Optimization Toolbox GUI that was launched by the command **optimtool** is no longer available in MATLAB, we can still use the commands in the toolbox to solve different types of optimization problems.

For the first example, let's solve the following problem: We want to minimize the cost (**C**) of producing two products, $\mathbf{x_1}$ and $\mathbf{x_2}$, subject to the following constraints:

a. The total production of both products must be at least 10 units.
b. The production of product $\mathbf{x_1}$ must be at least 4 units.
c. The production of product $\mathbf{x_2}$ must be at least 3 units.
d. The production of both products must not exceed 15 units combined.

The price of each unit of product $\mathbf{x_1}$ is \$5 and the price of each unit of product $\mathbf{x_2}$ is \$3, so the cost function will be:

$$C = 5x_1 + 3x_2 \tag{10.1}$$

First, we define the optimization problem using the command **optimproblem**, giving it a name and the optimization type (here, to minimize the cost function):

```
prob = optimproblem('ObjectiveSense', 'minimize');
```

Next, we define the optimization variables (here, as a vector with 2 elements), the type of these variables (continuous), and then any upper or lower bound on them:

```
x = optimvar('x', 2, 'Type', 'continuous', 'LowerBound', [4, 3]);
```

Then, we define the cost function:

```
cost = 5 * x(1) + 3 * x(2);
prob.Objective = cost;
```

And finally, we define the constraints:

```
cons1 = x(1) + x(2) >= 10;
cons2 = x(1) + x(2) <= 15;
prob.Constraints.cons1 = cons1;
prob.Constraints.cons2 = cons2;
```

Now it is time to solve the problem and show the results:

```
[sol, fval, exitflag, output] = solve(prob);
disp('Optimal Solution:');
disp(sol.x);
disp('Minimum Cost:');
disp(fval);
```

Since the cost function and constraints are all linear in terms of the variables, the solver that the toolbox selects is the **linprog** which stands for **linear programming**. The results are:

```
Optimal Solution:
4
6

Minimum Cost:
38
```

For the second example, let's solve a design problem for mass (or cost) minimization while keeping the stress below a given value (e.g. the yield strength): We have a cantilever beam with the following parameters:

a. Length, **L**
b. Width, **w**
c. Height, **h**
d. Density, **ρ**

The objective is to minimize the mass, **M**, of the beam while ensuring that the maximum stress σ does not exceed the allowable stress $\boldsymbol{\sigma_{max}}$.

The mass of the beam is given by: $\mathbf{M = \rho Lwh}$.

The stress in the beam due to a point load **P** at the free end is: $\mathbf{= 6PL/(wh^2)}$.

So, the constraints are:

a. The width, **w**, and height, **h**, must be within given bounds ($\mathbf{0.01m \leq w, h \leq 0.5m}$).
b. The maximum stress, **σ**, must not exceed $\boldsymbol{\sigma_{max}}$.

First, we define the problem, and the constant parameters involved in it:

```
L = 2.0; % Length of the beam in meters
rho = 7850; % Density of the beam material in kg/m^3
P = 1000; % Point load at the free end in Newtons
sigma_max = 250e6; % Allowable stress in Pascals (250 MPa)

prob = optimproblem('ObjectiveSense', 'minimize');
```

Next, we define the optimization variables (here, 2 separate scalars), the type of these variables (continuous), and the upper or lower bound on them:

```
w = optimvar('w', 'Type', 'continuous', 'LowerBound', 0.01, 'UpperBound', 0.5);
h = optimvar('h', 'Type', 'continuous', 'LowerBound', 0.01, 'UpperBound', 0.5);
```

Then, we define the cost/objective function:

```
mass = rho * L * w * h;
prob.Objective = mass;
```

And finally we define the constraints:

```
stress = (6 * P * L) / (w * h^2);
stress_constraint = stress <= sigma_max;
prob.Constraints.stress_constraint = stress_constraint;
```

Now, it is time to solve the problem and show the results. Note that the problem at hand is nonlinear, which requires initial guesses for the variables.

```
initialGuess.w = 0.1;
initialGuess.h = 0.1;
```

Also, while the Optimization Toolbox decides a solver based on the cost function and the constraint, i.e. whether they are linear or nonlinear and whether the problem is constrained or unconstrained, you can specify the solver too, but attention has to be paid so the solver selected is appropriate for the problem.

```
options = optimoptions('fmincon', 'Display', 'iter');
[sol, fval, exitflag, output] = solve(prob, initialGuess, 'Options', options);
disp('Optimal Solution:');
disp(['Width (w): ', num2str(sol.w), ' meters']);
disp(['Height (h): ', num2str(sol.h), ' meters']);
disp('Minimum Mass:');
disp([num2str(fval), ' kg']);
```

Since the cost function is nonlinear in terms of the variables, the solver that the toolbox would select is **fmincon**, which stands for **function minimization constrained**. The results are:

```
Optimal Solution:
Width (w): 0.010001 meters
Height (h): 0.069311 meters

Minimum Mass:
10.8826 kg
```

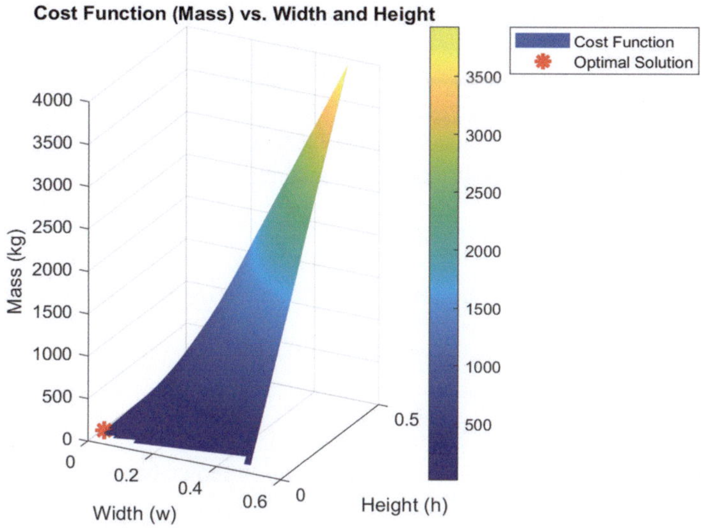


**Fig. 10.4** Visualization of the optimization results for the mass minimization of a cantilevered beam subjected to maximum stress and dimension bounds

The code below can be used to visualize the cost function versus the variables with the solution overlaid on the plot as can be seen in Fig. 10.4:

For our last example, let's solve a more complicated design problem with three variables that involves global optimization due to the potential existence of multiple local minima. We'll optimize the design of a spring by minimizing its weight while ensuring that it can support a given load without exceeding allowable stress limits.

The objective is to minimize the weight, **W**, of the spring subject to the following constraints:

a. The shear stress, $\boldsymbol{\tau}$, should not exceed the allowable shear stress, $\boldsymbol{\tau_{max}}$.
b. The spring must have a minimum required stiffness, $\mathbf{k_{min}}$.

The variables are:

a. Wire diameter, **d**
b. Mean coil diameter, **D**
c. Number of active coils, **N**

The weight of the spring is given by: $\mathbf{W = \rho\pi d^2DN}$

The shear stress is given by: $\boldsymbol{\tau} \mathbf{= 8FD/(\pi d^3)}$

The stiffness is given by: $\mathbf{k = Gd^4/(8D^3N_a)}$

where $\rho$ is the material density, **F** is the applied load, and **G** is the shear modulus of the material.

First, we define the problem, and the constant parameters involved in it:

```
rho = 7850; % Density of the material in kg/m^3
F = 1000; % Applied load in Newtons
tau_max = 400e6; % Allowable shear stress in Pascals (400 MPa)
G = 79.3e9; % Shear modulus in Pascals
k_min = 10000; % Minimum required stiffness in N/m
% Define the objective function (weight of the spring)

objective = @(x) rho * pi * x(1)^2 * x(2) * x(3);
```

Next, we define the upper and lower bounds on the optimization variables:

```
% Lower and upper bounds for the variables [d, D, N]
lb = [0.01, 0.1, 1];
ub = [0.1, 1.0, 20];
```

Then, we define the constraints using a function as shown below (here, no equality constraint exists):

```
% Define the nonlinear constraints
function [c, ceq] = springConstraints(x)
        d = x(1); % Wire diameter
        D = x(2); % Mean coil diameter
        N = x(3); % Number of active coils

        % Shear stress constraint
        tau = (8 * F * D) / (pi * d^3);
        c1 = tau - tau_max; % Shear stress should be less than tau_max

        % Stiffness constraint
        k = (G * d^4) / (8 * D^3 * N);
        c2 = k_min - k; % Stiffness should be greater than k_min

        % Nonlinear inequality constraints
        c = [c1, c2];

        ceq = [ ]; % No equality constraints
end
```

Finally, we need to define the options for the solver, solve the problem, and show the results. Note that the problem at hand is nonlinear, which requires initial guesses for the variables. Here, we use the **particle swarm** solver with 50 particles since it is a global optimization problem. We also opt to show the optimization iterations. The value "**3**" in the second line is the size of the variables array.

```
options = optimoptions('particleswarm', 'SwarmSize', 50, 'HybridFcn', @fmincon,
'Display', 'iter');

[x, fval] = particleswarm(objective, 3, lb, ub, options);

disp('Optimal Solution:');
disp(['Wire diameter (d): ', num2str(x(1)), ' meters']);
disp(['Mean coil diameter (D): ', num2str(x(2)), ' meters']);
disp(['Number of active coils (N): ', num2str(x(3))]);
disp('Minimum Weight:');
disp([num2str(fval), ' kg']);
```

The results are:

```
                             Best         Mean     Stall
Iteration    f-count         f(x)         f(x)   Iterations
    0           50          9.708        499.5       0
    1          100         0.7336        226.1       0
    2          150         0.2466        16.76       0
    3          200         0.2466        2.251       1
    4          250         0.2466       0.6172       2
    5          300         0.2466       0.3162       3
    6          350         0.2466       0.2671       4
    7          400         0.2466       0.2466       5
    8          450         0.2466       0.2466       6
    9          500         0.2466       0.2466       7
   10          550         0.2466       0.2466       8
   11          600         0.2466       0.2466       9
   12          650         0.2466       0.2466      10
   13          700         0.2466       0.2466      11
   14          750         0.2466       0.2466      12
   15          800         0.2466       0.2466      13
   16          850         0.2466       0.2466      14
   17          900         0.2466       0.2466      15

Optimal Solution:
Wire diameter (d): 0.01 meters
Mean coil diameter (D): 0.1 meters
Number of active coils (N): 1

Minimum Weight:
0.24662 kg
```

We must emphasize that the Optimization Toolbox covers a versatile range of constraints:

a. **Bound Constraints**: Define the upper and lower bounds for each variable. These are specified directly as vectors.
b. **Linear Equality Constraints**: Constraints of the form $\mathbf{A_{eq}x} = \mathbf{b_{eq}}$.
c. **Linear Inequality Constraints**: Constraints of the form $\mathbf{Ax} \leq \mathbf{b}$.
d. **Nonlinear Equality Constraints**: Constraints of the form $\mathbf{c_{eq}(x)} = \mathbf{0}$.
e. **Nonlinear Inequality Constraints**: Constraints of the form $\mathbf{c(x)} \leq \mathbf{0}$.

The code shown below is an example that includes all types of these constraints:

```
objective = @(x) x(1)^2 + x(2)^2;  % Objective function

% Bound constraints
lb = [0, 0];
ub = [5, 5];

% Linear inequality constraints
A = [1, 2; -1, 2];
b = [4; 1];

% Linear equality constraints
Aeq = [1, 1];
beq = 3;

% Nonlinear constraints
function [c, ceq] = nonlin_constraints(x)
        % Nonlinear inequality constraint: x1^2 + x2^2 - 4 <= 0
        c = x(1)^2 + x(2)^2 - 4;
        % Nonlinear equality constraint: x1^3 - x2 = 0
        ceq = x(1)^3 - x(2);
end

% Initial guess
x0 = [1, 1];

% Solve the problem
[x, fval, exitflag, output] = fmincon(objective, x0, A, b, Aeq, beq, lb, ub, @nonlin_constraints);
```

## 10.3 Controls Toolbox

The Control System Toolbox is a powerful tool for modeling, analyzing, and designing control systems. It provides a set of functions and graphical user interfaces (GUIs) that help engineers and scientists work with linear time-invariant (LTI) systems.

Key features and capabilities of the Control System Toolbox are:

i. **System Modeling**

- LTI Systems: The toolbox allows you to model linear systems in various forms, including transfer functions, state-space representations, and zero-pole-gain forms.
- Time and Frequency Domain Analysis: You can analyze system behavior in both the time domain (e.g., step response, impulse response) and frequency domain (e.g., Bode plots, Nyquist plots).

ii. **System Analysis**

- Stability Analysis: Tools for determining system stability using pole-zero maps, root locus plots, and other techniques.
- Performance Analysis: Functions to evaluate system performance metrics such as rise time, settling time, overshoot, and gain margin.

iii. **Control Design**

- PID Controllers: Tools for designing and tuning proportional-integral-derivative (PID) controllers.
- State-Space Design: Capabilities for designing controllers and observers using state-space methods, including pole placement and linear quadratic regulators (LQR).
- Frequency Domain Design: Functions for designing compensators using methods like lead-lag compensation and loop shaping.

iv. **Simulink Integration**

- The toolbox integrates with Simulink, allowing you to simulate control systems within a graphical environment. You can design controllers in MATLAB and then test them in Simulink.

v. **Automatic Tuning**

- PID Tuning: The toolbox includes tools for automatically tuning PID controllers based on performance criteria.
- Control System Tuning: You can tune SISO and MIMO controllers to meet specified time-domain and frequency-domain performance requirements.

vi. **Advanced Features**

- Robust Control: Tools for designing robust control systems that can handle model uncertainties and external disturbances.
- Model Predictive Control: Functions for implementing model predictive control (MPC), which is useful for systems with constraints on inputs and states.
- Linearization: Capabilities to linearize nonlinear systems around an operating point.

vii. **Graphical User Interfaces (GUIs)**

- The toolbox includes several GUIs that simplify the process of designing and analyzing control systems. For example, the Control System Designer app provides an interactive environment for tuning controllers and analyzing system responses.

viii. **Code Generation**

- You can generate C/C ++ code from your control algorithms, enabling deployment to embedded systems or integration with other software applications.

ix. **Toolbox Functions**

- The toolbox includes a wide range of MATLAB functions for tasks like **tf** (transfer function), **ss** (state-space model), **step** (step response), **bode** (Bode plot), and many others.

Below, we discuss the GUIs of the Control System Toolbox, their purposes and features, and how to access them:

1. **Control System Designer**

- **Purpose**: A comprehensive environment for designing and analyzing control systems.
- **Features**:
  - Interactive Bode, Nyquist, Nichols, and root locus plots.
  - Tools for tuning SISO (Single-Input, Single-Output) and MIMO (Multiple-Input, Multiple-Output) controllers.
  - Ability to design compensators using lead/lag, PID, or other methods.
  - Real-time updating of system responses as controller parameters are adjusted.
- **How to Access**: Use the **controlSystemDesigner** command in the MATLAB command window.

2. **SISO Design Tool**

- **Purpose**: A specialized GUI for the design and analysis of Single-Input, Single-Output (SISO) control systems.
- **Features**:
  - Interactive root locus, Bode, and Nichols plot editors.
  - Tools for adding poles, zeros, and adjusting gains.
  - Ability to fine-tune compensator parameters and see real-time effects on system behavior.
- **How to Access**: Use the **sisotool** command in the MATLAB command window.

3. **LTI Viewer**

   - **Purpose**: A GUI to visualize the time and frequency responses of Linear Time-Invariant (LTI) systems.
   - **Features**:

     – Simultaneously view multiple system responses such as step response, impulse response, Bode plots, and Nyquist plots.
     – Customize plots and compare responses from different systems.

   - **How to Access**: Use the **ltiview** command in the MATLAB command window.

4. **PID Tuner**

   - **Purpose**: A specialized tool for designing and tuning PID controllers.
   - **Features**:

     – Automatically tunes PID controller gains based on system dynamics.
     – Provides real-time simulation of step response to visualize the performance of the tuned controller.
     – Manually adjust PID parameters and instantly see the effects on system performance.

   - **How to Access**: Use the **pidTuner** command in the MATLAB command window.

5. **Linear System Analyzer**

   - **Purpose**: A GUI for analyzing linear systems in the time and frequency domains.
   - **Features**:

     – Allows for the plotting of step, impulse, Bode, Nyquist, and Nichols plots.
     – Provides tools for customizing the plots and analyzing multiple systems side by side.

   - **How to Access**: Use the **linearSystemAnalyzer** command in the MATLAB command window.

6. **Control System Tuner**

   - **Purpose**: GUI used for tuning control systems that are modeled in Simulink.
   - **Features**:

     – Works with Simulink models to tune control systems that may include multiple feedback loops.
     – Real-time simulation and validation within Simulink.

   - **How to Access**: Typically accessed from within Simulink when you are working with control systems, by right-clicking on a block and selecting the appropriate tuning option.

For our first example, we will look at a SISO LTI system of order 2 and type 0 (2 poles and no zeros) for which the transfer function is given by Eqn. 10.2:

$$\mathbf{G(s)} = \frac{\mathbf{10}}{\mathbf{s^2 + 2s + 0.1}} \tag{10.2}$$

We create this transfer function using the command **tf** in MATLAB and then open the Control System Design GUI for this transfer function using the command **controlSystemDesigner(G)**:

```
s = tf('s');
G = 10 / (s^2 + 2*s + 0.1); % or use G=tf([10],[1 2 0.1]) instead of these 2 lines
controlSystemDesigner(G)
```

Running these commands will open the Control System Designer GUI with the Bode plot, root locus and step response for the uncompensated system as shown in Fig. 10.5. More plots, such Nichols plot, Nyquist plot, and impulse response are available under "New Plot".

As can be seen, the uncompensated system is stable and it has an overshoot of more than 35%, a settling time of about 3.5 s, no steady-state error in the step response, an infinite gain margin, and a phase margin of 35.1 degrees. Since there is no steady-state error, the goal is to improve these **transient response characteristics** using a

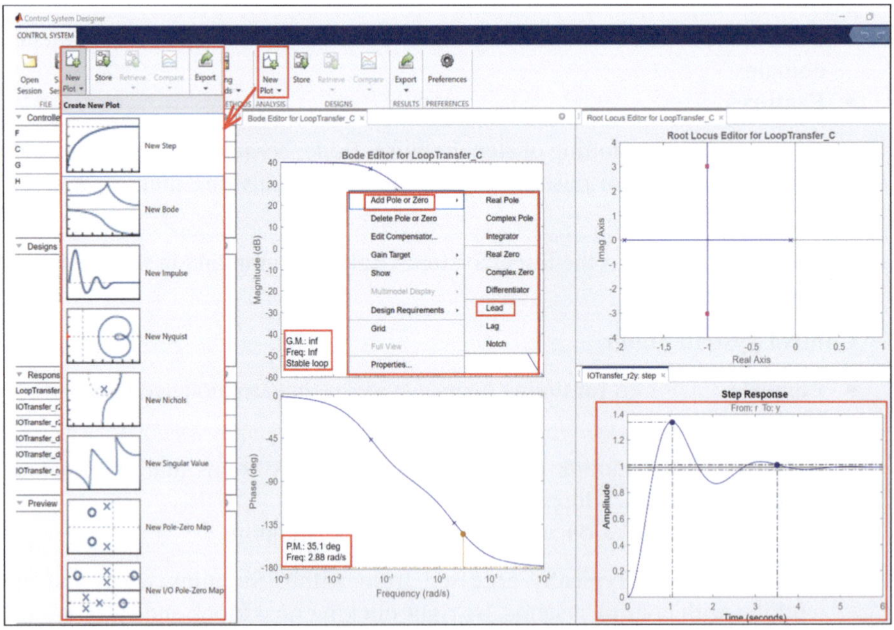


**Fig. 10.5** Control System Designer GUI for the uncompensated system

**lead compensator**. The desired maximum overshoot is less than 10% and the desired settling time is less than 2 s.

To show the overshoot and settling time responses on the step response plot, right-click on this plot and check them under the option "characteristics".

To add a lead compensator, right-click on the Bode plot, and choose "lead" from under the "Add Pole or Zero" option as shown in Fig. 10.5. A similar option exists for the root locus plot.

Now you can click on the Bode plot or root locus plot, and it will add a pole and a zero to both, which can be moved to shape the desired transient response characteristics as seen in Fig. 10.6.

The advantage of using this GUI is to visualize the effects of the lead compensator pole and zero locations on all response characteristics. For instance, in this system, we can see that by moving the lead compensator pole in the root locus plot to the left (shown by a cross mark on the plot), the maximum overshoot and the settling time will be reduced and vice versa.

Once the desired step response characteristics have been achieved by moving the lead compensator pole and zero as shown in Fig. 10.6, we can use the **Store** command in the GUI and add this compensator to the "Designs" menu and also send the tuned system (transfer function) to the workspace using the command **Export**.

Note that instead of moving the compensator pole and zero manually, we can use the **Optimization Based Tuning** under the "Tuning Methods" dropdown menu as can be seen in Fig. 10.7. In the launched window called "Response Optimization", first

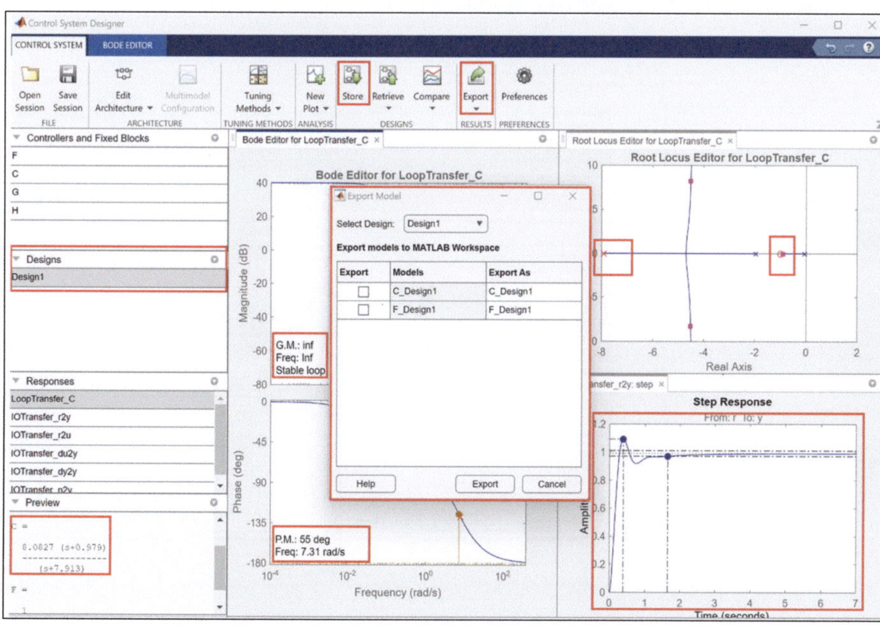


**Fig. 10.6** Response of the controlled system with a lead compensator manually designed

we click on the “Compensators” tab and check the **Lead Lag** compensator we chose. Next, we click on the “Design Requirements” tab, click on the “Add New Design Requirement” button, and enter the **Settling Time** and **Overshoot** values (Make sure to enter a value smaller than the settling time for the **Rise Time** too or you will get an error). Finally, we click on “Start Optimization” under the “Optimization” tab, and we can select the **Display Options** on the right during the optimization process. If a solution is found, the compensator will be added to the GUI and can be stored and exported as explained before. Furthermore, the design requirements will be added to the step response plot as shown in Fig. 10.7.

For our second example, we will look at a SISO LTI system of order 3 and type 0 for which the transfer function is given by Eqn. 10.3:

$$\mathbf{G(s)} = \frac{\mathbf{1}}{\mathbf{s(s+1)(s+2)}} \tag{10.3}$$

We create this transfer function using the command **tf** in MATLAB and then open the PID Tuner GUI for this transfer function using the command **pidTuner(G)**:

```
s = tf('s');
G = 1 / (s * (s + 1) * (s + 2));
pidTuner(G)
```

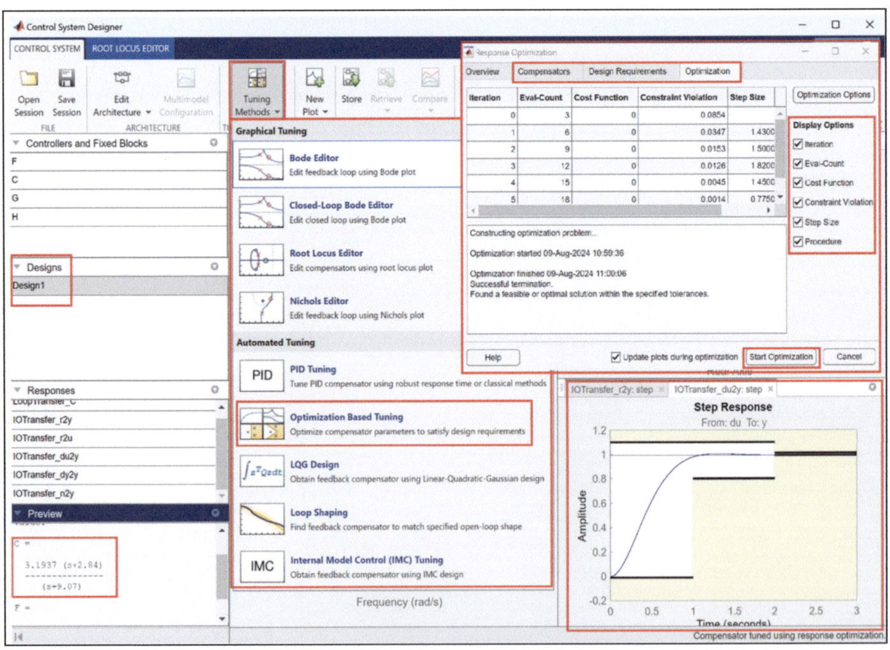


**Fig. 10.7** Response of the controlled system with a lead compensator automatically designed using optimization-based tuning

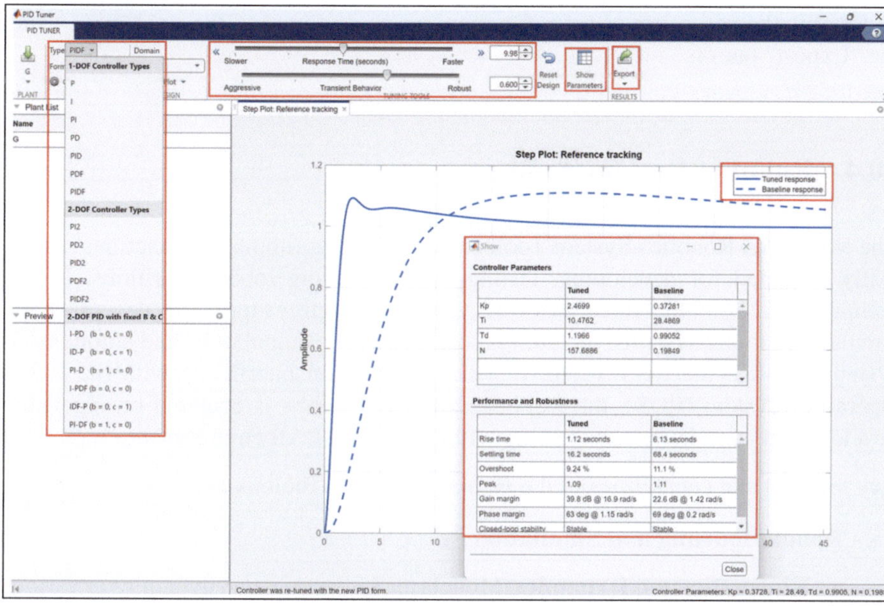


**Fig. 10.8** Response of the controlled system with a PIDF controller automatically tuned by providing the desired response time and transient response

In the launched window, you can see the step response of the system without the PID controller. By clicking on the "Export" command dropdown menu and selecting "Save as Baseline", we can see and compare the responses of the uncontrolled versus controlled system as can be seen in Fig. 10.8.

We can select the controller type via the "Type" dropdown menu (e.g., PI, PD, PID, or PIDF). The transfer function for a PID controller with a low-pass filter (PIDF) in the most general form that is suitable for systems with noisy signals is given by Eqn. 10.4:

$$\mathbf{C_{PIDF}(s)} = \mathbf{K_p}\left(\mathbf{1} + \frac{\mathbf{1}}{\mathbf{T_i s}} + \frac{\mathbf{T_d s}}{\mathbf{1} + \mathbf{T_f s}}\right) \tag{10.4}$$

Furthermore, the analysis can be done in the "Time" or "Frequency" domains (under the "Domain" dropdown menu) and the controller could be placed in "Parallel" or "Standard/Series" with the plant transfer function (under the "Form" dropdown menu). More response plots are available under the "Add Plot" dropdown menu.

Next, we can manually move the sliders for the **Response Time** and the **Transient Behavior** to left or right to make the former **Slower** or **Faster** and the latter **Aggressive** (more overshoot) or **Robust**.

Controller parameters ($\mathbf{K_p}$, $\mathbf{T_i}$, $\mathbf{T_d}$, $\mathbf{T_f}$) and the controlled system response would change accordingly and can be visualized in real time.

Finally, the controller transfer function can be exported to the workspace using the “Export” button.

## 10.4 Robotics System Toolbox

The MATLAB Robotics System Toolbox provides algorithms and functions specifically designed for developing, testing, and deploying robot algorithms. It is a comprehensive tool that integrates with MATLAB to offer support for designing and simulating robotic systems, including mobile robots, manipulators, and autonomous systems. Unlike the ROS Toolbox, which focuses on interfacing with the Robot Operating System (ROS), the Robotics System Toolbox is centered on algorithm development, simulation, and testing within the MATLAB environment.

Key features and capabilities of the Robotics System Toolbox are:

i. **Robot Modeling and Simulation**

- **Kinematics and Dynamics**: Models the kinematics and dynamics of various robotic systems, including manipulators and mobile robots. It supports forward and inverse kinematics, dynamics equations, and motion simulation.
- **URDF and Simulink Integration**: Imports robot models using URDF files and simulates them within MATLAB or Simulink. This integration allows for co-simulation with other MATLAB toolboxes and Simulink models.

ii. **Path Planning and Navigation**

- **Path Planning Algorithms**: Provides several path planning algorithms like A*, RRT (Rapidly-exploring Random Trees), PRM (Probabilistic Roadmaps), and more. These are used to compute collision-free paths for mobile robots and manipulators.
- **SLAM (Simultaneous Localization and Mapping)**: Includes algorithms for SLAM, which allow robots to build a map of an unknown environment while simultaneously keeping track of their location within that environment.
- **Localization and Mapping**: Supports algorithms for localization (such as Monte Carlo localization) and mapping, which are essential for autonomous navigation.

iii. **Control Algorithms**

- **Trajectory Generation**: Generates trajectories for robots using various polynomial, spline, and trapezoidal velocity profiles.
- **Motion Control**: Implements control strategies for robotic arms and mobile robots, including PID controllers, inverse dynamics control, and model predictive control.
- **Visual Servoing**: Supports visual servoing, which allows you to control robot motion based on visual feedback from cameras.

iv. **Sensor Integration and Simulation**

- **Lidar and Vision**: Simulates and processes data from Lidar sensors, stereo cameras, and depth sensors. The toolbox provides algorithms for 3D point cloud processing, object detection, and feature extraction.
- **IMU and GPS**: Integrates Inertial Measurement Unit (IMU) and GPS data for sensor fusion and state estimation, which is crucial for accurate robot localization.

v. **Robot Manipulators**

- **Manipulation Algorithms**: Develops algorithms for robot manipulators, including pick-and-place tasks, inverse kinematics, and control strategies for articulated robots.
- **Collision Detection and Avoidance**: Implements collision detection and avoidance mechanisms for manipulators, ensuring safe operation in complex environments.

vi. **Autonomous Mobile Robots**

- **Multi-Robot Systems**: Models and simulates multi-robot systems, enabling you to test cooperative behaviors, swarm robotics, and coordination strategies.
- **Custom Environments**: Creates custom simulation environments with different terrains, obstacles, and sensor placements to test robot navigation and interaction with the environment.

vii. **Code Generation and Deployment**

- **C/C ++ Code Generation**: Supports code generation for deploying algorithms to embedded systems, including microcontrollers and real-time operating systems.
- **Hardware Integration**: Integrates with hardware platforms like Arduino, Raspberry Pi, and other microcontrollers for real-time control and testing of robotic algorithms.

### Applications

The MATLAB Robotics System Toolbox is used across a wide range of robotics applications, including:

- Autonomous vehicle navigation.
- Industrial automation and robot manipulators.
- Mobile robot path planning and mapping.
- Multi-robot coordination and swarm robotics.
- Research and development in robotic systems and control algorithms.
- Educational purposes for teaching robotics concepts and control theory.

This toolbox is particularly valuable for researchers, engineers, and educators who are developing and testing complex robotic systems in a simulated environment before deploying them in real-world scenarios.

For our first example, we will look at a path planning demo for a mobile robot using the PRM algorithm via the built-in command **robotics.PRM**. The parameters to control are the number of nodes and the maximum distance allowed for two nodes to be connected. We create an occupancy grid map using the **robotics.OccupancyGrid**, which in the code below is loaded from an existing **.mat** file in the MATLAB installation folders. Once the start and end locations are set, we can create a **prm** object on which the **findpath** method can be applied using the **prm** start and end locations to find the **path**. Note that the path is not going to be optimal and will be random (different results every time you run the code).

In the code shown below, we use the number of nodes and the maximum connection distance to investigate their effects.

```
% Load a map file and create an occupancy grid

filePath = fullfile(fileparts(which('PathPlanningExample')),'data','exampleMaps.mat');

load(filePath)
map = robotics.OccupancyGrid(simpleMap,2);

% Create a roadmap with 100 nodes
rngState = rng;
prm = robotics.PRM(map,100)

% determine the start and end location
startLocation = [2 1];
endLocation = [12 1];

% calculate the path and show the map
figure(1);
path = findpath(prm,startLocation,endLocation);
show(prm)
title('PRM with 100 nodes and ConnectionDistance=inf');
```

In Fig. 10.9, we can see the results for the same start and end locations. Part (a) shows the resulting path using 100 nodes and an infinite maximum connecting distance; Part (b) shows the resulting path using 400 nodes and an infinite maximum connecting distance; and Part (c) shows the resulting path using 100 nodes and a maximum connecting distance of 3 units.

Our second example is about the Analytical Inverse Kinematics (IK) tool that can be applied to a variety of well-known robots. The list of these robots can be found in the following link:

https://www.mathworks.com/help/robotics/ref/loadrobot.html

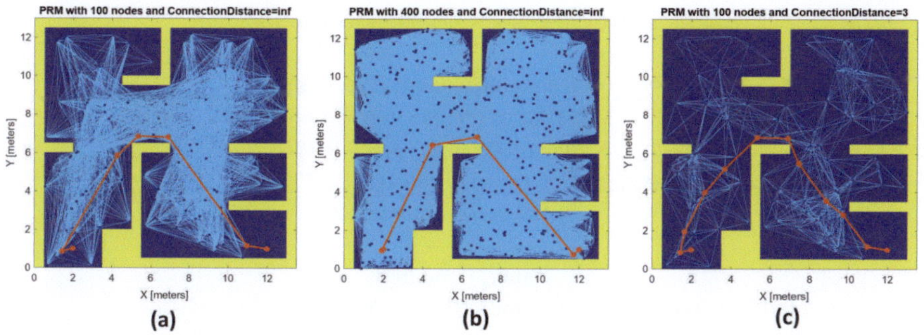


**Fig. 10.9** Path planning using PRM algorithm a) 100 nodes and an infinite maximum connecting distance b) 400 nodes and an infinite maximum connecting distance c) 100 nodes and a maximum connecting distance of 3 units

Note that in order guarantee the existence of an IK solution, the robot must be a six-DOF serial manipulator with a spherical wrist.

First, we load the robot object using the command **loadrobot**, an IK object using the command **analyticalInverseKinematics**, and an IK function using the command **generateIKFunction** (note the **IsValidGroupForIK: 1** indicates that this robot qualifies for an analytical IK solution):

```
% create the robot object
robot = loadrobot('abbIrb120','DataFormat','row')
```

```
robot = rigidBodyTree with properties:
NumBodies: 8
Bodies: {1×8 cell}
Base: [1×1 rigidBody]
BodyNames: {'base' 'link_1' 'link_2' 'link_3' 'link_4' 'link_5' 'link_6' 'tool0'}
BaseName: 'base_link'
Gravity: [0 0 0]
DataFormat: 'row'
```

```
% create the IK object and function
aik = analyticalInverseKinematics(robot)
showdetails(aik)
generateIKFunction(aik,'robotIK');
```

```
aik = analyticalInverseKinematics with properties:

            KinematicGroup: [1×1 struct]
             RigidBodyTree: [1×1 rigidBodyTree]
        KinematicGroupType: 'RRRSSS'
 KinematicGroupConfigIdx: [1 2 3 4 5 6]
          IsValidGroupForIK: 1
        --------------------
          Robot: (8 bodies)
 Index    Base Name   EE Body Name   Type          Actions
   -----  ---------  ------------  ----          -------
     1    base_link       link_6  RRRSSS  Use this kinematic group
     2    base_link        tool0  RRRSSS  Use this kinematic group
```

The first few lines of the **robotIK** function are shown below:

```
function qOpts = robotIK(eeTform, enforceJointLimits, sortByDistance,
referenceConfig)
%robotIK Function for generating closed-form inverse kinematics solutions to the DH
robot given by the parameters specified below
% $Revision: $ $Date: $
% Generated on 19-Aug-2024 14:06:10
dhParams = [0 -1.5707963267949 0.29 0;0.27 0 0 0;0.0700000000000001 -
1.5707963267949 0 0;0 1.5707963267949 0.302 0;0 -1.5707963267949 0 0;0 -
1.5707963267949 0.072 0];
thetaOffsets = [0 -1.5707963267949 0 0 0 -1.5707963267949];
lastThreeAxesSign = [1 -1 1];
jointLimits = [-2.87979 2.87979;-1.91986 1.91986;-1.91986 1.22173;-2.79253 2.79253;-
2.094395 2.094395;-6.98132 6.98132];
isJointRevolute = ([true true true true true true]);
```

Next, we define a desired pose (position and orientation) for the robot. It can be done using the command **trvec2tform** or the commands **homeConfiguration** and **getTransform** as shown below:

```
% manually set the eePose
eePosition = [0 0.5 0.5];
eePose = trvec2tform(eePosition);

% home configuration of the robot (joint values)
hmConfig = homeConfiguration(robot)
eePose = getTransform(robot,hmConfig,eeBodyName,baseName)
```

Now we can obtain the IK solutions using the command **robotIK** and plot them using the command **plotTransforms** (which requires the end-effector position and end-effector orientation in quaternions):

```
% IK solutions and their number
ikConfig = robotIK(eePose)
numSolutions = size(ikConfig,1);

% show all solutions
for ii = 1:numSolutions
        figure(ii);
        show(robot,ikConfig(ii,:));
        hold on;

        % plotTransforms(translations,rotations) plots 3-D transforms from
        % translations and rotations (in quaternions)
        plotTransforms(eePosition,tform2quat(eePose));
        title(['IK solution ' num2str(ii)]);
end
```

All 4 results for the home configuration of the **abbIrb120** robot can be seen in Fig. 10.10.

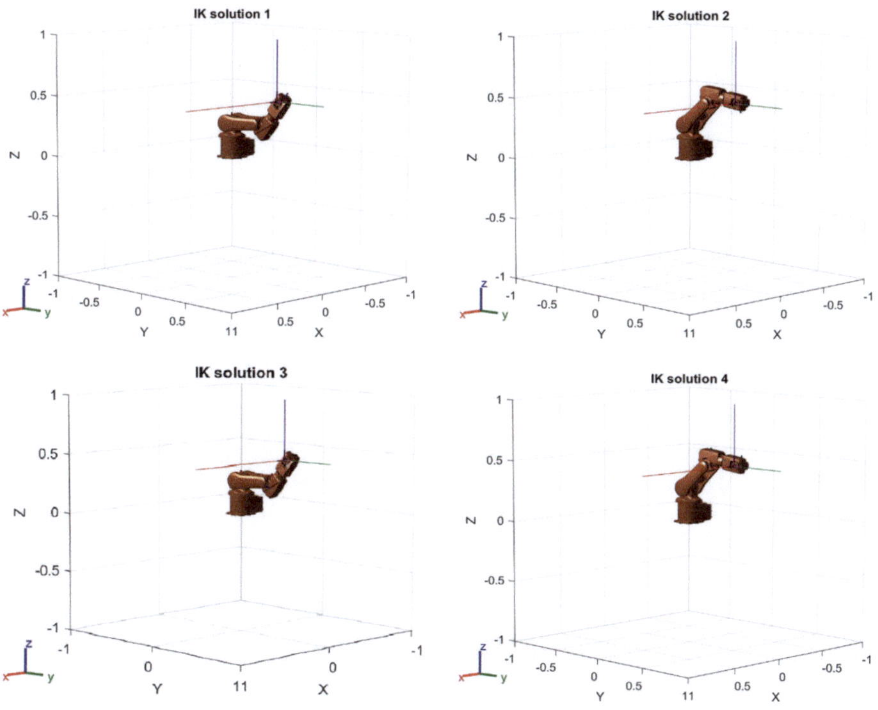


**Fig. 10.10** Analytical Inverse Kinematics for an abbIrb120 robot

For our last example, we will use the **inverseKinematicsDesigner** application GUI and the RRT path planning algorithm for a robotic manipulator given start and goal positions and extra constraints, such as mechanical joint limits.

First, we load a **universalUR5e** robotic manipulator using the function **loadrobot.**

```
% load the robot
ur5e_robot = loadrobot("universalUR5e");
```

Next, we create workspace objects to which collision must be avoided using the **collisionBox**, **collisionCylinder**, and **collisionSphere** commands and set their pose properties (we can also alter their orientations):

```
% collision box
obj1 = collisionBox(0.2,0.3,0.4); % size of the box
matZ = axang2tform([0 0 1 pi/2]); % rotate the box
matZ(1:3,4)=[0.3 -0.3 .05]'; % move the box
obj1.Pose = matZ; % pose is a 4x4 homogenous transformation metrix

% collision cylinder
obj2 = collisionCylinder(.1,0.4); % radius and length
obj2.Pose = trvec2tform([-.45 .05 0.2]);

% collision sphere
obj3 = collisionSphere(0.15);
obj3.Pose = trvec2tform([0.5 .3 -0.1]);
```

Next, we load the Inverse Kinematics Designer application GUI using the **inverseKinematicsDesigner** command.

Once the GUI is launched as can be seen in Fig. 10.11, we click on “New Session” to add the robot object as an **.stl** file from the workspace, and then click on “Add Collision Object” to load the collision objects as **.stl** files from the workspace.

We can also add other types of constraints (shown at the bottom of Fig. 10.11), such as joint limit constraints.

Now we need to define an initial point, a goal point, and mid points in the workspace for path planning. We can select the frame attached to any member from the “Maker Body” dropdown menu and translate or rotate that member (and the rest of the robot as a result) by dragging the marker frame that will be attached to it as shown in Fig. 10.12. Once the desired position and orientation is reached, the window on the right shows the pose for the frame attached to the member. Also, under “Check Collision”, we can ensure that the robot does not collide with obstacles in the workspace or with itself (self-collision can be ignored as can be seen in Fig. 10.12).

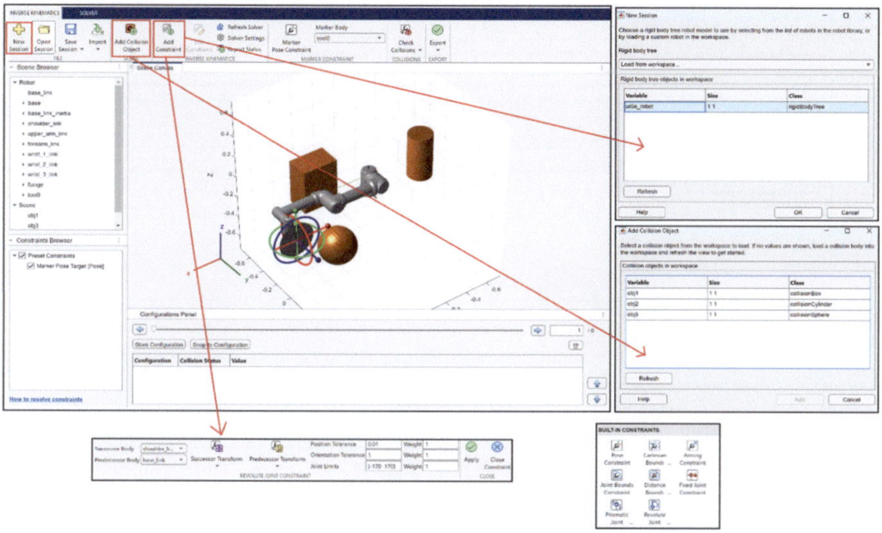

**Fig. 10.11** Using GUI for inverseKinematicsDesigner on a universalUR5e robot

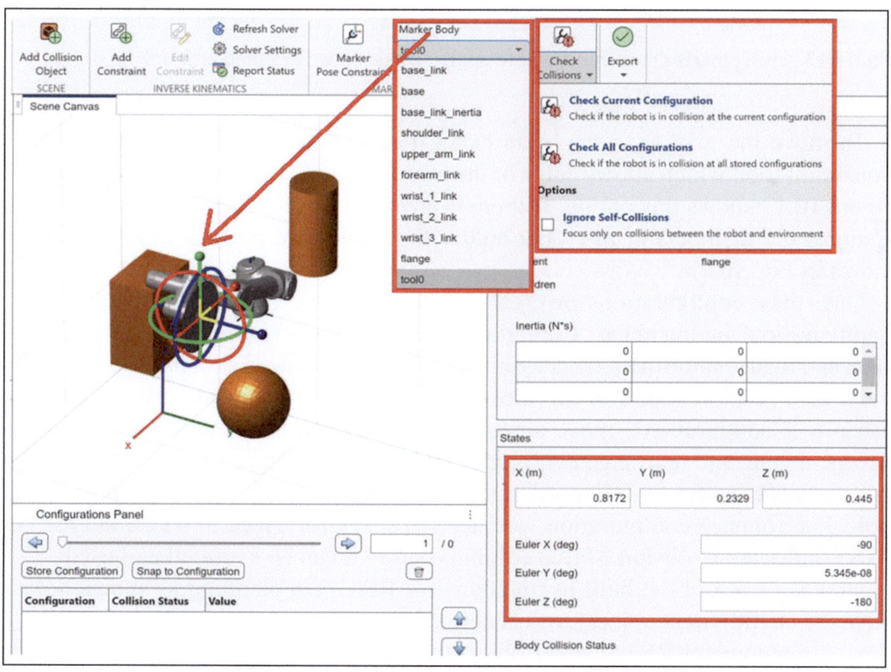


**Fig. 10.12** Moving a marker frame to a desired pose and checking collisions

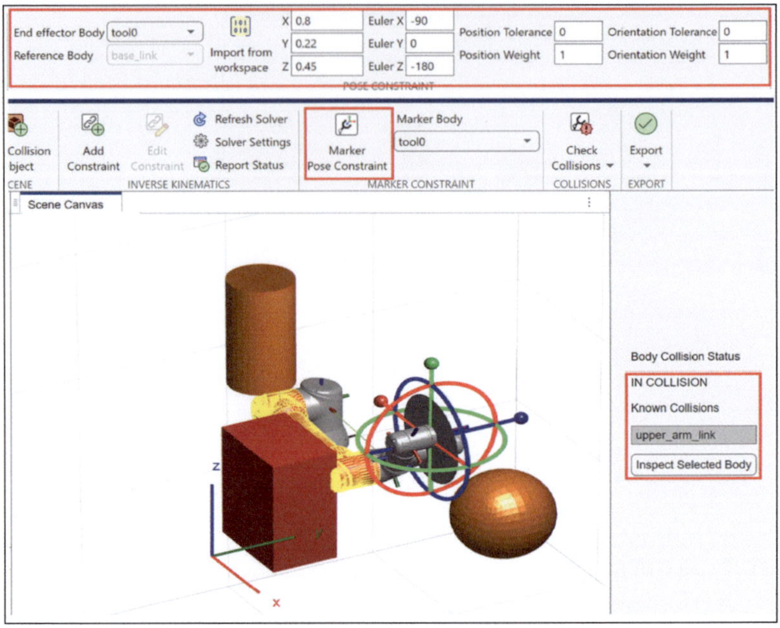


**Fig. 10.13** Using marker pose constraint to apply an exact pose to a marker frame

To move the marker frame to an exact pose, we need to use the Marker Pose Constraint tool, which allows entry of the exact numerical values in the top window. Figure 10.13 shows this option. If there is any collision in the current or all configurations, the members of the robot and/or the workspace obstacles become red as shown in Fig. 10.13.

Once these configurations are created, you can go to each one, click on the "Store Configuration" button in the "Configuration Panel" at the bottom of the GUI to store them and then rename them if necessary. Figure 10.14 shows a start point, a midpoint, and a goal point stored. You can move to a specific configuration using the button "Snap to Configuration", move among the configurations using the left and right arrows, delete, and rearrange configurations. The configurations can be exported to the workspace using the "Export" button under the variable name **waypointData**. Note that, for each configuration, you need to check for a lack of collision (you see **PASS** under the **Collision Status** column) before it can be exported and used.

Now we can use the built-in functions for RRT path planning as shown below. First, the environment object (**env**) is defined by joining the obstacles as a cell array. Next, a manipulator RRT planner object (**planner**) is defined by passing the robot object and the environment object to the command **manipulatorRRT**.

```
env = {obj1 obj2 obj3};
planner = manipulatorRRT(ur5e_robot,env);
```

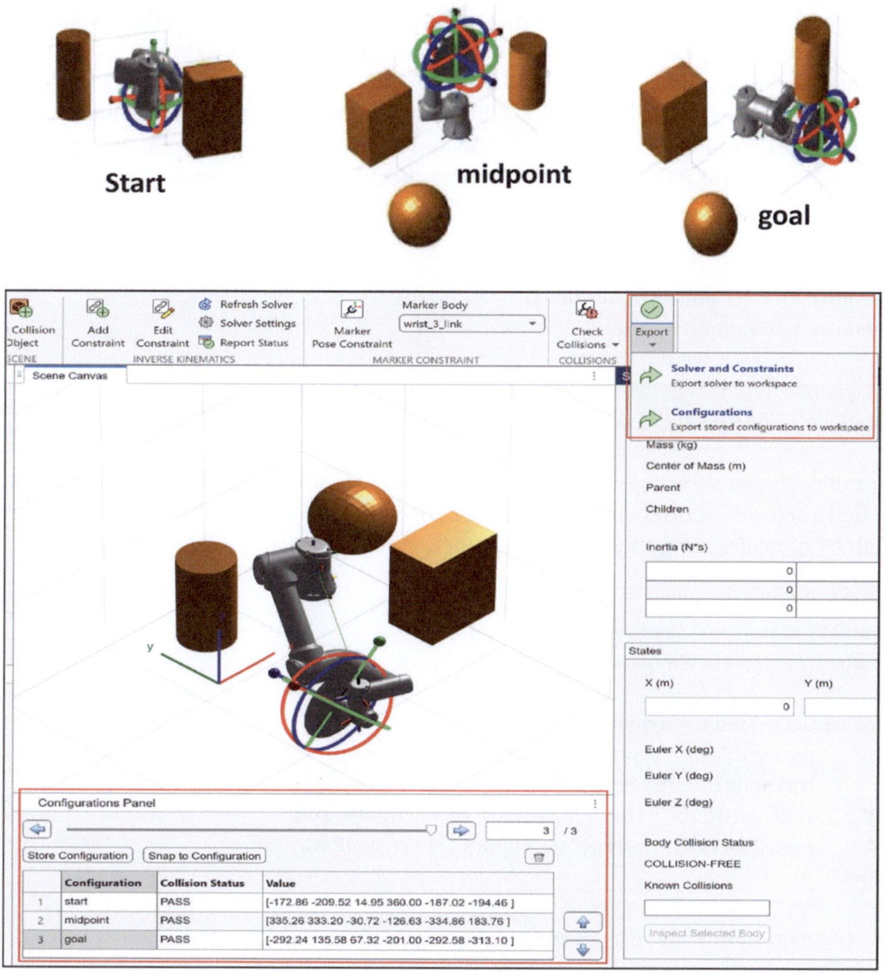


**Fig. 10.14** Storing and exporting configurations to workspace

The RRT algorithm has several parameters that can be tuned, including **MaxConnectionDistance**, **EnableConnectHeuristic** (ignoring the maximum connection distance)**, ValidationDistance** (distance between intermediate points for a path segment ensuring no collision along that path segment), **IgnoreSelfCollision**, and **SkippedSelfCollisions** (**adjacent** means only skip checking collision among adjacent members and **parent** means only skip checking collision among members and their predecessors):

```
% planner.MaxConnectionDistance = 0.5;
planner.EnableConnectHeuristic=1;
planner.ValidationDistance = 0.05;

% planner.IgnoreSelfCollision = 0;
planner.SkippedSelfCollisions = "adjacent";
```

For each pair of waypoints, we run the commands **plan**, **shorten**, and **interpolate** to find a smooth, shortened path segment between the points. We then make the entire path by concatenating these path segments.

```
numPts = 30;
numWaypoints = size(waypointData,1);
paths = cell(1,numWaypoints);

for segIdx = 1:numWaypoints
        plannedPath = plan(planner, waypointData(segIdx,:),
        waypointData(mod(segIdx,numWaypoints)+1,:));
        shortenedPath = shorten(planner, plannedPath, 10);
        paths{segIdx} = interpolate(planner, shortenedPath, 10);
End

% complete path is made by combining the segments
totalSegs = vertcat(paths{:});
```

Finally, the robot, environment, and the path can be visualized and animated:

```
ax = show(ur5e_robot);
hold all
for ii = 1:numel(env)
        env{ii}.show("Parent", ax);
end
view(-33,14);

% Display the figure window the animation
pathFig = ancestor(ax, 'figure');
set(pathFig, "Visible", "on")

% Set up timing and configure robot
r = rateControl(50);
tvec = linspace(1,numWaypoints,numWaypoints*numPts);
ur5e_robot.DataFormat = "row";

% Animate all path segments
for pathSegIdx = 1:numel(paths)
        path = paths{pathSegIdx};
        % For each path segment, step through all the configurations
        for configIdx = 1:size(path,1)
                show(ur5e_robot, path(configIdx,:), "FastUpdate",true,
                "PreservePlot",false,"Parent",ax);
                waitfor(r);
        end
        % Hold the pose and update the title each time a waypoint is reached
        title(sprintf('Segment %i completed', pathSegIdx), "Parent",ax);
        pause(0.1);
end
```

## 10.5 Signal Processing Toolbox

The Signal Processing Toolbox in MATLAB is a comprehensive set of functions and apps designed to analyze, design, and process signals. It provides engineers, scientists, and researchers with the tools needed to work on both analog and digital signals, enabling a wide range of tasks in signal processing, including filtering, spectral analysis, and time–frequency analysis.

Key features and capabilities of the Signal Processing Toolbox are:

i. **Filter Design and Analysis**

   - Design various types of filters, including FIR (Finite Impulse Response) and IIR (Infinite Impulse Response) filters.

- Analyze filter responses, such as magnitude, phase, group delay, and impulse response.
- Implement adaptive filters and multi-rate filters (like decimators and interpolators).

ii. **Spectral Analysis**

- Perform spectral estimation using methods like FFT (Fast Fourier Transform), Welch's method, and Multi-taper method.
- Compute power spectral densities, spectrograms, and periodograms.
- Analyze signals in the frequency domain for better understanding and manipulation.

iii. **Time–Frequency Analysis**

- Perform time–frequency analysis using wavelets, Wigner-Ville distributions, and other techniques.
- Analyze how the frequency content of a signal evolves over time, useful for non-stationary signals.

iv. **Signal Generation**

- Generate signals for testing and simulation, such as sine waves, chirps, and random noise.
- Create multi-channel signals and perform operations like modulation and demodulation.

v. **Signal Transformation**

- Apply various transforms like Fourier, Hilbert, and wavelet transforms.
- Perform short-time Fourier transforms (STFT) for analyzing signal components over time.

vi. **Filtering and Smoothing**

- Implement and apply filters to smooth data, reduce noise, or extract specific components from signals.
- Use moving average filters, Savitzky-Golay filters, and more advanced methods.

vii. **Wavelet Analysis**

- Perform discrete and continuous wavelet transforms.
- Use wavelets for signal compression, denoising, and feature extraction.

viii. **Analog and Digital Modulation/Demodulation**

- Simulate analog and digital modulation schemes such as AM, FM, PSK, and QAM.
- Analyze and process modulated signals for communication system design.

ix. **Multi-rate Signal Processing**

- Design and implement systems that operate at multiple sampling rates.
- Perform operations like up-sampling, down-sampling, and fractional delay filtering.

x. **Measurement and Analysis**

- Measure and analyze signal characteristics like signal-to-noise ratio (SNR), harmonic distortion, and peak analysis.
- Use tools for statistical analysis of signals, including correlation and covariance.

xi. **Interactive Tools and Apps**

- Design, analyze, and visualize signals without needing to write extensive code using interactive apps like Filter Designer and Signal Analyzer.
- Use the Signal Analyzer app to visualize and explore time-domain and frequency-domain characteristics of signals.

**Applications**

- **Communications**: Design and simulate communication systems, including modulation, demodulation, and signal filtering.
- **Audio Processing**: Analyze and process audio signals for applications like noise reduction, equalization, and effects processing.
- **Biomedical Signal Processing**: Work with ECG, EEG, and other biomedical signals for diagnostics and research.
- **Vibration Analysis**: Analyze mechanical vibrations and noise, useful in predictive maintenance and structural health monitoring.
- **Radar and Sonar**: Process radar and sonar signals for target detection and tracking.

For our first example, we look at an audio processing example in which we create an audio file using an existing MATLAB file, read its content using the command **audioread**, apply a Butterworth low-pass filter of order 10 with a cut-off frequency of 500 Hz using the commands **butter** and **filter** to remove high-frequency noise, visualize the results both in the time domain and in the time/frequency domain together with a spectrogram using the command **spectrogram**, listen to the original and modified audio contents, and finally write the modified **.wav** file to the local folder using the **audiowrite** command.

```
% create the wav file from an existing mat file
load handel.mat;
filename = 'handel.wav';
audiowrite(filename,y,Fs);
clear y Fs;

% Load the audio file
[audioIn, Fs] = audioread('handel.wav'); % Replace 'example.wav' with the audio file

% Define the duration to zoom in on (e.g., first 0.1 seconds)
duration = 0.1; % 0.1 seconds
samples = round(duration * Fs); % Number of samples to plot

% Plot the original audio signal (zoomed in)
t = (0:samples-1)/Fs; % Time vector for the zoomed-in plot
figure;
subplot(3,1,1);
plot(t, audioIn(1:samples));
title('Original Audio Signal (Zoomed In)');
xlabel('Time (s)');
ylabel('Amplitude');

% Design a low-pass filter
Fc = 500; % Cutoff frequency of 1000 Hz
n = 10; % Order of the filter
[b, a] = butter(n, Fc/(Fs/2), 'low'); % Butterworth low-pass filter design

% Apply the filter to the audio signal
audioOut = filter(b, a, audioIn);

% Plot the filtered audio signal (zoomed in)
subplot(3,1,2);
plot(t, audioOut(1:samples));
title('Filtered Audio Signal (Low-Pass, Fc = 500 Hz, Order = 10, Zoomed In)');
xlabel('Time (s)');
ylabel('Amplitude');

% Perform and plot the frequency analysis (spectrogram) of the original signal
subplot(3,1,3);
spectrogram(audioIn, 256, [], [], Fs, 'yaxis');
title('Spectrogram of Original Audio Signal');
% Listen to the original and filtered audio signals
sound(audioIn, Fs); % Play original audio
pause(length(audioIn)/Fs + 1); % Wait for the audio to finish
sound(audioOut, Fs); % Play filtered audio

% Save the filtered audio to a new file
audiowrite('filtered_example.wav', audioOut, Fs); % Replace 'filtered_example.wav'
with your desired output file name
```

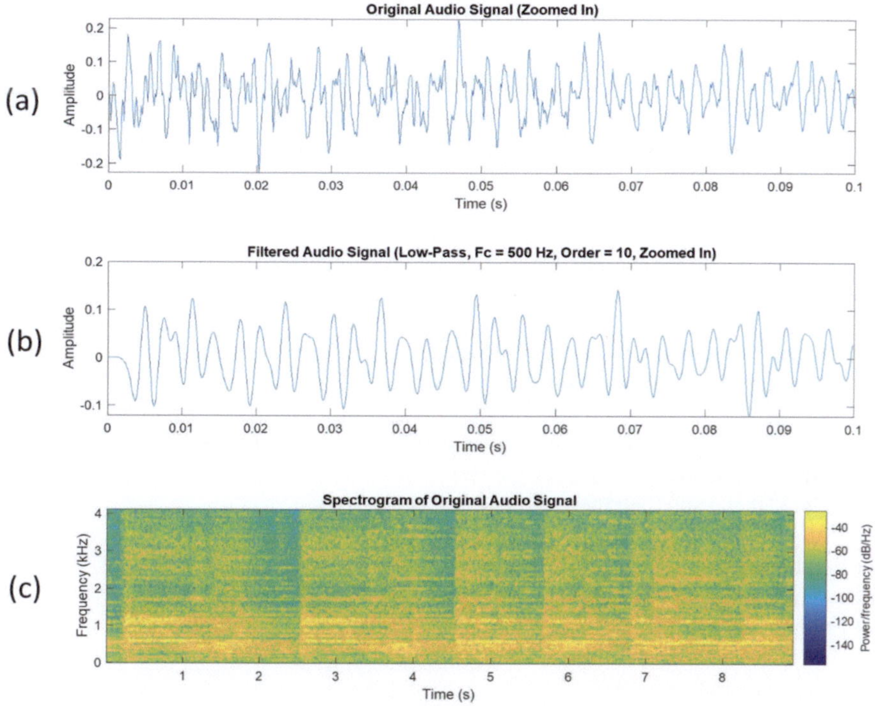


**Fig. 10.15** Audio Processing (filtering high-frequency content) using MATLAB Signal Processing Toolbox: a) original audio file zoomed in, b) modified audio file zoomed in, and c) spectrogram of the original audio file

Results of the code above can be seen in Fig. 10.15, where part (a) shows the original audio, part (b) shows the modified audio where the high-frequency noise is removed, and part (c) shows the spectrogram of the original audio.

For our second example, we look at a vibration analysis problem. Let's say you are tasked with analyzing the vibration data from a rotating machine to identify any abnormal behavior or potential mechanical faults. The machine operates at a known rotation speed, and you suspect that there might be an imbalance or misalignment causing additional vibration components at harmonic frequencies.

### Given

- Vibration data sampled at 10 kHz (sampling frequency).
- The machine operates at a fundamental frequency of 50 Hz.
- The vibration signal is stored in a file called **vibration_data.mat** containing a variable **vibrationSignal**.

Note that in this example, we artificially create the vibration data, but real data can be downloaded from websites such as kaggle.com. The code to generate this data is given below:

```
Fs = 10000; % Sampling frequency (10 kHz)
t = 0:1/Fs:10-1/Fs; % Time vector for 10 seconds

% Simulate a vibration signal with a fundamental frequency of 50 Hz
f0 = 50; % Fundamental frequency (50 Hz)
vibrationSignal = sin(2*pi*f0*t);

% Add harmonics
vibrationSignal = vibrationSignal + 0.5*sin(2*pi*2*f0*t); % 2nd harmonic
vibrationSignal = vibrationSignal + 0.3*sin(2*pi*3*f0*t); % 3rd harmonic

% Add some random noise
vibrationSignal = vibrationSignal + 0.1*randn(size(t));

% Save the simulated data to a .mat file
save('vibration_data.mat', 'vibrationSignal');
```

### Tasks

- Load and visualize the vibration signal in the time domain.
- Perform spectral analysis to identify the frequency components using the command **FFT** (Fast Fourier Transform).
- Apply a 4th-order Butterworth band-pass filter with a frequency range of [40 200] Hz to isolate the frequency components around the fundamental frequency and its harmonics using commands **butter** and **filter**.
- Plot the filtered signal in the time domain and compare it with the original signal.

To performs these tasks, we use the code below:

```
% Load the simulated vibration signal
load('vibration_data.mat'); % Load the data file containing 'vibrationSignal'
Fs = 10000; % Sampling frequency (10 kHz)
t = (0:length(vibrationSignal)-1)/Fs; % Time vector
% Design a band-pass filter to isolate the fundamental frequency and harmonics
Fc1 = 40; % Lower cutoff frequency (40 Hz)
Fc2 = 200; % Upper cutoff frequency (200 Hz)
[b, a] = butter(4, [Fc1 Fc2]/(Fs/2), 'bandpass'); % 4th order Butterworth band-pass
filter

% Apply the band-pass filter to the vibration signal
filteredSignal = filter(b, a, vibrationSignal);

% Compute the FFT of the filtered signal
n = length(filteredSignal);
f = (0:n-1)*(Fs/n); % Frequency vector
filteredSignalFFT = abs(fft(filteredSignal)/n); % Compute FFT of the filtered signal
filteredSignalFFT = filteredSignalFFT(1:floor(n/2)); % Keep only the positive
frequencies
f = f(1:floor(n/2)); % Adjust frequency vector

% Plot the data
figure;
% Subplot 1: Original Signal (Zoomed In)
subplot(3,1,1);
plot(t, vibrationSignal);
title('Original Vibration Signal (Zoomed In)');
xlabel('Time (s)');
ylabel('Amplitude');
xlim([0 0.1]); % Zoom in on the first 0.1 seconds
% Subplot 2: Filtered Signal (Zoomed In)
subplot(3,1,2);
plot(t, filteredSignal);
title('Filtered Vibration Signal (Zoomed In)');
xlabel('Time (s)');
ylabel('Amplitude');
xlim([0 0.1]); % Zoom in on the first 0.1 seconds
% Subplot 3: FFT of Filtered Signal
subplot(3,1,3);
plot(f, filteredSignalFFT);
title('Frequency Spectrum of Filtered Signal');
xlabel('Frequency (Hz)');
ylabel('Amplitude');
```

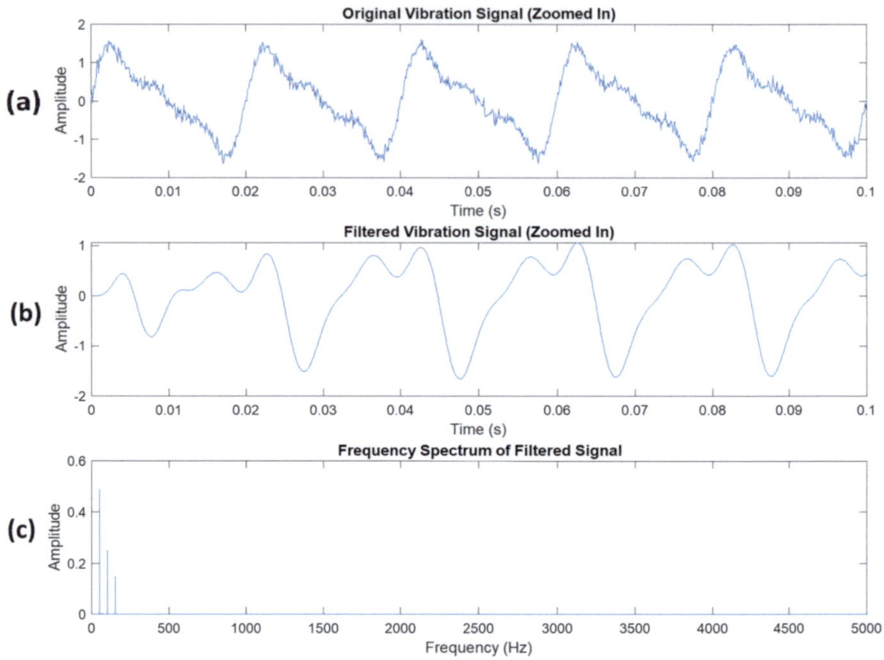


**Fig. 10.16** Vibration analysis (applying a bandpass filter to look for harmonics) using MATLAB Signal Processing Toolbox: a) original signal, b) filtered signal, and c) FFT of the filtered signal

Results of the code above can be seen in Fig. 10.16, where part (a) shows the original vibration signal, part (b) shows the filtered signal, and part (c) shows the FFT of the filtered signal, where the harmonics of the fundamental frequency are detected.

## 10.6 Image Processing and Computer Vision Toolboxes

MATLAB offers powerful toolboxes for image processing and computer vision, which are widely used in various applications ranging from basic image manipulation to advanced computer vision tasks. Here's an overview of what you can do with them:

1. Image Processing Toolbox

The Image Processing Toolbox in MATLAB provides a comprehensive set of functions for image processing, analysis, visualization, and algorithm development. Some key capabilities include:

i. **Image Filtering and Enhancement**: Functions for noise reduction, sharpening, edge detection, and contrast enhancement.

ii. **Geometric Transformations**: Tools to perform scaling, rotation, translation, and affine transformations.
iii. **Image Segmentation**: Techniques for dividing an image into meaningful regions, including thresholding, edge-based, region-based, and clustering methods.
iv. **Morphological Operations**: Tools for dilating, eroding, opening, closing, and other morphological operations used to process binary images.
v. **Image Registration**: Tools to align multiple images taken from different viewpoints, sensors, or times.
vi. **Feature Detection and Extraction**: Functions to detect edges, corners, blobs, and other features.
vii. **Image Analysis**: Tools to measure object properties, such as area, perimeter, and centroid, as well as to perform statistical analysis on image data.
viii. **Image Visualization**: Functions to display images in various formats, apply color maps, and create interactive tools for exploring image data.
ix. **Image Arithmetic**: Tools to perform operations like addition, subtraction, multiplication, and division on images.

2. Computer Vision Toolbox

The Computer Vision Toolbox extends MATLAB's capabilities to include more specialized functions for computer vision. Some of its key features are:

i. **Object Detection and Recognition**: Algorithms for detecting and recognizing objects, faces, and other patterns in images and video.
ii. **Feature Matching**: Tools for finding correspondences between features in different images.
iii. **3D Vision**: Functions for 3D reconstruction, structure from motion, and stereo vision, allowing the creation of 3D models from images.
iv. **Video Processing**: Techniques for video stabilization, tracking, and motion analysis, including optical flow.
v. **Camera Calibration**: Tools to estimate camera parameters, rectify images, and perform un-distortion.
vi. **Deep Learning for Vision**: Integration with deep learning frameworks for tasks like image classification, semantic segmentation, and object detection using pre-trained models or custom networks.
vii. **Simultaneous Localization and Mapping (SLAM)**: Functions for mapping environments and tracking the position of a moving camera in real-time.
viii. **Gesture and Motion Analysis**: Tools for analyzing human gestures and motion for applications like sign language recognition and activity monitoring.
ix. **Automated Driving and Robotics**: Tools to simulate and design vision-based perception systems for automated vehicles and robots.

**Applications**

These toolboxes are used across a wide range of industries and research areas, including:

- **Medical Imaging**: For enhancing and analyzing medical images such as MRI, CT scans, and X-rays.
- **Automated Inspection**: In manufacturing, to detect defects or irregularities in products.
- **Remote Sensing**: Analyzing satellite or aerial imagery for environmental monitoring, agriculture, and urban planning.
- **Surveillance**: For face recognition, people counting, and anomaly detection in security systems.
- **Automotive**: In the development of advanced driver-assistance systems (ADAS) and autonomous vehicles.
- **Augmented Reality**: To overlay digital content on real-world images and videos.
- **Robotics**: For vision-based navigation, object manipulation, and human–robot interaction.

In this section, we will not show examples with deep learning, instead leaving them for Sect. 10.8.

For our first example, we look at automated recognition of US bills using Speeded-Up Robust Features (SURF) features. Pictures for the front of US bills (templates) and some test images in which the image is rotated, scaled, or cluttered can be downloaded from the following link: https://drive.google.com/file/d/1qWcMpctt0MdzUB1E_ZlwYnTh63hK6HZt/view

Note that the image to be recognized (the **billImage** variable) is directly read into MATLAB using the command **imread**, but as can be seen in lines 4 to 7 of the following code (commented out), it can also be read from a USB webcam (if the support package for USB webcams is installed) using the command **snapshot** and a webcam that was detected and defined using the commands **webcamlist** and **webcam(n),** respectively (**n** is the webcam number if more than one exists).

Next, pictures of the template bills are read, converted to grayscale, and SURF features are detected for the template images and test image using the command **detectSURFFeatures.**

The SURF feature descriptors are obtained for all these images using the command **extractFeatures** and the best match from the templates for the test image is found using the command **matchFeatures.** A selection refinement is done using an inlier analysis, where we remove the matched images for which the number of matched SURF points is below a threshold (20).

Finally, the test image and best match found are shown side by side with the SURF features and their matches overlaid using the command **showMatchedFeatures** as can be seen in Fig. 10.11 for the US 1$ bill.

```
% Load the US dollar bill image
billImage = imread('us_bill1.jpg');

% use webcam capture
% webcamlist
% cam = webcam(2)
% billImage = snapshot(cam);
% Define templates for different denominations
templates = {
imread('one_dollar_template.jpg'),
imread('five_dollar_template.jpg'),
imread('ten_dollar_template.jpg'),
imread('twenty_dollar_template.jpg'),
imread('fifty_dollar_template.jpg'),
imread('hundred_dollar_template.jpg')
};
% Create a SURF object
surfObj = detectSURFFeatures(rgb2gray(billImage));
% Initialize variables to store the best match
bestMatchIndex = 0;
maxInlierCount = 0;

% Iterate through templates and perform feature matching
for ii = 1:numel(templates)
        template = templates{ii};

        % Detect SURF features in the template image
        templatePoints = detectSURFFeatures(rgb2gray(template));

        % Extract feature descriptors
        [templateFeatures, templatePoints] = extractFeatures(rgb2gray(template),
        templatePoints);
        [billFeatures, surfObj] = extractFeatures(rgb2gray(billImage), surfObj);

        % Match features between the bill image and the template
        indexPairs = matchFeatures(templateFeatures, billFeatures);
```

```
        % Calculate the number of inliers (good matches)
        numInliers = length(indexPairs);
        % Update if this template has more inliers
                if numInliers > maxInlierCount
                        bestMatchIndex = ii;
                        maxInlierCount = numInliers;
                end
end

% Define a threshold for considering a match
minInlierCountThreshold = 20; % Adjust as needed
% Check if the best match exceeds the threshold
if maxInlierCount > minInlierCountThreshold
        % Display the recognized denomination
        denominations = {'1 Dollar', '5 Dollars', '10 Dollars', '20 Dollars', '50 Dollars',
        '100 Dollars'};
        recognizedDenomination = denominations{bestMatchIndex};
        fprintf('Recognized US Bill: %s\n', recognizedDenomination);
else
        fprintf('No US bill recognized.\n');
end

% Highlight inliers (good matches)
figure;
showMatchedFeatures(rgb2gray(billImage), rgb2gray(templates{bestMatchIndex}), ...
surfObj.Location(indexPairs(:, 2), :), templatePoints(indexPairs(:, 1), :),'montage');
```

As can be seen in Fig. 10.17, recognition is done correctly. If the goal is to help people with visual disabilities, we can add a speaking component to the code to say the recognized bill:

```
NET.addAssembly('System.Speech');
obj = System.Speech.Synthesis.SpeechSynthesizer;
obj.Volume = 100;
Speak(obj, recognizedDenomination);
```

**Fig. 10.17** Recognition of US bills using SURF features: left image is the test image rotated and scaled down; right image is the best match; the matching SURF features are overlaid

For the second example, we look at finding round objects in an image and determining their diameters in millimeters if intrinsic parameters of the capturing camera (focal length and pixel size) and the distance from the camera to the scene are known.

In Fig. 10.18, a quarter coin exists in the scene along with other objects (part of a notebook and a pen). To find circular objects with a radius in a range **[$\mathbf{R_{min}}$ $\mathbf{R_{max}}$]** in an image, we use the command **imfindcircle** which works based on the **circular Hough Transform**. This command has a parameter **sensitivity** in the range [0 1]. As you increase the sensitivity factor, **imfindcircles** detects more circular objects, including weak and partially obscured circles. Outputs of this command are the center location and the radius of detected circles.

The diameter of the circular object in millimeters is given by Eqn. 10.5:

$$D(mm) = 1000 \times D(pixels) \times pixel \times size\left(\frac{m}{picxel}\right) \times \frac{dist}{f} \tag{10.5}$$

In this equation, **dist** is the distance from camera to the scene and**f** is the focal length of the camera in meters.

Command **viscircle** is used to show a circle overlaid on the detected coin in Fig. 10.18.

Finally, the detected diameter of the coin is displayed on the command window, which is very close to its actual size of 24.26 mm.

```
% Read the image
image = imread('test_image.png');

% Convert the image to grayscale
grayImage = rgb2gray(image);
% Detect circles in the image
[centers, radii] = imfindcircles(grayImage, [20 100], 'Sensitivity', 0.8);

% Check if a circle was found
if isempty(radii)
        disp('No circles were detected.');
else
        % Calculate the diameter of the detected circle in pixels
        diameterInPixels = 2 * radii(1);

        % Known Camera Parameters
        focalLength = 5e-3; % Focal length in meters (e.g., 5mm)
        pixelSize = 1.4e-6; % Pixel size in meters (e.g., 1.4 micrometers)
        distanceToObject = 0.6; % Distance to object in meters (e.g., 60 cm)

        % Calculate the diameter in the real world (in meters)
        diameterInMeters = (diameterInPixels * pixelSize * distanceToObject) /
        focalLength;

        % Convert diameter to millimeters
        diameterInMillimeters = diameterInMeters * 1000;

        % Display the result
        fprintf('The diameter of the round object is %.2f mm.\n',
        diameterInMillimeters);

        % Display the image with the detected circle
        imshow(image);
        hold on;
        viscircles(centers, radii, 'EdgeColor', 'b');
        hold off;
end
```

```
The diameter of the round object is 24.24 mm.
```

For our last example, we will look at face detection and tracking in the two images shown in Fig. 10.19, where the face has moved in the images (as a result of camera motion). To detect faces in an image, we use the Haar Cascade classifier. First, we create a **detector** object with the command **vision.CascadeObjectDetector()** and then use the **step** command to detect the faces. The output of this command is a bounding box, which will be used to draw a rectangle around the face in the first image and define a Region of Interest (ROI) in it for feature detection and tracking.

**Fig. 10.18** Diameter calculation for round objects detected in an image

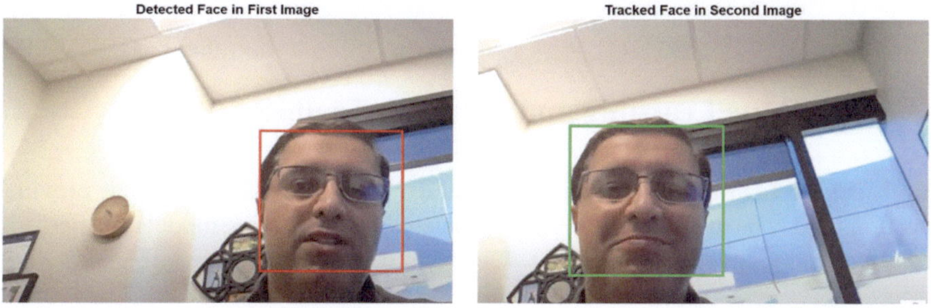


**Fig. 10.19** Face detection and tracking

The command **detectMinEigenFeatures** returns a **cornerPoints** object that contains information about the corner features in this ROI using the minimum eigenvalue algorithm.

Then the commands **vision.PointTracker** and **initialize** are used to track these interest corner points using the **Kanade-Lucas-Tomasi (KLT)** feature-tracking algorithm.

Once the interest points are tracked in the second image using the **step** command, a geometric transformation is calculated using the locations of these points in both images via the command **estimateGeometricTransform,** which in turn will be used to find the new shape of the ROI of the first image in the second image.

Finally, the detected ROIs in both images are highlighted and shown in Fig. 10.19.

```
% Load the first and second images
image1 = imread('face_image1.jpg');
image2 = imread('face_image2.jpg');

% Detect the face in the first image
faceDetector = vision.CascadeObjectDetector();
bbox1 = step(faceDetector, image1);

% Check if a face was detected in the first image
if isempty(bbox1)
        disp('No face detected in the first image.');
else
        % Initialize point tracker
        roi = bbox1(1,:); % Region of interest
        points1 = detectMinEigenFeatures(rgb2gray(image1), 'ROI', roi);
        pointTracker = vision.PointTracker('MaxBidirectionalError', 2);
        initialize(pointTracker, points1.Location, image1);

        % Track the points in the second image
        [points2, isFound] = step(pointTracker, image2);
        visiblePoints1 = double(points1.Location(isFound, :));
        visiblePoints2 = double(points2(isFound, :));
```

## 10.7 Machine Learning and Statistics Toolbox

The Machine Learning and Statistics Toolbox is a powerful toolkit that provides functions and apps to create, train, and evaluate machine learning models and perform statistical analyses. Here's an overview of what it includes:

i. **Machine Learning**

- **Model Training and Evaluation**: Algorithms for supervised learning (e.g., classification, regression) and unsupervised learning (e.g., clustering, dimensionality reduction). It supports decision trees, support vector machines (SVMs), k-nearest neighbors (KNN), ensemble learning methods (e.g., random forests, boosting), and neural networks.
- **Hyperparameter Optimization**: Tools to automatically optimize hyperparameters using techniques like grid search and Bayesian optimization to improve model performance.
- **Feature Selection and Extraction**: Tools for selecting the most relevant features for your models and for extracting new features through techniques like Principal Component Analysis (PCA).

- **Model Exporting**: Trained models can be exported to other environments or deployed as part of a larger system. MATLAB allows you to convert models into C/C++ code, or even generate standalone executables.

ii. **Statistical Analysis**

- **Descriptive Statistics**: Functions to calculate a wide range of descriptive statistics such as mean, median, variance, and standard deviation.
- **Statistical Testing**: Functions for performing hypothesis testing, including t-tests, chi-square tests, ANOVA, and non-parametric tests.
- **Regression Analysis**: Multiple types of regression analysis, including linear regression, logistic regression, and generalized linear models.
- **Multivariate Statistics**: Tools for multivariate data analysis, such as principal component analysis (PCA), factor analysis, and cluster analysis.
- **Probability Distributions**: Functions to work with a variety of probability distributions, perform distribution fitting, and generate random samples from these distributions.

iii. **Data Preprocessing**

- **Data Cleaning**: Functions for handling missing data, outlier detection, and data transformation.
- **Data Transformation**: Techniques such as normalization, scaling, and encoding categorical variables.
- **Time Series Analysis**: Tools for analyzing and forecasting time series data, including functions for autocorrelation, spectral analysis, and smoothing.

iv. **Visualization**

- **Interactive Visualization**: Interactive apps for visualizing data and model performance. This includes plots for confusion matrices, ROC curves, and decision boundaries.
- **Custom Plots**: Customized plots for visualizing the results of statistical analyses or machine learning models.

v. **Apps and Live Scripts**

- **Apps for Easy Model Development**: Interactive apps like the Classification Learner and Regression Learner apps, which allow you to quickly prototype and compare different models without writing code.
- **Live Scripts**: Live scripts that let you combine code, output, and formatted text in an interactive notebook environment, making it easier to document and share your analyses.

vi. **Integration with Other MATLAB Toolboxes**

- The Machine Learning and Statistics Toolbox integrates seamlessly with other MATLAB toolboxes, allowing for complex workflows that might involve signal processing, image processing, or control systems, among others.

For our first example, we perform vibration data analysis for a rotating loaded shaft to detect whether it has faulty components (e.g. gears or bearings). The 1000 data points are synthetic and made up of two features: the Root Mean Square of vibration velocity amplitude (mm/s) and the peak frequency (Hz), where the training data for normal and faulty conditions (500 data points each) are generated using random values from two normal distributions with different means and standard deviations using the **randn** command. Labels of 0 and 1 are considered for normal and faulty conditions, respectively, and the data is shuffled using the command **randperm**.

The code for this part is given in the following:

```
% Parameters
num_samples = 1000; % Number of samples
```

```
% Generate synthetic data for 'normal' condition
% RMS amplitude: mean 1 mm/s, small variance
% Peak frequency: mean 50 Hz, small variance
RMS_normal = 1 + 0.1 * randn(num_samples/2, 1); % in mm/s
Freq_normal = 50 + 5 * randn(num_samples/2, 1); % in Hz
% Generate synthetic data for 'faulty' condition
% RMS amplitude: mean 3 mm/s, larger variance
% Peak frequency: mean 100 Hz, larger variance
RMS_faulty = 2 + 0.5 * randn(num_samples/2, 1); % in mm/s
Freq_faulty = 75 + 20 * randn(num_samples/2, 1); % in Hz

% Combine the data
X = [RMS_normal, Freq_normal; RMS_faulty, Freq_faulty];

% Labels: 0 for normal, 1 for faulty
y = [zeros(num_samples/2, 1); ones(num_samples/2, 1)];

% Shuffle the dataset
data = [X y];
data = data(randperm(num_samples), :);
```

Now, we preprocess the synthetic data by separating them into features and labels, normalizing the feature values using the **normalize** command, and splitting the entire data into training data (70% of all data) and testing data (30% of all data) using the **cvpartition** command:

```
% Split features and labels
X = data(:, 1:end-1);
y = data(:, end);

% Normalize the features
X = normalize(X);

% Split the Data into Training and Testing Sets
cv = cvpartition(y, 'HoldOut', 0.3);
X_train = X(training(cv), :);
y_train = y(training(cv), :);
X_test = X(test(cv), :);
y_test = y(test(cv), :);
```

Since this is a classification problem with only two classes and we want to find optimal decision boundaries (that have the maximum width), we will use the **Support Vector Machines (SVM)** method. First, we fit a SVM model (variable **mdl**) to the training features and labels using **Radial Basis Functions (RBF)** as kernels (command **fitcsvm**). Then we predict the labels of the testing data using the fitted model and the **predict** command. Finally, we measure the accuracy of our model by comparing the predicted labels for testing data and their actual labels:

```
% Train the SVM classifier
mdl = fitcsvm(X_train, y_train, 'KernelFunction', 'rbf', 'Standardize', true);

% Predict on the test set
y_pred = predict(mdl, X_test);

% Compute accuracy
accuracy = sum(y_pred == y_test) / length(y_test);
```

Now we can plot the **confusion matrix**, showing the number of True and False positives and negatives, as shown in Fig. 10.20:

```
% Confusion Matrix
confMat = confusionmat(y_test, y_pred);

% Visualization
figure;
confusionchart(confMat);
title('Confusion Matrix for SVM Predictive Maintenance Model');
```

To visualize the results of SVM better, we can plot the data, where the classes, support vectors, and decision boundaries are plotted in Fig. 10.21 using the code below. Note that we used the property **mdl.SupportVectors** to access the support vectors of the fitted model.

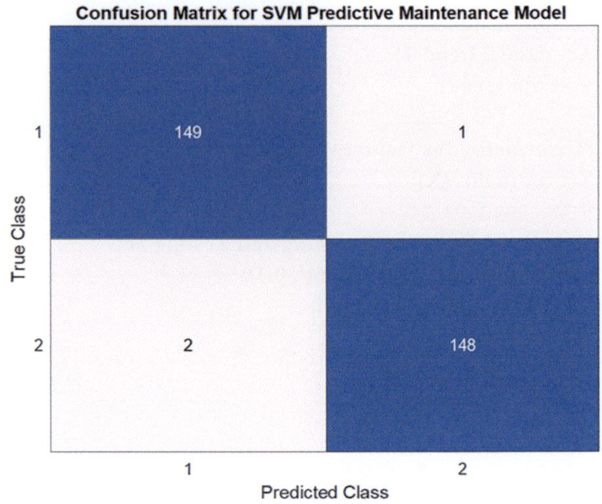


**Fig. 10.20** Confusion matrix for fault detection in vibration data (feature 1 is the RMS velocity amplitude and feature 2 is the peak frequency) using SVM

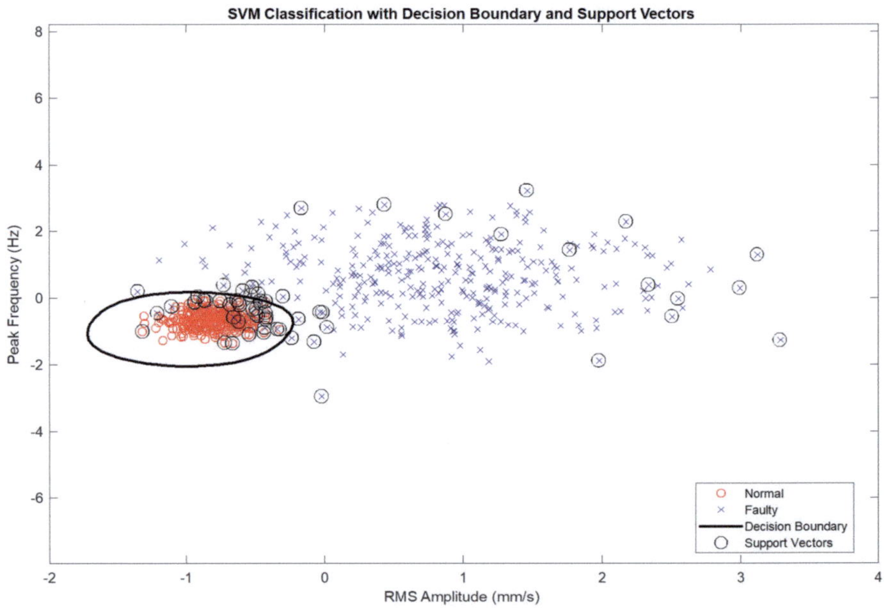


**Fig. 10.21** Vibration data with classes rendered in different colors and markers, support vectors rendered as circles, and the decision boundary rendered as a black line

```
% Define grid for plotting decision boundary
x1_range = linspace(min(X(:,1))-0.5, max(X(:,1))+0.5, 100);
x2_range = linspace(min(X(:,2))-5, max(X(:,2))+5, 100);
[X1, X2] = meshgrid(x1_range, x2_range);
X_grid = [X1(:), X2(:)];

% Predict over the grid
[~, scores] = predict(mdl, X_grid);
decision_boundary = reshape(scores(:,2), size(X1));

% Plot the original data
figure;
gscatter(X_train(:,1), X_train(:,2), y_train, 'rb', 'ox');
hold on;

% Plot the decision boundary
contour(X1, X2, decision_boundary, [0 0], 'k', 'LineWidth', 2);

% Plot support vectors
sv = mdl.SupportVectors;
plot(sv(:,1), sv(:,2), 'ko', 'MarkerSize', 10);

% Labels and title
xlabel('RMS Amplitude (mm/s)');
ylabel('Peak Frequency (Hz)');
title('SVM Classification with Decision Boundary and Support Vectors');
legend('Normal', 'Faulty', 'Decision Boundary', 'Support Vectors');
```

We can also check the prediction for new vibration data:

```
% Predict on New Data (For Completeness)
% Example of new vibration data
new_vibration_data = [2.5, 90]; % Example new data point with RMS Amplitude in mm/s
and Peak Frequency in Hz
% Manually normalize the new data using the mean and std from the training set
new_vibration_data_mean = mean(X_train);
new_vibration_data_std = std(X_train);
new_vibration_data_normalized = (new_vibration_data - new_vibration_data_mean) ./ new_vibration_data_std;


% Predict the condition of the new data
prediction = predict(mdl, new_vibration_data_normalized);

% Display prediction
if prediction == 1
        disp('The machinery is predicted to be faulty.');
else
        disp('The machinery is predicted to be normal.');
end
```

The result in the command window is:

```
The machinery is predicted to be faulty.
```

For our second example, we will use statistical analysis tools. Consider an engineer analyzing the effect of different material types on the tensile strength of a mechanical component. He has data from experiments conducted on three different materials (Material A, Material B, and Material C). Each material was tested multiple times under the same conditions to measure its tensile strength. The goal is to use Analysis of Variance (ANOVA) to determine if there is a statistically significant difference in the mean tensile strength between the three materials.

Once again, the data is synthetic and generated by the following code:

```
% Parameters
num_samples_per_group = 30; % Number of samples for each material
material_types = {'Material A', 'Material B', 'Material C'}; % Labels for materials

% Mean tensile strength for each material (in MPa)
mean_strength_A = 500; % Material A
mean_strength_B = 520; % Material B
mean_strength_C = 490; % Material C
% Standard deviation for each material
std_dev = 15; % Assumed the same for simplicity

% Generate random tensile strength data for each material
strength_A = mean_strength_A + std_dev * randn(num_samples_per_group, 1);
strength_B = mean_strength_B + std_dev * randn(num_samples_per_group, 1);
strength_C = mean_strength_C + std_dev * randn(num_samples_per_group, 1);

% Combine data into a single array
tensile_strength = [strength_A; strength_B; strength_C];

% Create a grouping variable corresponding to the material types
group = [repmat({'Material A'}, num_samples_per_group, 1);
repmat({'Material B'}, num_samples_per_group, 1);
repmat({'Material C'}, num_samples_per_group, 1)];
```

Now we perform ANOVA using the **anova1** command, which returns the **p** value, ANOVA table (variable **tbl**), and ANOVA statistics (variable **stats**). The following code shows the p-value, determines whether a significant difference among the groups was detected, and visualizes the results as shown in Fig. 10.22 using the code below:

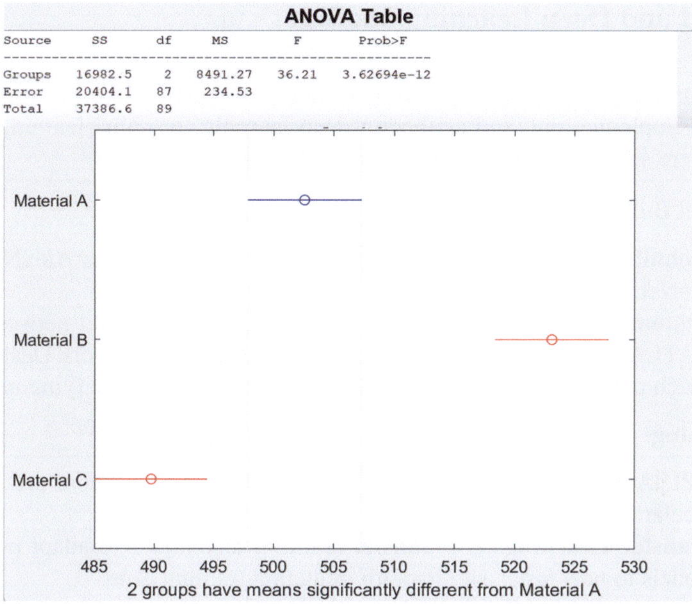

ANOVA Table

| Source | SS | df | MS | F | Prob>F |
|---|---|---|---|---|---|
| Groups | 16982.5 | 2 | 8491.27 | 36.21 | 3.62694e-12 |
| Error | 20404.1 | 87 | 234.53 | | |
| Total | 37386.6 | 89 | | | |



**Fig. 10.22** ANOVA analyzing the effect of different material types on the tensile strength of a mechanical component

```
% Perform ANOVA
[p, tbl, stats] = anova1(tensile_strength, group);

% Display p-value
disp(['p-value from ANOVA: ', num2str(p)]);

% Post-Hoc Analysis (if ANOVA shows significant results)
if p < 0.05
        disp('Significant differences found between groups.');
        c = multcompare(stats);
else
        disp('No significant differences found between groups.');
end
```

The results displayed in the command window are:

```
p-value from ANOVA: 3.6269e-12
Significant differences found between groups.
```

## 10.8 AI and Deep Learning Toolbox

The AI and Deep Learning Toolbox provides a comprehensive environment for designing, implementing, and analyzing deep learning, machine learning, and AI models. Here's a breakdown of the features and capabilities:

i. **Model Building**:
   - **Prebuilt Models**: Use and customize prebuilt models like AlexNet, VGG, ResNet, YOLO, and others.
   - **Custom Networks**: Design and implement custom neural networks using MATLAB's high-level commands or using the Deep Network Designer app, which provides a graphical interface for building and modifying networks.

ii. **Training**:
   - **GPU Acceleration**: Leverage NVIDIA GPUs and multicore processors to accelerate the training process.
   - **Transfer Learning**: Use transfer learning techniques to adapt pre-trained models to new tasks, significantly reducing training time.
   - **Hyperparameter Tuning**: Use tools for automatic tuning of hyperparameters like learning rates, batch sizes, and network architectures.

iii. **Data Handling**:
   - **Preprocessing**: Use functions for data augmentation, image preprocessing, and handling of large datasets.
   - **Data Importing**: Import data from various formats, including images, videos, and custom formats.

iv. **Training Algorithms**:
   - **Optimizers**: Use various optimization algorithms such as SGD, Adam, and RMSProp.
   - **Custom Training Loops**: Create custom training loops for more control over the training process.
   - **Distributed Training**: Train models on clusters or in the cloud for handling large-scale datasets.

v. **Visualization**:
   - **Network Architecture**: Visualize network architectures and inspect layer properties.
   - **Training Progress**: Monitor the progress of training with live plots showing accuracy, loss, and other metrics.

- **Activation and Feature Maps**: Visualize and analyze feature activations, gradients, and output of layers to understand model behavior.

vi. **Deployment**:

- **Code Generation**: Generate C/C++ or CUDA code automatically for embedded systems, allowing for deployment on hardware like NVIDIA GPUs and ARM processors.
- **MATLAB Production Server**: Deploy models on the server for integration with enterprise systems.

### Application Areas

- **Computer Vision**: Implement deep learning for object detection, image segmentation, and recognition tasks.
- **Natural Language Processing**: Use for text classification, sentiment analysis, and sequence-to-sequence models.
- **Time Series and Signal Processing**: Apply deep learning to time-series forecasting, anomaly detection, and signal classification.
- **Reinforcement Learning**: Design and train reinforcement learning agents using deep Q-networks, policy gradients, and other methods.

For our first example, we will utilize the Deep Learning Toolbox to detect cars in an image captured by a surveillance camera. We will use the pretrained YOLO3 object detector network with the **yolov3ObjectDetector** command to create a detector object and the method **detect** from this object. Note that in order to use this network, the Computer Vision Toolbox Model for YOLO v3 support package must be installed.

First, we load the test image and the pretrained network:

```
% Load the image
img = imread('test_image.png');

% Load a pre-trained YOLOv3 object detector using the darknet53 backbone
detector = yolov3ObjectDetector('darknet53-coco');
```

Then, we detect the cars in the image:

```
% Detect objects in the image
[bboxes, scores, labels] = detect(detector, img);

% Filter for car detections
jj = 1;
carIdx = []; % Initialize carIdx array
for ii = 1:length(labels)
        if strcmp(char(labels(ii)), 'car')
                carIdx(jj) = ii;
                jj = jj + 1;
        end
end
```

Next, we crop the image region corresponding to the largest car and annotate it on the original image with a red box.

To further narrow the search region for the license plate number, we have to either train YOLO3 with a new label/category “license plate”, which can take a lot of time and training data, or use the command **detectTextCRAFT**, which detects text in images using the character region awareness for text detection (CRAFT) deep learning model. To use this command, we should first install the Computer Vision Toolbox Model for Text Detection support package.

Now we crop the license plate number region using the bounding box provided by **detectTextCRAFT** and also annotate it on the original image with a green box.

Then we pass it to the command **ocr** (Optical Character Recognition algorithm) for text recognition. Note that for the **ocr** command to work properly, the text must have a significant contrast to its background.

The code below shows all these operations while also handling the scenarios in which no car or no text on the car is detected:

```
% Get bounding boxes of cars
carBboxes = bboxes(carIdx, :);

% Find the largest car based on area
if ~isempty(carBboxes)
        carAreas = carBboxes(:, 3) .* carBboxes(:, 4);
        [~, largestCarIdx] = max(carAreas);
        largestCarBbox = carBboxes(largestCarIdx, :);

        % Annotate the largest car on the image
        detectedImg = insertObjectAnnotation(img, 'rectangle', largestCarBbox, 'Car',
        'LineWidth', 3, 'Color', 'red');

        % Extract the largest car area
        largestCarImg = imcrop(img, largestCarBbox);

        % Detect text within the largest car bounding box using CRAFT
        textRegions = detectTextCRAFT(largestCarImg);

        if ~isempty(textRegions)
                % Adjust the position of the detected text bounding boxes to fit the
                original image
                for k = 1:size(textRegions, 1)
                textRegions(k, 1:2) = textRegions(k, 1:2) + largestCarBbox(1:2);
                end

                % Annotate the detected text regions (assumed to be license plate)
                for k = 1:size(textRegions, 1)
                        detectedImg = insertObjectAnnotation(detectedImg, 'rectangle',
                        textRegions(k, :), 'License Plate', 'LineWidth', 3, 'Color', 'green');
                end

                % Crop and apply OCR to each detected text region (we assume only one
                license plate)
                licensePlateImg = imcrop(img, textRegions(1, :)); % Crop the first
                detected text region

                ocrResults = ocr(licensePlateImg);
                licensePlateText = strtrim(ocrResults.Text); % Trim to remove extra
                spaces
```

```
            % Set the final image title with the detected license plate number
            titleText = sprintf('Detected License Plate: %s', licensePlateText);
        else
            titleText = 'No text detected in the largest car.';
        end
        % Show the final annotated image with the title as the detected license plate
        number
        figure;
        imshow(detectedImg);
        title(titleText);
else
        titleText = 'No cars detected in the image.';
        figure;
        imshow(img);
        title(titleText);
end
```

The result can be seen in Fig. 10.23, where the largest car and its license plate are annotated with red and green bounding boxes, respectively, and the license plate number is added to the figure title.

For our second example, we will utilize the Deep Learning Toolbox to estimate body pose in a picture using the pretrained, open-source, and state-of-the-art **Open-Pose** neural network developed by researchers at Carnegie Melon University. This network has 2 branches and 6 stages. Its top branch predicts **heatmaps** (18 body key points for the body plus one key point for the background) and its bottom branch predicts **Part Affinity Fields (PAFs)** for body parts (38 2D vector field showing the limbs and the direction of connection between 2 key points) in an input image. It has 6 stages that improve predictions gradually by utilizing the image features, heatmaps, and PAFs from previous stages (the first stage only receives the image features). For more information about this network, watch the video in the following link:

https://www.youtube.com/watch?v=qr9065bxslc

The pretrained network files (zipped folder) must be downloaded from the following URL and placed in the current directory:

https://ssd.mathworks.com/supportfiles/vision/data/human-pose-estimation.zip

First, we extract the OpenPose network's layers and remove the last layers that send out the name of the detected heatmaps and PAFs not needed in this example:

```
% extract and import the OpenPose network
dataDir=cd;
unzip(fullfile(dataDir,'human-pose-estimation.zip'),dataDir);
modelfile = fullfile(dataDir,'human-pose-estimation.onnx');
layers = importONNXLayers(modelfile,"ImportWeights",true);

% Remove the unused output layers
layers = removeLayers(layers,layers.OutputNames);
net = dlnetwork(layers);
```

Fig. 10.23 Car detection, text detection, and text recognition for license plate number recognition using YOLO3, CRAFT, and OCR

Next, we load the test image shown in Fig. 10.24a and preprocess the image so it is useable as the network input:

```
% load the test image
im = imread('test_image_body_pose_estimation.png');
imshow(im)

% pre-processing the input for the network
% The network expects image data of data type single in the range [-0.5, 0.5].
netInput = im2single(im)-0.5;

% The network expects the color channels in the order blue, green, red.
netInput = netInput(:,:,[3 2 1]);

% Store the image data as a dlarray (deep learning array)
netInput = dlarray(netInput,"SSC");
```

Now we can pass this test image to the network and predict the heatmaps and PAFS. We can also visualize some results, such as the heatmap corresponding to the key point with **ID = 0** (nose) as shown in Fig. 10.24b, and the x components of the PAFs associated with the limb that connects the neck to the right side of the hip joint as can be seen in Fig. 10.24c:

```
% Predict the heatmaps and PAFs
[heatmaps,pafs] = predict(net,netInput);

% extract the heatmaps and show the body parts associated with them
heatmaps = extractdata(heatmaps);

% Remove the background heatmap
heatmaps = heatmaps(:,:,1:end-1);

% show the heat maps associated with ID=0 (nose)
idx = 1;
hmap = heatmaps(:,:,idx);
hmap = imresize(hmap,size(im,[1 2]));
figure;
imshowpair(hmap,im);
```

Finally, we can use the utility functions developed by MathWorks to draw body poses over the test image using lines with different colors as shown in Fig. 10.24d.

```
% post-processing to identify and connect the individual poses of the people in the
image using the heatmaps and PAFs
params = getBodyPoseParameters; % helper function that returns parameters and
constants for OpenPose pose estimation algorithm
poses = getBodyPoses(heatmaps,pafs,params); % helper function that returns the pose
of bodies in the image
renderBodyPoses(im,poses,size(heatmaps,1),size(heatmaps,2),params); % helper
function that plots the poses over the original image
```

All the required utility files, the live script containing the code above, and the network zipped folder can be downloaded from the following link:

https://drive.google.Com/File/D/1Y1PU_M1oze4jlsgtq2tbdi3nyo81mudx/View

For the last part of this section, we will look at the important functions of MATLAB's Deep Learning Toolbox, including their syntax and applications. Different types of layers are shown in Table 10.1:

Other functions used for visualizing the network, splitting the data, training options, network training, prediction and classification, manually training networks, and creating custom networks are shown in Table 10.2 along with their syntax and applications.

If we want to create a feedforward regression network, a CNN network, or an RNN network, we can create them using different layers in Table 10.1 with the codes shown in Fig. 10.25, parts (a) to (c), respectively, where visualizations of these networks are also shown next to their codes.

To create a custom network (e.g. with parallel branches), we use the **layerGraph, addLayers, connectLayers** commands to create a layer graph for the network structure, add previously made layers to the graph, and then connect layers in a custom manner, respectively, as shown in Fig. 10.26:

One common task in deep learning is to split the entire data into training and validation. To do this, we can use the built-in function **cvpartition** and the associated

**Table 10.1** Layers in MATLAB's Deep Learning Toolbox, including their syntax and applications

| Function | Description | Syntax |
|---|---|---|
| Featureinputlayer | Define an input layer for features in a network | **layer = featureInputLayer(numFeatures, 'Normalization')** |
| Fullyconnectedlayer | Create a fully connected layer (dense layer) for a neural network | **layer = fullyConnectedLayer(numUnits)** |
| Relulayer | Add a ReLU activation function layer to the network | **layer = reluLayer** |
| Regressionlayer | Add a regression output layer to the network | **layer = regressionLayer** |
| tanhLayer | Add a hyperbolic tangent activation function layer to the network | **layer = tanhLayer** |
| leakyReluLayer | Add a leaky ReLU activation function layer to the network | **layer = leakyReluLayer(scale)** |
| clippedReluLayer | Add a clipped ReLU activation function layer to the network | **layer = clippedReluLayer** |
| swishLayer | Add a Swish activation function layer to the network | **layer = swishLayer** |
| batchNormalizationLayer | Normalize the activations across a batch of data | **layer = batchNormalizationLayer** |
| softmaxLayer | Add a softmax layer for multi-class classification | **layer = softmaxLayer** |
| classificationLayer | Add a classification output layer to the network | **layer = classificationLayer** |
| flattenLayer | Flatten the input by collapsing all dimensions into one | **layer = flattenLayer** |

(continued)

**Table 10.1** (continued)

| Function | Description | Syntax |
|---|---|---|
| dropoutLayer | Add a dropout layer to reduce overfitting during training | **layer = dropoutLayer(rate)** |
| averagePooling1dLayer | Apply average pooling to 1D input data | **layer = averagePooling1dLayer(poolSize,Padding = m,Stride = n)** |
| averagePooling2dLayer | Apply average pooling to 2D input data | **layer = averagePooling2dLayer(poolSize,Padding = m,Stride = n)** |
| maxPooling1dLayer | Apply max pooling to 1D input data | **layer = maxPooling1dLayer(poolSize,Padding = m,Stride = n)** |
| maxPooling2dLayer | Apply max pooling to 2D input data | **layer = maxPooling2dLayer(poolSize,Padding = m,Stride = n)** |
| imageInputLayer | Define an input layer for images in a network | **layer = imageInputLayer([height, width, channels])** |
| convolution1dLayer | Apply 1D convolution to the input data | **layer = convolution1dLayer(filterSize, numFilters)** |
| convolution2dLayer | Apply 2D convolution to the input data | **layer = convolution2dLayer(filterSize, numFilters)** |
| sequenceInputLayer | Define an input layer for sequences in a network | **layer = sequenceInputLayer(inputSize)** |
| lstmLayer | Add an LSTM layer for sequence learning | **layer = lstmLayer(numHiddenUnits)** |
| additionLayer | Combine inputs from multiple layers using addition | **layer = additionLayer(numInputs)** |

commands **training** and **test** as shown below (**y** is the labels variable and **x** is the features variable):

```
c = cvpartition(length(y), 'HoldOut', 0.2); % Hold out 20% of the data for validation
% Split data into training and validation sets
xTrain = x(training(c));
yTrain = y(training(c));
xVal = x(test(c));
yVal = y(test(c));
```

**Table 10.2** Important Functions in MATLAB's Deep Learning Toolbox, including their syntax and applications

| Function | Description | Syntax |
|---|---|---|
| Analyzenetwork | Analyze the architecture of the network | **analyzeNetwork(layers)** |
| cvpartition | Partition data into training and testing sets | **c = cvpartition(n, 'Holdout', p)** |
| Training | Extract indices of the training data from a partition | **idx = training(c)** |
| Test | Extract indices of the testing data from a partition | **idx = test(c)** |
| Trainingoptions | Specify training options for the network | **options = trainingOptions('sgdm', 'MaxEpochs', 30)** |
| Trainnetwork | Train the neural network with the specified data and options | **net = trainNetwork(XTrain, YTrain, layers, options)** |
| Predict | Make predictions using the trained network | **YPred = predict(net, XTest)** |
| Classify | Classify input data using the trained network | **YPred = classify(net, XTest)** |
| dlnetwork | Create a deep learning network for custom training | **net = dlnetwork(layers)** |
| dlarray | Convert data to a deep learning array for use with automatic differentiation | **dlX = dlarray(data, 'CB')** |
| dlfeval | Evaluate the gradients of a function for custom training | **[gradients, loss] = dlfeval(@modelGradients, net, dlX, dlY)** |
| layerGraph | Create a layer graph for a network | **lgraph = layerGraph(layers)** |
| addLayers | Add layers to a layer graph | **lgraph = addLayers(lgraph, layer)** |
| connectLayers | Connect layers within a layer graph | **lgraph = connectLayers(lgraph, layer1Name, layer2Name)** |

```
% extract the PAFs
pafs = extractdata(pafs);

% show the PAFs connecting the first and second heat map IDs (backbone)
idx = 1;
impair = horzcat(im,im);
pafpair = horzcat(pafs(:,:,2*idx-1),pafs(:,:,2*idx));
pafpair = imresize(pafpair,size(impair,[1 2]));
figure;
imshowpair(pafpair,impair);
```

After creating the network, we can define the training options using the **trainingOptions** command with a general syntax shown below to define the solver (e.g.

**Fig. 10.24** Body pose estimation using OpenPose neural network, a) Test image, b) Detected key points corresponding to nose, c) Detected right side (PAF) of the trunk, and d) Body pose laid over the test image using lines with different colors

**'adam'**), define the initial learning rate and learning rate scheduling (changing), define the momentum in gradient methods, define the maximum number of epochs and size of minimum batches, define the evaluation metrics, define the validation data, display the results on the command window (verbose), and plot the results.

```
options = trainingOptions('solver', LearnRateSchedule='piecewise','InitialLearnRate',
0.001,'Momentum', 0.9,'MaxEpochs',20,'MiniBatchSize',64, ...
Metrics = 'accuracy',ValidationData={XTest,labelsTest},...
'Verbose', false,'Plots', 'training-progress');
```

Simple standard networks can be trained automatically using the command **trainNetwork**, which takes the training data, the network layers, the loss function, and the training options:

```
net = trainNetwork(xTrain, yTrain, layers, lossFcn, options);
```

If a network has dynamic data that changes size during the training or we want to provide custom functions for the gradient, loss, or learning rate scheduling functions,

**Fig. 10.25** Creation and visualization of neural networks in MATLAB, a) A feedforward regression network, b) A CNN network used for image classification, and c) An RNN network used for data sequence classification

we have to manually train the network. To do so, first, we need to convert the layers into a **deep learning network** using the **dlnetwork** command:

```
net = dlnetwork(layers);
```

We also need to convert all data passed to the network (training, validation, and test) into **deep learning arrays** using the command **dlarray**:

```
% Transpose to match [1, batchSize] format:    C ----> channel ,    B -----> batch
dlX = dlarray(XMiniBatch', "CB");
```

Then, use a **for** loop to perform the loss function, gradient, and parameter update calculations and display the progress (verbose) of the neural network using the custom utility functions:

```
for epoch = 1:numEpochs
    % Shuffle the data
    idx = randperm(size(x, 1));
    x = x(idx);
    y = y(idx);

    % Loop over mini-batches
    for ii = 1:miniBatchSize:size(x, 1)

        % Get the mini-batch data
        idxMiniBatch = ii:min(ii + miniBatchSize - 1, size(x, 1));
        XMiniBatch = x(idxMiniBatch, :);
        YMiniBatch = y(idxMiniBatch, :);

        % Convert mini-batch data to dlarray with correct dimensions
        dlX = dlarray(XMiniBatch', "CB"); % Transpose to match [1, batchSize] format:
        C-> channel , B-> batch
        dlY = dlarray(YMiniBatch', "CB"); % Transpose to match [1, batchSize] format

        % Evaluate the model gradients and loss using the custom function
        [gradients, loss] = dlfeval(@modelGradients, net, dlX, dlY);

        % Update the network parameters using the gradients
        net = sgdmUpdate(net, gradients, learningRate);
    end

    % Store the loss for this epoch
    losses(epoch) = loss;

    % Display progress
    if mod(epoch, 100) == 0
    disp(['Epoch ' num2str(epoch) ': Loss = ' num2str(loss)]);
    end
end
```

In the code above, utility functions **modelGradients** and **sgdmUpdate** are custom-written and shown below:

```
function [gradients, loss] = modelGradients(net, dlX, dlY)
        % Forward pass
        dlYPred = forward(net, dlX);
        % Compute the mean squared error loss
        loss = mse(dlYPred, dlY);
        % Compute the gradients of the loss with respect to the network parameters
        gradients = dlgradient(loss, net.Learnables);
end

function net = sgdmUpdate(net, gradients, learnRate)
        % Update the learnable parameters using SGD with momentum
        learnables = net.Learnables;
        for ii = 1:size(learnables, 1)
                paramName = learnables.Parameter(ii);
                paramGrad = gradients.Value{ii};
                net.Learnables.Value{i} = net.Learnables.Value{ii} - learnRate *
                paramGrad;
        end
end
```

Finally, when the network is trained, we can use the network to predict or classify the test data using commands **predict** and **classify**:

```
yPred = predict(net, xTest);
yPred = classify(net, xTest);
```

To put everything together, the following code performs approximation for a sine function using a feedforward neural network made up of a feature input layer, 3 sets of fully connected (dense) layers + Relu layers, and finally a dense layer + regression layer. The network was shown in Fig. 10.25a. The 3 hidden dense layers each have 20 neurons but the last layer only has one neuron as the output must be a scalar. The solver is "**adam**", and is assigned a max epochs of 200 and an initial learning rate of 0.001. The solver will not output data to the command window but will plot the training process. The data is split into 80% for training and 20% for validation. The test data is randomly created in the range of training data and used to predict the y values and compare against the actual y values for the test data.

```
% Generate Training and Validation Data
x = linspace(-2*pi, 2*pi, 5000)';
y = sin(x); % Corresponding sine values
% Partition the data
c = cvpartition(length(y), 'HoldOut', 0.2); % Hold out 20% of the data for validation
xTrain = x(training(c));
yTrain = y(training(c));
xVal = x(test(c));
yVal = y(test(c));

% Define the Network Architecture
layers = [
featureInputLayer(1, 'Normalization', 'none', 'Name', 'input')
fullyConnectedLayer(20, 'Name', 'fc1')
reluLayer('Name', 'relu1')
fullyConnectedLayer(20, 'Name', 'fc2')
reluLayer('Name', 'relu2')
fullyConnectedLayer(20, 'Name', 'fc3')
reluLayer('Name', 'relu3')
fullyConnectedLayer(1, 'Name', 'fc4')
regressionLayer('Name', 'output')];

% plot and analyze the network
analyzeNetwork(layers);

% Specify Training Options with validation data
options = trainingOptions('adam', ...
'MaxEpochs', 200, ...
'InitialLearnRate', 0.001, ...
'ValidationData', {xVal, yVal}, ...
'ValidationFrequency', 50, ...
'Verbose', false, ...
'Plots', 'training-progress');

% Train the Network
net = trainNetwork(xTrain, yTrain, layers, options);

% Evaluate the Network
xTest = linspace(-2*pi, 2*pi, 100)'; % Test data
yPred = predict(net, xTest); % Predict using the trained network
% Compare predicted values with the actual sine function
figure;
plot(xTest, sin(xTest), 'r', 'LineWidth', 2); % Actual sine function
hold on;
plot(xTest, yPred, 'b--', 'LineWidth', 2); % Predicted values
legend('Actual Sine Function', 'Network Approximation');
title('Function Approximation using Feedforward Neural Network');
xlabel('x');
ylabel('y');
```

Results of the function approximation can be seen in Fig. 10.27, where the network could achieve a relatively accurate approximation of the function as shown in part

(a) due to the loss function convergence to very small values, as provided in part (b). To improve the results, smoother cost functions, such as **tanh** or **swash** can be used instead of Relu, more training data can be used, and more neurons and more hidden layers can be used. We must avoid overfitting when increasing the neurons and layers by checking the loss function over the validation data and using dropout layers if needed.

### Practice Problems

1. Write a script that analyzes the stress–strain behavior of a new material that exhibits nonlinear elastic behavior. After performing a tensile test, you obtain a set of experimental stress–strain data (here, artificially made with the code shown below). Use the Curve-Fitting Toolbox to fit a suitable model to the data. Use a third order polynomial and plot the original stress versus strain data as well as the fitted curve. Show the coefficients of the fitted polynomial and the goodness of fit statistics in the command window.

```
input1 = imageInputLayer([28 28 1], 'Name', 'input1');
input2 = imageInputLayer([28 28 1], 'Name', 'input2');
conv1 = convolution2dLayer(3, 8, 'Padding', 'same', 'Name', 'conv1');
conv2 = convolution2dLayer(3, 8, 'Padding', 'same', 'Name', 'conv2');
addLayer = additionLayer(2, 'Name', 'add');
fc = fullyConnectedLayer(10, 'Name', 'fc');
softmax = softmaxLayer('Name', 'softmax');
output = classificationLayer('Name', 'output');
```

```
lgraph = layerGraph(input1);
lgraph = addLayers(lgraph, input2);
lgraph = addLayers(lgraph, conv1);
lgraph = addLayers(lgraph, conv2);
lgraph = addLayers(lgraph, addLayer);
lgraph = addLayers(lgraph, fc);
lgraph = addLayers(lgraph, softmax);
lgraph = addLayers(lgraph, output);
```

```
lgraph = connectLayers(lgraph, 'input1', 'conv1');
lgraph = connectLayers(lgraph, 'input2', 'conv2');
lgraph = connectLayers(lgraph, 'conv1', 'add/in1');
lgraph = connectLayers(lgraph, 'conv2', 'add/in2');
lgraph = connectLayers(lgraph, 'add', 'fc');
lgraph = connectLayers(lgraph, 'fc', 'softmax');
lgraph = connectLayers(lgraph, 'softmax', 'output');
```

```
analyzeNetwork(lgraph);
```

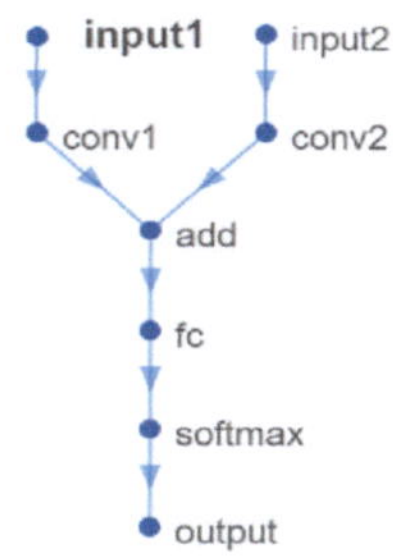


**Fig. 10.26** Creation and visualization of a custom neural network in MATLAB

```
E = 200;   % Young's Modulus (MPa) - linear elastic region
yield_strain = 0.01;  % Yield strain
yield_stress = E * yield_strain; % Yield stress (MPa)
hardening_modulus = 50;  % Post-yield hardening modulus (MPa)
strain_max = 0.05;  % Maximum strain value
n_points = 100;  % Number of data points
noise_level = 0.5;  % Noise level (MPa)

% Generate strain values
strain = linspace(0, strain_max, n_points);

% Generate stress values with a bilinear model and add noise
stress = zeros(size(strain));
for ii = 1:length(strain)
  if strain(ii) <= yield_strain
    % Elastic region (Hooke's Law)
    stress(ii) = E * strain(ii);
  else
    % Plastic region (Linear hardening)
    stress(ii) = yield_stress + hardening_modulus * (strain(ii) - yield_strain);
  end
end

% Add some random noise to simulate experimental data
stress = stress + noise_level * randn(size(stress));
```

2. Write a script using the Optimization Toolbox to minimize the total weight of the truss shown in Fig. 10.28 by finding the optimal cross-sectional areas for each member. Your code needs to return the optimum weight, return the optimal cross sections, and show the optimization iterations.

   **Cost function**: The weight of the truss **W** is given by:

$$W = \sum_{i=1}^{n} \rho L_i A_i \tag{10.6}$$

   where $\rho$ is the material density and $\mathbf{L}_i$ and $\mathbf{A}_i$ are the length and cross section of member **i**.

   **Variables**: The cross-sectional areas $\mathbf{A}_i$ of the truss members **i = 1, 2,..., n** where **n** is the total number of members in the truss.

   **Constraints**: The design must satisfy both stress constraints (to ensure that no member exceeds the yield stress of the material) and geometric constraints (to keep the cross-sectional areas within realistic limits):

$$\sigma_i = \frac{F_i}{A_i} \leq \sigma_{yield}, \; A_{min} \leq A_i \leq A_{max} \tag{10.7}$$

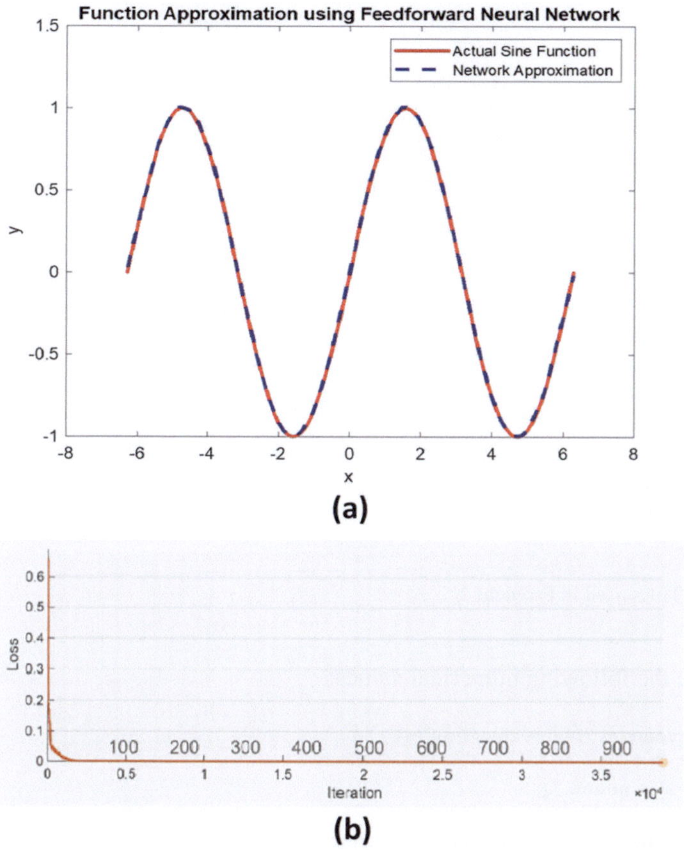


**Fig. 10.27** Function approximation using a feedforward neural network, a) Predicted values versus actual values for the test data, and b) Loss function versus iteration

In the above relationships, $\mathbf{F_i}$ is the load applied to member **i**, $\sigma_{\mathbf{yield}}$ is the yield strength of the material chosen, and $\mathbf{A_{min}}$ and $\mathbf{A_{max}}$ are the minimum and maximum allowed cross sections for all members.

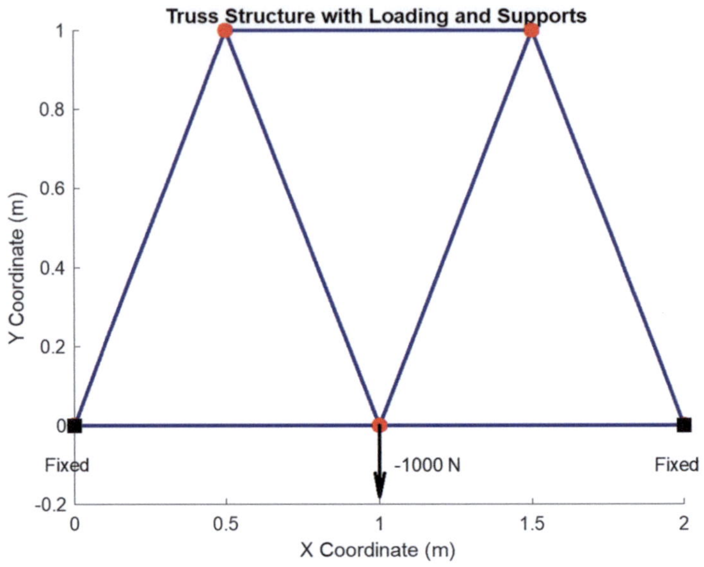


**Fig. 10.28** Truss used in Problem 2

Use the following numerical values:

```
% Truss geometry (nodes coordinates)
nodes = [0, 0; 1, 0; 2, 0; 0.5, 1; 1.5, 1];
n_nodes = size(nodes, 1);

% Truss connectivity (which nodes are connected by which members)
members = [1, 2; 2, 3; 1, 4; 2, 4; 2, 5; 3, 5; 4, 5];
n_members = size(members, 1);

% External load (applied to nodes)
load = zeros(n_nodes, 2);  % No load initially
load(2, 2) = -1000;        % Apply a downward load of 1000 N at node 2 (middle bottom
node)

% Fixed nodes (supports)
fixed_nodes = [1, 3];

% Material properties
E = 210e9;             % Young's Modulus (Pa)
density = 7850;    % Density of steel (kg/m^3)
yield_stress = 250e6; % Yield stress of steel (Pa)
```

3. Write a script to design a feedback control system to regulate the temperature of a chemical reactor using a Proportional-Integral-Derivative (PID) controller (in series with the plant) using the Control Toolbox (use commands **tf**, **series**, and **feedback** to form the plant transfer function). The temperature of the reactor

must be maintained at a specific setpoint despite disturbances such as varying ambient temperatures or changes in the exothermic reaction rate.

The reactor is modeled as a first-order system with a transfer function given by Eqn. 10.7, where **K** is the process gain and $\boldsymbol{\tau}$ is the time constant:

$$G(s) = \frac{K}{\tau s + 1} \tag{10.8}$$

Use the following numerical values:

```
K = 2;      % Process gain
tau = 5;    % Time constant (seconds)
```

Plot the step response of the controlled system and display the time-domain specifications of this response, such as rise time, settling time, and overshoot on the command window using the **stepinfo** function.

4. Write a script using the Robotics Toolbox to find the Inverse Kinematics (IK) solution of a 6-DOF robot with all rotary joints. Use the code below to define the lengths of the members and the pose of the end-effector. Display the robot details and the results of IK on the command window. Apply Forward Kinematics on the resulting angles to show the end-effector pose can indeed be reached and show the results on the command window. Plot the desired end-effector pose and show the robot reaching it as can be seen in Fig. 10.29. Use the code below to construct the robot.

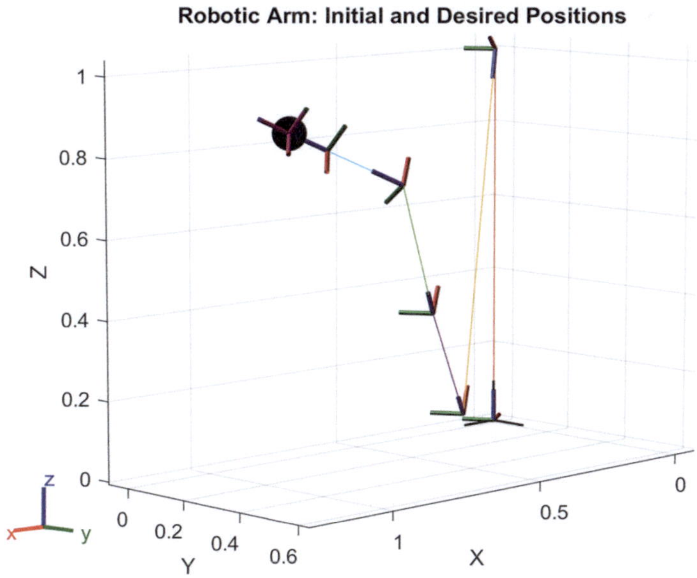


**Fig. 10.29** Robotic arm in Problem 4

```
L1 = 1; % Length of link 1
L2 = 1; % Length of link 2
L3 = 0.5; % Length of link 3
L4 = 0.5; % Length of link 4 (added for 6-DOF)
L5 = 0.2; % Length of link 5 (added for 6-DOF)
L6 = 0.1; % Length of link 6 (added for 6-DOF)

desiredPosition = [1.2; 0.5; 0.9]; % Desired position (x, y, z) in meters
desiredOrientation = eul2quat([pi/4, pi/6, pi/3]); % Set a desired orientation

% Create a rigid body tree model for the robot
robot = rigidBodyTree('DataFormat', 'column', 'MaxNumBodies', 6);

% Create and add first link (base to link 1)
body1 = rigidBody('body1');
jnt1 = rigidBodyJoint('jnt1', 'revolute');
setFixedTransform(jnt1, trvec2tform([0 0 0])); % No offset for the first joint
jnt1.JointAxis = [0 0 1];
body1.Joint = jnt1;
addBody(robot, body1, 'base');

% Create and add second link (link 1 to link 2)
body2 = rigidBody('body2');
jnt2 = rigidBodyJoint('jnt2', 'revolute');
setFixedTransform(jnt2, trvec2tform([0 0 L1]));
jnt2.JointAxis = [0 1 0];
body2.Joint = jnt2;
addBody(robot, body2, 'body1');

% Create and add third link (link 2 to link 3)
body3 = rigidBody('body3');
jnt3 = rigidBodyJoint('jnt3', 'revolute');
setFixedTransform(jnt3, trvec2tform([0 0 L2]));
jnt3.JointAxis = [0 1 0];
body3.Joint = jnt3;
addBody(robot, body3, 'body2');

% Create and add fourth link (link 3 to link 4)
body4 = rigidBody('body4');
jnt4 = rigidBodyJoint('jnt4', 'revolute');
setFixedTransform(jnt4, trvec2tform([0 0 L3]));
jnt4.JointAxis = [0 1 0];
body4.Joint = jnt4;
addBody(robot, body4, 'body3');

% Create and add fifth link (link 4 to link 5)
body5 = rigidBody('body5');
jnt5 = rigidBodyJoint('jnt5', 'revolute');
setFixedTransform(jnt5, trvec2tform([0 0 L4]));
jnt5.JointAxis = [1 0 0];
body5.Joint = jnt5;
addBody(robot, body5, 'body4');

% Create and add sixth link (link 5 to link 6)
body6 = rigidBody('body6');
jnt6 = rigidBodyJoint('jnt6', 'revolute');
setFixedTransform(jnt6, trvec2tform([0 0 L5]));
jnt6.JointAxis = [0 0 1];
body6.Joint = jnt6;
addBody(robot, body6, 'body5');

% Add end-effector
tool = rigidBody('tool');
setFixedTransform(tool.Joint, trvec2tform([0 0 L6]));
addBody(robot, tool, 'body6');
```

5. You are tasked with analyzing the harmonics in the voltage signal of an electrical system. Harmonics can cause distortion in the power supply, leading to inefficiency and potential damage to electrical equipment. You will use the Signal Processing Toolbox to identify the harmonic components in the voltage signal, quantify their contributions, and assess the Total Harmonic Distortion (THD), which is the ratio of the sum of the powers of all harmonic components to the power of the fundamental frequency. Your code must plot the identified harmonics, and display their amplitudes and the calculated THD value.

   Use the code below to create an artificial voltage signal:

```
Fs = 5000;          % Sampling frequency (Hz)
T = 1/Fs;           % Sampling period (seconds)
L = 5000;           % Length of the signal
t = (0:L-1) * T;    % Time vector

% Fundamental frequency (e.g., 60 Hz)
f_fundamental = 60;
fundamental_signal = 230 * sin(2 * pi * f_fundamental * t);

% Add 3rd harmonic (180 Hz) and 5th harmonic (300 Hz)
f_h3 = 3 * f_fundamental; % 3rd harmonic
f_h5 = 5 * f_fundamental; % 5th harmonic
harmonic_signal_3 = 30 * sin(2 * pi * f_h3 * t); % 3rd harmonic
harmonic_signal_5 = 20 * sin(2 * pi * f_h5 * t); % 5th harmonic

% Combined voltage signal
voltage_signal = fundamental_signal + harmonic_signal_3 + harmonic_signal_5;
```

6. Write a script using the Image Processing Toolbox for automatic vegetable color consistency inspection. The code must first read a reference image and acquire the mean values of **A** and **B** channels for the LAB color space version. Then the code must read a test image and apply the same operations on it. Finally, the code must check whether the test image has color consistency (matching with the reference image) by comparing the distance between these mean values against a threshold of 10. Test your code on the two peppers in Fig. 10.30.
7. You want to write a script that can perform predictive maintenance for industrial machines. The goal is to use sensor data from these machines (temperature, vibration, pressure) to predict when a machine is likely to fail so that maintenance can be performed proactively. You will use the Machine Learning Toolbox to build and train a model (here, a **decision tree**) that predicts machine failures based on historical sensor data and a failure category (label). Your code must plot the decision tree and evaluate the model's performance using metrics such as accuracy, precision, recall, and the F1-score. Use the command **cvpartition** to separate 30% of the entire data for testing.

   Use the following code to generate the artificial data for this supervised learning problem:

**Fig. 10.30** Test images for Problem 6

```
num_samples = 1000;
rng('default'); % For reproducibility

% Simulated features (e.g., temperature, vibration, pressure)
temperature = 60 + 15 * randn(num_samples, 1); % Mean = 60, Std = 15
vibration = 0.5 + 0.2 * randn(num_samples, 1); % Mean = 0.5, Std = 0.2
pressure = 100 + 10 * randn(num_samples, 1);  % Mean = 100, Std = 10

% Simulated labels (0 = no failure, 1 = failure)
failure = (temperature > 75 | vibration > 0.7 | pressure > 115);

% Combine into a table
data = table(temperature, vibration, pressure, failure);
```

8. Write a script to develop a predictive model for estimating the Remaining Useful Life (RUL) of aircraft engines based on sensor data. The goal is to build a regression model that can predict how much longer an engine will operate before it needs maintenance or replacement. You will use the Deep Learning Toolbox to design, train, and evaluate a simple feedforward Regression network for this time series regression problem.

   The dataset features include temperature, vibration, and pressure, along with the actual RUL values. Keep 20% of the entire data for testing.

   Your code must calculate the Root Mean Square Error (RMSE) to evaluate the model's performance and plot the true versus predicted RUL values for visual inspection.

In this case, we use the following code to generate artificial data:

```
numEngines = 500;
numTimeSteps = 200;
numFeatures = 3; % Assume 3 sensor readings

% Simulate better correlated sensor data
time = linspace(0, 1, numTimeSteps); % Normalized time

% Feature 1: Temperature (decreases as the engine wears out)
temperature = repmat(100 - 50 * time, numEngines, 1) + 0.25 * randn(numEngines,
numTimeSteps);

% Feature 2: Vibration (increases as the engine wears out)
vibration = repmat(0.1 + 0.9 * time, numEngines, 1) + 0.02 * randn(numEngines,
numTimeSteps);

% Feature 3: Pressure (slight decrease over time, indicating wear)
pressure = repmat(200 - 20 * time, numEngines, 1) + 0.3 * randn(numEngines,
numTimeSteps);

% Combine the features into a single dataset
sensorData = cat(3, temperature, vibration, pressure);

% Simulated Remaining Useful Life (RUL) values
RUL_baseline = linspace(100, 0, numTimeSteps); % Baseline RUL for one engine
RUL = repmat(RUL_baseline, numEngines, 1);%+ 0.1 * randn(numEngines,
numTimeSteps);

% Flatten the data so each time step is treated as a separate data point
X = reshape(sensorData, [], numFeatures); % Flatten into (NumSamples x Features)
Y = RUL(:); % Flatten into (NumSamples x 1)
```

# Appendix I

Complete code for the Truss Calculator example:

```
classdef TrussCalculator < MATLAB.apps.AppBase

    % Properties that correspond to app components
    properties (Access = public)
        TrussCalculatorUIFigure             MATLAB.ui.Figure
        InfoButton                          MATLAB.ui.control.Button
        DeformationAmplificationSlider      MATLAB.ui.control.Slider
        DeformationAmplificationSliderLabel MATLAB.ui.control.Label
        TabGroup                            MATLAB.ui.container.TabGroup
        VerticesTab                         MATLAB.ui.container.Tab
        DeleteVertexButton                  MATLAB.ui.control.Button
        AddVertexButton                     MATLAB.ui.control.Button
        UITable                             MATLAB.ui.control.Table
        BeamsTab                            MATLAB.ui.container.Tab
        DeleteBeamButton                    MATLAB.ui.control.Button
        AddBeamButton                       MATLAB.ui.control.Button
        UITable_2                           MATLAB.ui.control.Table
        Results                             MATLAB.ui.container.Tab
        ReactionForcesNTextArea             MATLAB.ui.control.TextArea
        ReactionForcesNTextAreaLabel        MATLAB.ui.control.Label
        AxialForcesNTextArea                MATLAB.ui.control.TextArea
        AxialForcesNTextAreaLabel           MATLAB.ui.control.Label
        CalculateButton                     MATLAB.ui.control.Button
        UIAxes                              MATLAB.ui.control.UIAxes
    end
```



M. Andalibi and S. A. Cherkauer, *Advanced MATLAB for Engineers*, https://doi.org/10.1007/978-3-032-10677-3

```
properties (Access = private)
        U % Displacement
end

methods (Access = private)

        function updatePlot(app)

                % Read new data
                vertexTable = app.convert(app.UITable.Data);
                beamTable = app.convert(app.UITable_2.Data);

                % Clear the plot
                app.U = [];
                cla(app.UIAxes)

                % If there are any valid vertices to plot
                if ~isempty(vertexTable)

                        % Switch plot between 2D and 3D
                        if all(vertexTable(:, 3) == 0) && all(vertexTable(:, 6) == 0)
                                view(app.UIAxes, 2)
                        else
                                view(app.UIAxes, 3)
                        end

                        % Get max applied force to scale appropriately
                        maxF = 0;
                        for index = 1:1:height(vertexTable)
                                maxF = max([maxF, norm(vertexTable(index, 4:6))]);
                        end
                        if maxF == 0
                                maxF = 1;
                        end

                        % Get distance of corner to opposite corner of the truss's
                        % bounding box to scale appropriately
                        scale = 0.25 * norm([max(vertexTable(:, 1)) -
min(vertexTable(:, 1)), max(vertexTable(:, 2)) - min(vertexTable(:, 2)), max(vertexTable(:,
3)) - min(vertexTable(:, 3))]);

                        % Replot all vertices, force arrows, and support triangles
                        hold(app.UIAxes, 'on');
                        for index = 1:1:height(vertexTable)
                                row = vertexTable(index, :);

                                % Replot all vertices
                                plot3(app.UIAxes, row(1), row(2), row(3), 'Marker', 'o',
'MarkerSize', 4, 'MarkerFaceColor', 'k', 'MarkerEdgeColor', 'k');

                                % Replot all arrows
                                app.drawArrow(row(1:3), row(1:3) + (scale*row(4:6) /
maxF))

                                % Replot all supports
                                app.drawTriangle(row(1:3), scale/10, row(7:9))
                        end
                        % If there are any valid beams to plot
                        if ~isempty(beamTable)
                                for index = 1:1:height(beamTable)
```

```
                                        row = beamTable(index, :);

                                        % Replot all beams
                                        plot3(app.UIAxes, [vertexTable(row(1), 1),
vertexTable(row(2), 1)], [vertexTable(row(1), 2), vertexTable(row(2), 2)],
[vertexTable(row(1), 3), vertexTable(row(2), 3)], 'k', 'LineWidth', 2);
                                  end
                            end
                            hold(app.UIAxes, 'off');
                      end
                end

                function drawArrow(app, startPos, endPos)
                      quiver3(app.UIAxes, startPos(1), startPos(2), startPos(3), endPos(1) -
startPos(1), endPos(2) - startPos(2), endPos(3) - startPos(3), 'r', 'LineWidth', 2)
                end

                function drawTriangle(app, pos, scale, planes)
                      sidelength = scale * 2/sqrt(3);

                      % Fixed in X
                      if planes(1)
                            fill3(app.UIAxes, [pos(1), pos(1) - scale, pos(1) - scale, pos(1)],
[pos(2), pos(2) + sidelength/2, pos(2) - sidelength/2, pos(2)], [pos(3), pos(3), pos(3), pos(3)],
[0 0.4470 0.7410], 'LineStyle', "none")
                            fill3(app.UIAxes, [pos(1), pos(1) - scale, pos(1) - scale, pos(1)],
[pos(2), pos(2), pos(2), pos(2)], [pos(3), pos(3) + sidelength/2, pos(3) - sidelength/2, pos(3)],
[0 0.4470 0.7410], 'LineStyle', "none")
                      end

                      % Fixed in Y
                      if planes(2)
                            fill3(app.UIAxes, [pos(1), pos(1) + sidelength/2, pos(1) -
sidelength/2, pos(1)], [pos(2), pos(2) - scale, pos(2) - scale, pos(2)], [pos(3), pos(3), pos(3),
pos(3)], [0 0.4470 0.7410], 'LineStyle', "none")
                            fill3(app.UIAxes, [pos(1), pos(1), pos(1), pos(1)], [pos(2), pos(2)
- scale, pos(2) - scale, pos(2)], [pos(3), pos(3) + sidelength/2, pos(3) - sidelength/2, pos(3)], [0
0.4470 0.7410], 'LineStyle', "none")
                      end

                      % Fixed in Z
                      if planes(3)
                            fill3(app.UIAxes, [pos(1), pos(1) + sidelength/2, pos(1) -
sidelength/2, pos(1)], [pos(2), pos(2), pos(2), pos(2)], [pos(3), pos(3) - scale, pos(3) - scale,
pos(3)], [0 0.4470 0.7410], 'LineStyle', "none")
```

```
                    fill3(app.UIAxes, [pos(1), pos(1), pos(1), pos(1)], [pos(2), pos(2)
+ sidelength/2, pos(2) - sidelength/2, pos(2)], [pos(3), pos(3) - scale, pos(3) - scale, pos(3)], [0
0.4470 0.7410], 'LineStyle', "none")
                end
            end

            function updateDropdowns(app)

                % Make the 2 vertices of the beam table dropdowns (dropdowns can
                % only contain strings)
                options = cellstr(string(1:1:height(app.UITable.Data)));
                app.UITable_2.ColumnFormat = {options options 'char' 'char'};
            end

            function plotDeformation(app, scaleFactor)
                vertexTable = app.convert(app.UITable.Data);
                beamTable = app.convert(app.UITable_2.Data);

                % Coord, Con, New vertices
                X = app.DeformedTruss(vertexTable(:, 1:3)', beamTable(:, 1:2)',
scaleFactor*app.U);

                % Replot the truss and scaled deformation
                cla(app.UIAxes)
                hold(app.UIAxes, 'on');

                % Deformation
                plot3(app.UIAxes, X(:,4), X(:,5), X(:,6), 'm', 'Marker', 'o',
'MarkerSize', 4, 'MarkerFaceColor', 'm', 'MarkerEdgeColor', 'm', 'LineWidth', 2);

                % Original
                plot3(app.UIAxes, X(:,1), X(:,2), X(:,3), 'k', 'Marker', 'o',
'MarkerSize', 4, 'MarkerFaceColor', 'k', 'MarkerEdgeColor', 'k', 'LineWidth', 2);
                hold(app.UIAxes, 'off');
            end
        end
        methods (Access = private, Static = true)

            % Solve the truss using FEA
            function [F, U, R] = TrussSolver(Coord, Con, Re, Load, E, A)
                w = size(Re);
                S = zeros(3*w(2));
                U = 1-Re;
                f = find(U);

                for i = 1:size(Con,2)
```

```
                H = Con(:,i);
                C = Coord(:,H(2)) - Coord(:,H(1));
                Le = norm(C);
                T = C/Le;
                s = T*T';
                G = E(i)*A(i)/Le;
                Tj(:,i) = G*T;
                e = [3*H(1)-2:3*H(1),3*H(2)-2:3*H(2)];
                S(e,e) = S(e,e)+G*[s -s;-s s];
            end
            U(f) = S(f,f)\Load(f);
            F = sum(Tj.*(U(:,Con(2,:))-U(:,Con(1,:))));
            R = reshape(S*U(:),w);
            R(f) = 0;
        end

        function X = DeformedTruss(Coord, Con, U)
            C = [Coord; Coord+U];
            e = Con(1,:);
            f = Con(2,:);
            for i=1:6
                M = [C(i,e); C(i,f); NaN(size(e))];
                X(:,i) = M(:);
            end
        end

        function [array, unused] = convert(cells)

            % UniformOutput converts invalid inputs in empty cells
            array = cellfun(@str2num, string(cells), 'UniformOutput', false);

            % Remove all rows that have empty cells
            array(any(cellfun(@isempty, array), 2), :) = [];

            % Convert cells to a number matrix
            array = cell2mat(array);

            % Value indicating that not every row is being used
            unused = height(cells) - height(array);
        end
    end

    % Callbacks that handle component events
    methods (Access = private)
```

```
        % Code that executes after component creation
        function startupFcn(app)

            % Make the 3 DOF columns of the vertice table checkboxes. Make
            % the rest character cells so that numeric functions like sqrt
            % and ^ can be used.
            app.UITable.ColumnFormat = {'char' 'char' 'char' 'char' 'char'
'char' 'logical' 'logical' 'logical'};
        end

        % Cell edit callback: UITable
        function UITableCellEdit(app, event)
            app.updatePlot()
        end

        % Button pushed function: AddVertexButton
        function AddVertexButtonPushed(app, event)

            % Add a new table row
            if isempty(app.UITable.Data)
                app.UITable.Data = {'0' '0' '0' '0' '0' '0' 1 1 1};
            else
                app.UITable.Data(end + 1, :) = {'0' '0' '0' '0' '0' '0' 0 0 1};
            end

            app.updateDropdowns()
            app.updatePlot()
        end

        % Button pushed function: DeleteVertexButton
        function DeleteVertexButtonPushed(app, event)
            app.UITable.Data(app.UITable.Selection, :) = [];

            app.updateDropdowns()
            app.updatePlot()
        end

        % Button pushed function: AddBeamButton
        function AddBeamButtonPushed(app, event)

            % Only add another row if another beam can be added (Can only
            % have n(n-1)/2 beams, where n is the number of vertices)
            if ((height(app.UITable.Data)*(height(app.UITable.Data)-1))/2 -
height(app.UITable_2.Data)) > 0
                if isempty(app.UITable_2.Data)
                    app.UITable_2.Data = {'Choose' 'Choose' '1' '1'};
```

```
                else
                        app.UITable_2.Data(end + 1, :) = [{'Choose' 'Choose'}
app.UITable_2.Data(end, 3:4)];
                end
            end

            app.updatePlot()
        end

        % Button pushed function: DeleteBeamButton
        function DeleteBeamButtonPushed(app, event)

            % Clear all rows that are currently selected
            app.UITable_2.Data(app.UITable_2.Selection, :) = [];

            app.updatePlot()
        end

        % Cell edit callback: UITable_2
        function UITable_2CellEdit(app, event)
            app.updatePlot()
        end

        % Button pushed function: CalculateButton
        function CalculateButtonPushed(app, event)
            [vertexTable, unusedVerts] = app.convert(app.UITable.Data);
            [beamTable, unusedBeams] = app.convert(app.UITable_2.Data);

            % Raise an alert if there are empty option boxes
            if unusedBeams > 0 || unusedVerts > 0
                uialert(app.TrussCalculatorUIFigure, 'Cannot have empty or
invalid inputs.', 'Incorrect Data')

            % Raise an alert if there are repeat vertices
            elseif height(vertexTable(:, 1:3)) > height(unique(vertexTable(:, 1:3),
'rows'))
                uialert(app.TrussCalculatorUIFigure, 'Cannot have multiple
vertices in the same location.', 'Incorrect Data')

            else
                try

                        % Coord, Con, Re, Load, E, A
                        [F, app.U, R] = app.TrussSolver(vertexTable(:, 1:3)',
beamTable(:, 1:2)', vertexTable(:, 7:9)', vertexTable(:, 4:6)', beamTable(:, 4)'.*1e6,
beamTable(:, 3)');
```

```
                        % Convert the string arrays into a single string for
                        % the text areas
                        if ~isempty(R)
                            app.ReactionForcesNTextArea.Value =
sprintf("[%d]\t" + strjoin(repmat("%.4f", 1, height(R)), ', \t') + "\n", [1:1:width(R); R]);
                        end
                        if ~isempty(F)
                            app.AxialForcesNTextArea.Value =
sprintf("[%d]\t" + strjoin(repmat("%.4f", 1, height(F)), ', \t') + "\n", [1:1:width(F); F]);
                        end

                        % Update the plot to show the deformation

app.plotDeformation(app.DeformationAmplificationSlider.Value)
                    catch

                        % Catch any errors in solving to prevent the app from
                        % crashing
                        uialert(app.TrussCalculatorUIFigure, 'An unknown
error occured. Check that all inputs are correct.', 'Unknown Error')
                    end
                end
            end

            % Button pushed function: InfoButton
            function InfoButtonPushed(app, event)
                uialert(app.TrussCalculatorUIFigure, "This application analyzes 2D
and 3D trusses. Insert new truss vertices and beams on the respective tabs using the 'Add
Vertex' and 'Add Beam' buttons. New beams copy the area and Young's modulus from the
previous beam. Select table rows and click the delete button to remove the corresponding
vertices and beams. All numerical inputs can be decimals or any mathematical functions.
Beam vertices are selected using dropdown menus, and the numbers correspond to the
vertex row numbers. Vertex DOFs are selected using checkboxes (A checkmark means
than axis is fixed). Outputs on the results page are displayed with the corresponding vertex
or beam number.", 'App Information', 'Icon', 'info');
            end

            % Value changing function: DeformationAmplificationSlider
            function DeformationAmplificationSliderValueChanging(app, event)

                % Update the deformation plot if no vertex or beam inputs have been
                % changed
                if ~isempty(app.U)
                    app.plotDeformation(event.Value)
                end
```

```
        end
    end

    % Component initialization
    methods (Access = private)

        % Create UIFigure and components
        function createComponents(app)

            % Create TrussCalculatorUIFigure and hide until all components are
            % created
            app.TrussCalculatorUIFigure = uifigure('Visible', 'off');
            app.TrussCalculatorUIFigure.Position = [100 100 640 480];
            app.TrussCalculatorUIFigure.Name = 'Truss Calculator';
            app.TrussCalculatorUIFigure.Icon = 'TrussCalculatorIcon.png';
            app.TrussCalculatorUIFigure.Scrollable = 'on';

            % Create UIAxes
            app.UIAxes = uiaxes(app.TrussCalculatorUIFigure);
            xlabel(app.UIAxes, 'x (m)')
            ylabel(app.UIAxes, 'y (m)')
            zlabel(app.UIAxes, 'z (m)')
            app.UIAxes.DataAspectRatio = [1 1 1];
            app.UIAxes.PlotBoxAspectRatio = [1 1 1];
            app.UIAxes.XLimitMethod = 'padded';
            app.UIAxes.YLimitMethod = 'padded';
            app.UIAxes.ZLimitMethod = 'padded';
            app.UIAxes.XGrid = 'on';
            app.UIAxes.YGrid = 'on';
            app.UIAxes.ZGrid = 'on';
            app.UIAxes.Position = [304 194 337 287];

            % Create TabGroup
            app.TabGroup = uitabgroup(app.TrussCalculatorUIFigure);
            app.TabGroup.Position = [1 1 304 480];

            % Create VerticesTab
            app.VerticesTab = uitab(app.TabGroup);
            app.VerticesTab.Title = 'Vertices';

            % Create UITable
            app.UITable = uitable(app.VerticesTab);
            app.UITable.ColumnName = {'x (m)'; 'y (m)'; 'z (m)'; 'Fx (N)'; 'Fy
(N)'; 'Fz (N)'; 'Fixed X'; 'Fixed Y'; 'Fixed Z'};
            app.UITable.SelectionType = 'row';
            app.UITable.ColumnEditable = true;
```

```
                    app.UITable.CellEditCallback = createCallbackFcn(app,
@UITableCellEdit, true);
                    app.UITable.Tooltip = {''};
                    app.UITable.Position = [1 39 303 417];

                    % Create AddVertexButton
                    app.AddVertexButton = uibutton(app.VerticesTab, 'push');
                    app.AddVertexButton.ButtonPushedFcn = createCallbackFcn(app,
@AddVertexButtonPushed, true);
                    app.AddVertexButton.Tooltip = {'Adds a new row to the table.'};
                    app.AddVertexButton.Position = [33 8 100 22];
                    app.AddVertexButton.Text = 'Add Vertex';

                    % Create DeleteVertexButton
                    app.DeleteVertexButton = uibutton(app.VerticesTab, 'push');
                    app.DeleteVertexButton.ButtonPushedFcn = createCallbackFcn(app,
@DeleteVertexButtonPushed, true);
                    app.DeleteVertexButton.Tooltip = {'Removes all selected rows from
the table.'};
                    app.DeleteVertexButton.Position = [171 8 100 22];
                    app.DeleteVertexButton.Text = 'Delete Vertex';

                    % Create BeamsTab
                    app.BeamsTab = uitab(app.TabGroup);
                    app.BeamsTab.Title = 'Beams';

                    % Create UITable_2
                    app.UITable_2 = uitable(app.BeamsTab);
                    app.UITable_2.ColumnName = {'Vertex 1'; 'Vertex 2'; 'A (m²)'; 'E
(MPa)'};
                    app.UITable_2.SelectionType = 'row';
                    app.UITable_2.ColumnEditable = true;
                    app.UITable_2.CellEditCallback = createCallbackFcn(app,
@UITable_2CellEdit, true);
                    app.UITable_2.Tooltip = {''};
                    app.UITable_2.Position = [1 39 303 417];

                    % Create AddBeamButton
                    app.AddBeamButton = uibutton(app.BeamsTab, 'push');
                    app.AddBeamButton.ButtonPushedFcn = createCallbackFcn(app,
@AddBeamButtonPushed, true);
                    app.AddBeamButton.Tooltip = {'Adds a new row to the table.'};
                    app.AddBeamButton.Position = [33 8 100 22];
                    app.AddBeamButton.Text = 'Add Beam';

                    % Create DeleteBeamButton
```

```
            app.DeleteBeamButton = uibutton(app.BeamsTab, 'push');
            app.DeleteBeamButton.ButtonPushedFcn = createCallbackFcn(app,
@DeleteBeamButtonPushed, true);
            app.DeleteBeamButton.Tooltip = {'Removes all selected rows from the
table.'};
            app.DeleteBeamButton.Position = [171 8 100 22];
            app.DeleteBeamButton.Text = 'Delete Beam';

            % Create Results
            app.Results = uitab(app.TabGroup);
            app.Results.Title = 'Results';

            % Create CalculateButton
            app.CalculateButton = uibutton(app.Results, 'push');
            app.CalculateButton.ButtonPushedFcn = createCallbackFcn(app,
@CalculateButtonPushed, true);
            app.CalculateButton.Tooltip = {'Analyzes the truss.'};
            app.CalculateButton.Position = [8 424 288 22];
            app.CalculateButton.Text = 'Calculate';

            % Create AxialForcesNTextAreaLabel
            app.AxialForcesNTextAreaLabel = uilabel(app.Results);
            app.AxialForcesNTextAreaLabel.HorizontalAlignment = 'right';
            app.AxialForcesNTextAreaLabel.Position = [32 393 91 22];
            app.AxialForcesNTextAreaLabel.Text = 'Axial Forces (N)';

            % Create AxialForcesNTextArea
            app.AxialForcesNTextArea = uitextarea(app.Results);
            app.AxialForcesNTextArea.Editable = 'off';
            app.AxialForcesNTextArea.WordWrap = 'off';
            app.AxialForcesNTextArea.Tooltip = {'Axial force in each truss
member. Each row shows the axial force in the corresponding beam on the "Beams" tab.'};
            app.AxialForcesNTextArea.Position = [8 8 140 377];

            % Create ReactionForcesNTextAreaLabel
            app.ReactionForcesNTextAreaLabel = uilabel(app.Results);
            app.ReactionForcesNTextAreaLabel.HorizontalAlignment = 'right';
            app.ReactionForcesNTextAreaLabel.Position = [169 393 113 22];
            app.ReactionForcesNTextAreaLabel.Text = 'Reaction Forces (N)';

            % Create ReactionForcesNTextArea
            app.ReactionForcesNTextArea = uitextarea(app.Results);
            app.ReactionForcesNTextArea.Editable = 'off';
            app.ReactionForcesNTextArea.WordWrap = 'off';
```

```
            app.ReactionForcesNTextArea.Tooltip = {'Reaction force on each
truss vertex. Each row shows the x, y, and z components of the reaction force on the
corresponding vertex on the "Vertices" tab.'};
            app.ReactionForcesNTextArea.Position = [156 8 140 377];

            % Create DeformationAmplificationSliderLabel
            app.DeformationAmplificationSliderLabel =
uilabel(app.TrussCalculatorUIFigure);
            app.DeformationAmplificationSliderLabel.HorizontalAlignment =
'right';
            app.DeformationAmplificationSliderLabel.Position = [397 140 142
22];
            app.DeformationAmplificationSliderLabel.Text = 'Deformation
Amplification';

            % Create DeformationAmplificationSlider
            app.DeformationAmplificationSlider =
uislider(app.TrussCalculatorUIFigure);
            app.DeformationAmplificationSlider.Limits = [1 1000];
            app.DeformationAmplificationSlider.MajorTicks = [1 250 500 750
1000];
            app.DeformationAmplificationSlider.ValueChangingFcn =
createCallbackFcn(app, @DeformationAmplificationSliderValueChanging, true);
            app.DeformationAmplificationSlider.Tooltip = {'Changes the scaling
of the truss''s resulting deformation.'};
            app.DeformationAmplificationSlider.Position = [358 128 220 3];
            app.DeformationAmplificationSlider.Value = 1;

            % Create InfoButton
            app.InfoButton = uibutton(app.TrussCalculatorUIFigure, 'push');
            app.InfoButton.ButtonPushedFcn = createCallbackFcn(app,
@InfoButtonPushed, true);
            app.InfoButton.Icon = 'InfoIcon.png';
            app.InfoButton.IconAlignment = 'center';
            app.InfoButton.BackgroundColor = [0.902 0.902 0.902];
            app.InfoButton.FontColor = [0.902 0.902 0.902];
            app.InfoButton.Tooltip = {'Open a help information window.'};
            app.InfoButton.Position = [310 456 21 22];
            app.InfoButton.Text = '';

            % Show the figure after all components are created
            app.TrussCalculatorUIFigure.Visible = 'on';
        end
    end

    % App creation and deletion
```

```
    methods (Access = public)

        % Construct app
        function app = TrussCalculator

            % Create UIFigure and components
            createComponents(app)

            % Register the app with App Designer
            registerApp(app, app.TrussCalculatorUIFigure)

            % Execute the startup function
            runStartupFcn(app, @startupFcn)

            if nargout == 0
                clear app
            end
        end

        % Code that executes before app deletion
        function delete(app)

            % Delete UIFigure when app is deleted
            delete(app.TrussCalculatorUIFigure)
        end
    end
end
```

MIX
Papier aus verantwortungsvollen Quellen
Paper from responsible sources
FSC® C105338

If you have any concerns about our products,
you can contact us on
**ProductSafety@springernature.com**

In case Publisher is established outside the EU,
the EU authorized representative is:
**Springer Nature Customer Service Center GmbH**
**Europaplatz 3, 69115 Heidelberg, Germany**

Printed by Libri Plureos GmbH
in Hamburg, Germany